MIR PUBLISHERS

Проф. Ю. М. ЛАХТИН

Основы металловедения

ИЗДАТЕЛЬСТВО «МЕТАЛЛУРГИЯ»
МОСКВА

Prof. Y. LAKHTIN

ENGINEERING PHYSICAL METALLURGY

Translated from the Russian

by

Nicholas Weinstein

Mir Publishers Moscow

CBS Publishers & Distributors Pvt. Ltd.

New Delhi • Bengaluru • Chennai • Kochi • Kolkata • Mumbai

Hyderabad • Uttarakhand • Nagpur • Patna • Pune • Jharkhand

ISBN: 81-239-0602-1

First Indian Edition: 1998

This edition has been published in India by arrangement with Butterworth Heinemann Limited, UK.

Published by **Satish Kumar Jain** and produced by **Varun Jain** for
CBS Publishers & Distributors Pvt. Ltd.,
4819/XI Prahlad Street, 24 Ansari Road, Daryaganj, New Delhi - 110002
delhi@cbspd.com, cbspubs@airtelmail.in • www.cbspd.com
Ph.: 23289259, 23266861, 23266867 • Fax: 011-23243014

Corporate Office: 204 FIE, Industrial Area, Patparganj, Delhi - 110 092
Ph: 49344934 • Fax: 011-49344935
E-mail: publishing@cbspd.com • publicity@cbspd.com

Branches:

- ***Bengaluru:*** 2975, 17th Cross, K.R. Road, Bansankari 2nd Stage, Bengaluru - 70 • Ph: +91-80-26771678/79 • Fax: +91-80-26771680 E-mail: cbsbng@gmail.com, bangalore@cbspd.com
- ***Chennai:*** No. 7, Subbaraya Street, Shenoy Nagar, Chennai - 600030 Ph: +91-44-26681266, 26680620 • Fax: +91-44-42032115 E-mail: chennai@cbspd.com
- ***Kochi:*** Ashana House, 39/1904, A.M. Thomas Road, Valanjambalam, Ernakulum, Kochi • Ph: +91-484-4059061-65 Fax: +91-484-4059065 • E-mail: cochin@cbspd.com
- ***Kolkata:*** 6-B, Ground Floor, Rameshwar Shaw Road, Kolkata - 700014 Ph: +91-33-22891126/7/8 • E-mail: kolkata@cbspd.com
- ***Mumbai:*** 83-C, Dr. E. Moses Road, Worli, Mumbai - 400018 Ph: +91-9833017933, 022-24902340/41 • E-mail: mumbai@cbspd.com

Representatives:

- Hyderabad: 0-9885175004
- Patna: 0-9334159340
- Jharkhand: 0-9811541605
- Nagpur: 0-9021734563
- Pune: 0-9623451994
- Uttarakhand: 0-9716462459

Printed at:
J.S. Offset Printers, Delhi (India)

CONTENTS

PREFACE

This book should be of particular aid to new engineering personnel, only recently engaged in industry, in coordinating their theoretical knowledge with the actual engineering practice they encounter and should also help them to better understand special treatises on physical metallurgy and heat treatment. More experienced engineers may use it to renew their theoretical knowledge on the subject.

The book may be expediently employed as supplementary reading by students of metallurgical and mechanical engineering institutes and technical schools attending courses in general physical metallurgy.

One aim of the author was to elucidate the latest developments made in engineering physical metallurgy, both in the Soviet Union and in other countries. Principal attention is given to the physical nature of the phenomena described.

The author does not claim to have made a complete exposition of all the aspects of physical metallurgy. His intention was merely to set forth the fundamentals of physical metallurgy and heat treatment of steel, cast iron, and nonferrous metals in a consecutive and easily understandable manner.

LIST OF SYMBOLS

RN—rate of nucleation
RG—rate of crystal growth
a, b, c—lattice parameters
R_0—equilibrium interatomic distance
F—free energy
H—total energy (enthalpy)
T—absolute temperature, degrees Kelvin
S—entropy
ΔT—degree of supercooling
T_s—equilibrium freezing temperature
T_k—actual freezing temperature
$\perp$—positive dislocation
$\top$—negative dislocation
σ_u—ultimate strength
Bhn—Brinell hardness number
δ—relative elongation
Ψ—relative reduction of area
S_T—cohesive strength
τ_k—shear strength
T_r—recrystallisation temperature
T_m—melting point in the absolute temperature scale

ρ—electrical resistivity
α—temperature coefficient of resistivity
μ—magnetic permeability
B—magnetic induction
H—magnetising force
K—magnetic susceptibility
I—intensity of magnetisation
B_r—residual magnetic induction
σ_p—proportional limit
σ_s—yield point
σ_e—elastic limit
ln—natural logarithm
R_c—Rockwell hardness, Scale C
R_B—Rockwell hardness, Scale B
VHN—Vickers hardness number
a_k—impact strength
σ_w—fatigue limit
V_c—critical cooling rate
f—current frequency
D—diffusion coefficient
R—gas constant
σ_{-1}—endurance limit for complete stress reversal

Chapter 1

CRYSTAL STRUCTURE OF METALS

1-1. Atomic Structure of Metals

Physical metallurgy is that branch of science which deals with the general relationship between the composition, structure, and properties of metals and alloys, as well as the changes brought about by thermal, chemical, and mechanical treatment.

The aim of physical metallurgy, as a science, is to establish the physical laws governing the structure of an alloy and its properties and to find the best possible composition, manufacturing techniques, and treatment of the alloy to obtain the required physical and mechanical properties.

All metals and their alloys are crystalline solids. They differ from amorphous solids, in which the atoms are arranged chaotically, in that crystalline solids have a definite and orderly internal structure repeated in three dimensions. If the centres of the atoms are considered to be connected together by straight lines, then a system will be obtained comprising a great number of equal parallelepipeds. This system is known as the *space or crystal lattice*. The smallest parallelepiped which could be transposed in three coordinate directions to build the space lattice is called the *unit cell*. The space lattices of various substances differ in the size and shape of their unit cells.

The space lattice of a crystal is described by means of a three-dimensional coordinate system in which the coordinate axes coincide with any three edges of the crystal that intersect at one point and do not lie in a single plane.

Seven different coordinate systems of *reference axes* are employed to describe crystal structure:

1) triclinic (Fig. 1, *a*) in which $a \neq b \neq c$ and $\alpha \neq \beta \neq \gamma$;
2) monoclinic (Fig. 1, *b*) in which $a \neq b \neq c$ and $\alpha = \gamma = 90° \neq \beta$;
3) orthorhombic (Fig. 1, *c*) in which $a \neq b \neq c$ and $\alpha = \beta = \gamma = 90°$;
4) hexagonal (Fig. 1, *d*) in which $a = b \neq c$, $\alpha = \beta = 90°$, and $\gamma = 120°$;
5) rhombohedral (Fig. 1, *e*) in which $a = b = c$ and $\alpha = \beta = \gamma \neq 90°$;
6) tetragonal (Fig. 1, *f*) in which $a = b \neq c$ and $\alpha = \beta = \gamma = 90°$; and
7) cubic (Fig. 1, *g*) in which $a = b = c$ and $\alpha = \beta = \gamma = 90°$.

Complex crystal lattices, most frequently encountered in metals, may be considered as comprised of several primitive translation lattices, displaced in relation to each other. The crystals of most metals have a highly symmetrical structure with close-packed atoms. The most common types of space lattices are the following (Fig. 2):

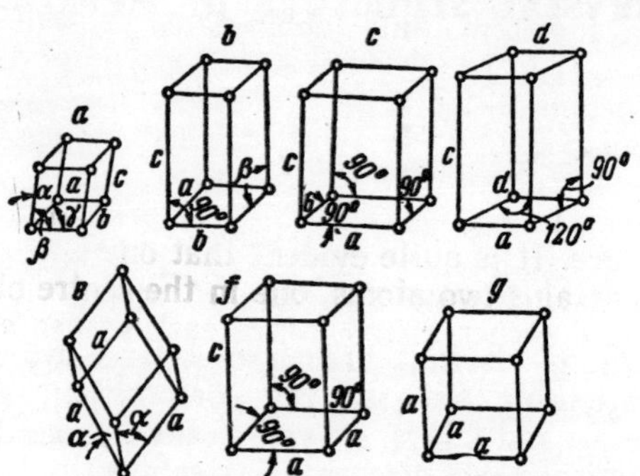

Fig. 1. Various types of unit cells:
a—triclinic, *b*—monoclinic, *c*—orthorhombic, *d*—hexagonal, *e*—rhombohedral, *f*—tetragonal *g*—cubic

Body-centred cubic space lattice (Fig. 2, *a*) in which atoms are located at the corners of the cube and one atom at its centre. This type of lattice is found, for example, in the following metals: Li, Na, K, Ba, β-Ti, V, Nb, Ta, α-Cr, Mo, α-W, α-Fe, and others.

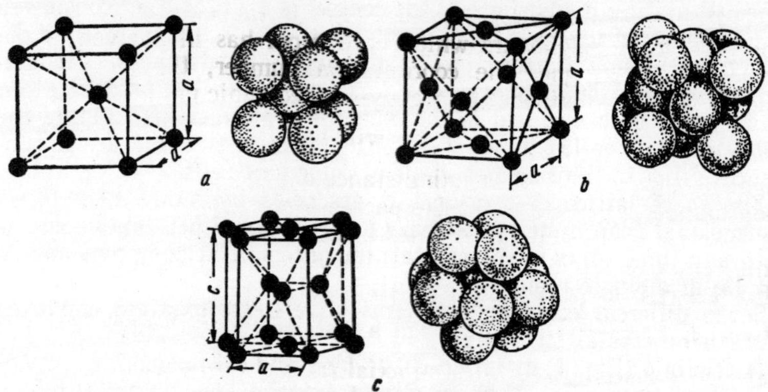

Fig. 2. Types of crystal lattices of metals and packing of atoms:
a—body-centred cubic, *b*—face-centred cubic, *c*—hexagonal

Face-centred cubic space lattice (Fig. 2, *b*) in which atoms are located at the corners of the cube and at the centres of each face. This type is typical of the metals: Cu, Ag, Au, Ca, Sr, Al, Pb, γ-Fe, β-Co, β-Ni, Rh, Pd, Ir, Pt, etc.

Close-packed hexagonal space lattice (Fig. 2, *c*) is found in such metals as: Be, Mg, Ca, Zn, Cd, γ-Ti, and others.

The dimensions of a space lattice are specified by lattice constants or parameters which represent the distances between parallel planes of the atoms comprising the unit cell (Fig. 2). The lattice constants are measured in angstrom units (abbreviated Å, and 1Å=10^{-8} cm) or in kilo-X units (1 kXU=1.00202×10^{-8} cm).

The lattice constant (*a*) for metals which solidify into a cubic system ranges from 2.86 to 6.07 Å. In the hexagonal system *a* ranges from 2.28 to 3.98 Å and *c* from 3.57 to 6.52 Å.

Atoms belonging to a given unit cell comprise the so-called *basic atoms* of the lattice. It is quite evident that one unit cell of a body-centred lattice contains two atoms, one in the centre of the cube and one comprising the corner atoms. Since each corner atom is shared by eight cells, only $^1/_8$ of the mass of this atom is referred to the given unit cell but $^1/_8 \times 8 = 1$ for the whole cell. One unit cell of a face-centred cubic lattice contains four atoms $[(^1/_2 \times 6)+(^1/_8 \times 8)=4]$ and the unit cell of the close-packed hexagonal lattice contains six atoms $[3+ +(^1/_6 \times 12)+(^1/_2 \times 2)=6]$. The positions of the basic atoms are expressed in a coordinate system of which the axes are edges of the unit cell, and the units are its dimensions. The coordinates of the basic atoms, for example, in a body-centred cubic space lattice will be: first atom (0, 0, 0), second atom ($^1/_2$, $^1/_2$, $^1/_2$). For a face-centred cubic lattice they will be—(0, 0, 0), ($^1/_2$, $^1/_2$, 0)($^1/_2$, 0, $^1/_2$), and (0, $^1/_2$, $^1/_2$).

The packing of a crystal lattice, i.e., the volume occupied by the atoms (which may be assumed as hard spheres) (Fig. 2), is characterised by the *coordination number*. This is the number of equally-spaced nearest neighbours which each atom has in a given crystal structure. The higher the coordination number, the more closely packed the atoms will be. In a body-centred cubic unit cell, the minimum distance between the atoms will be $\frac{a\sqrt{3}}{2}$. The given atom has eight neighbours spaced at this distance and, therefore, the coordination number will be 8 (C8). The packing factor, which is determined as the ratio of the volume occupied by the atoms to the volume of the unit cell, will be 68 per cent in this case. The coordination number of a face-centred crystal lattice will be 12 (C12). In this lattice, each atom has 12 nearest neighbours at a distance of $\frac{a\sqrt{2}}{2}$. A close-packed hexagonal space lattice, with an axial ratio of $c/a=1.633$, has a coordination number of 12 (H12), which also corresponds to the most closely packed atom structure. Most metals which crystallise into a hexagonal system have axial ratios c/a from 1.57 to 1.64, i.e., they deviate only slightly from a close-packed structure.

If the axial ratio (c/a) differs considerably from 1.633 (as in Zn and Cd), the coordination number of the hexagonal lattice will equal 6.

Face-centred cubic and hexagonal lattices (in which $c/a=1.633$) are the most compact types and have a packing factor of 74 per cent.

If the coordination number is reduced to 6 the packing factor will be about 50 per cent. It will be only about 25 per cent for a coordination number of 4.

Certain metals have more complex lattices than those mentioned above. The space lattice for bismuth (Fig. 3), for example, also characteristic of arsenic and antimony, differs from typical metal structures. This is usually called a layer- or arsenic-type structure. In each layer, the atom is surrounded by three equally distant nearest neighbours.

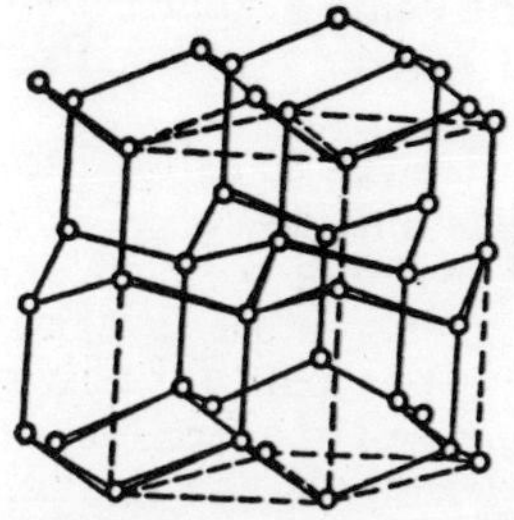

Fig. 3. Crystal lattice of Bi

Typical structures of certain nonmetallic materials are illustrated in Fig. 4. The first is the diamond structure with a coordination number of 4. Silicon, germanium, and grey tin (α-Sn) have similar lattices. The second, layer-type lattice (Fig. 4, *b*) is characteristic of graphite.

In all cases, the coordination number of the crystal structure for nonmetallic elements and metalloids in Groups IV, V, and VI of the Mendeleyev Periodic Table will equal 8—N, in which N is the number of the group to which the element belongs.

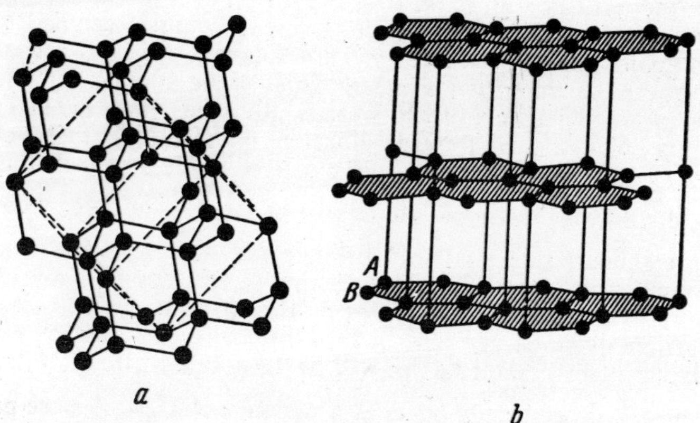

Fig. 4. Crystal structures of carbon:
a—diamond, *b*—graphite

The type of crystal lattice and the properties of a substance are determined by the interatomic bonding forces which depend on the structure of the atoms concerned.

The special type of bond peculiar to metals is often called the metallic bond.

It is known that an atom consists of the nucleus, carrying a positive electrostatic charge, about which negatively charged electrons revolve. The number of electrons equals the positive charge of the nucleus and corresponds to the atomic number of the element in the periodic table. In its "stationary" or lowest-energy condition, every atom is electrically neutral. Electrons belonging to an atom may be classified as "outer" or valence electrons, which are least firmly bound by the atomic nucleus, and "inner" or core electrons, more firmly bound. According to the present theory of the structure of metals, valence electrons are so weakly bound that they are no longer associated with particular atoms and belong to the aggregate of atoms as a whole. This means that they form an electron cloud or gas and may be considered to move freely. Valence electrons occupy a definite energy level and move about freely in the crystal lattice of the metal.

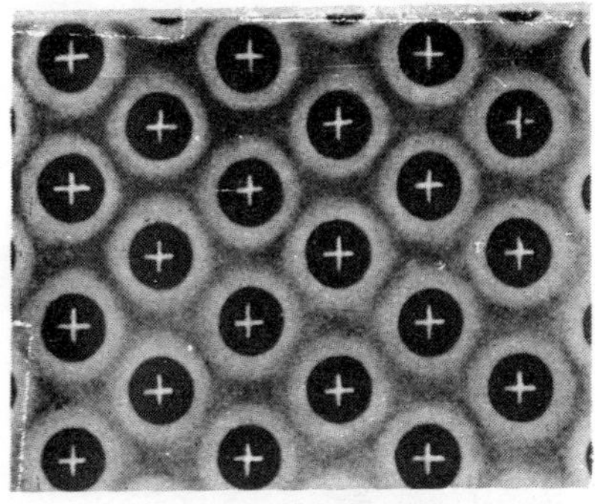

Fig. 5. Structure of a metal. Positively-charged ions are located at the points of the lattice and are surrounded by a cloud of free electrons

Therefore, a metal crystal should be considered to consist of positive ions, located at the points of the lattice, and surrounded by a singular plasma of electron gas (Fig. 5).

Metallic ions continuously oscillate at a frequency peculiar to each metal and with an amplitude depending on the temperature. It is, in fact, the midpoints in space of their centres of gravity that are represented in depicting simple crystal lattices.

A metallic bond is due to interaction of the ions and electrons. Free electrons are simultaneously attracted by several positive ions to form the bond between them.

The forces holding atoms of a metal together are determined by the repulsion between ions and electrons and the attraction between ions and free electrons and these forces are not of a sharply defined directional nature.

Atoms tend to maintain a definite distance from each other at which the energy of interaction will be at a minimum (Fig. 6, *a*).

As evident from Fig. 6, this condition is satisfied by the distance R_0. The value of R_0 for certain elements is indicated in the chart of Fig. 7.

It will be necessary to do a definite amount of work against the forces of repulsion and attraction to bring the atoms (ions) nearer than the distance R_0 or farther apart than this equilibrium spacing.

Therefore, atoms (ions) in a state of equilibrium are arranged in a definite order, forming a regular crystal lattice which corresponds to the minimum energy of interaction between the atoms.

A model of a metallic crystal is shown in Fig. 6, *b*. Variation of the potential energy of the atoms in the lattice is shown graphically

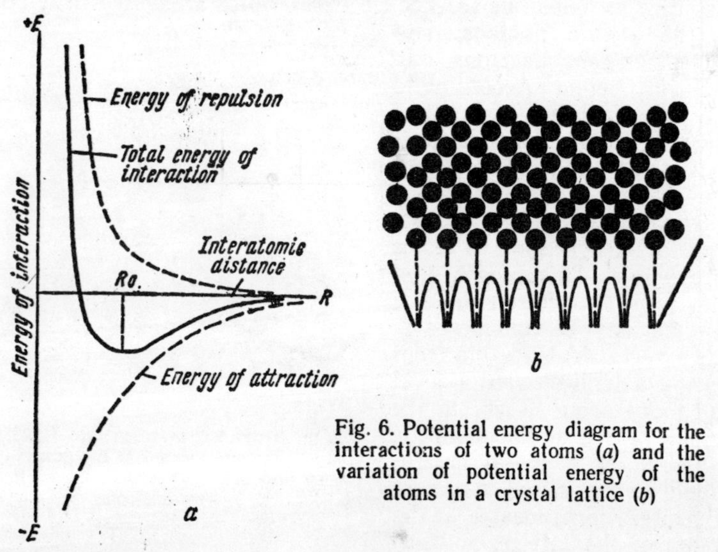

Fig. 6. Potential energy diagram for the interactions of two atoms (*a*) and the variation of potential energy of the atoms in a crystal lattice (*b*)

below the model. The atoms (ions) occupy positions corresponding to minimum potential energy. Atoms on the surface possess a higher potential energy.

Due to the spherical symmetry of the fields of positive ions, metallic crystals have multiple-fold symmetry and closely packed atoms (C12, C8, and H12).

1-2. Allotropy

Existence of a given metal in two or more stable but different crystal structures (modifications), depending on the conditions of temperature or pressure, is called allotropy. The essence of allotropic transformations is that the atoms of a crystallic solid are converted from one crystalline form to another, i.e., they form a new crystal lattice.

Modifications, stable at lower temperatures, are designated by the Greek letter α (alpha); β (beta) designates a second form of the same material that is stable at some higher temperatures; γ (gamma) at still higher temperatures; etc.

The allotropy of iron is of especial practical importance. Iron exists in two allotropic forms: α-iron, with a body-centred cubic lattice, stable at temperatures up to 910°C and over 1,401°C, and

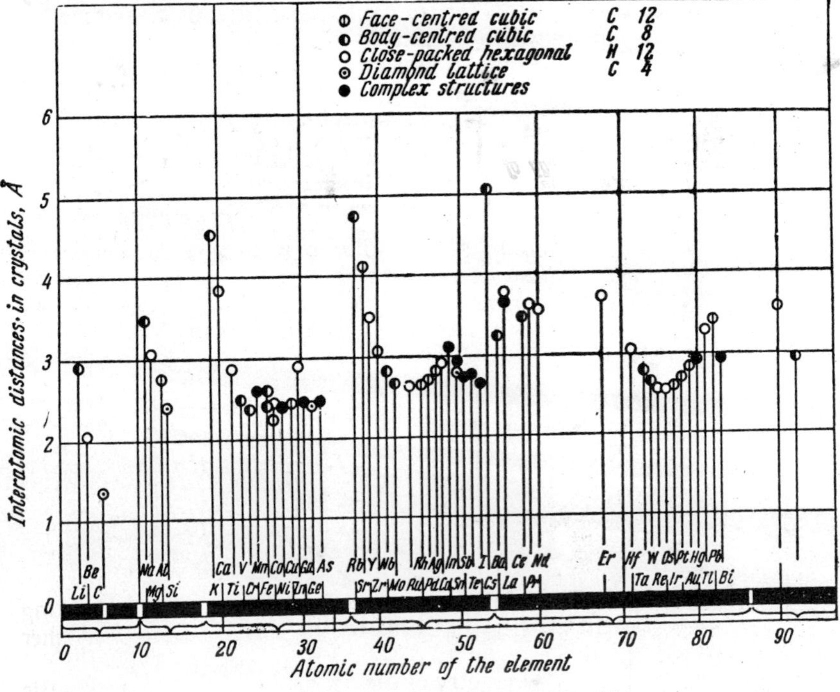

Fig. 7. Interatomic distances in crystals

γ-iron, with a face-centred cubic lattice, stable in the range from 910° to 1,401°C.

There are also three allotropic modifications of manganese (α-Mn, β-Mn, and γ-Mn) with complex crystal lattices, two modifications of cobalt (α-Co and β-Co), two—of tin (α-Sn and β-Sn), two—of titanium (α-Ti and β-Ti), two—of zirconium (β-Zr and α-Zr), and two—of tellurium (α-Te and β-Te).

Changes in the packing density of the crystal lattice, in conversion from one allotropic form to another, lead to changes in the volume of the material as well.

For example, the packing of γ-iron (C12) is 1.05 per cent denser than that of α-iron (C8) and its specific volume is correspondingly less. This is of great importance, as we shall see further on, in the heat treatment of steel.

In the transformation of white tin (β-Sn) to grey tin (α-Sn), the volume increases by 26 per cent. This increase in volume hinders the allotropic transformation β-Sn⇄α-Sn, which takes place only on the surface of the metal. On the other hand, it causes asymmetric stresses in the metal which results in cracking of the brittle α-Sn and in its transformation into a grey crystallic powder (tin pest or tin plague).

The high brittleness of α-Sn is a result of its less-closely packed crystallic structure (see Fig. 4, *a*), typical of nonmeta'ʼs.

1-3. Crystallographic Notation of Atomic Planes

The positions of atomic planes (planes passing through atoms) in crystal space lattices are determined by indices (h, k, l) which are three whole rational values, reciprocals of the intercepts. The

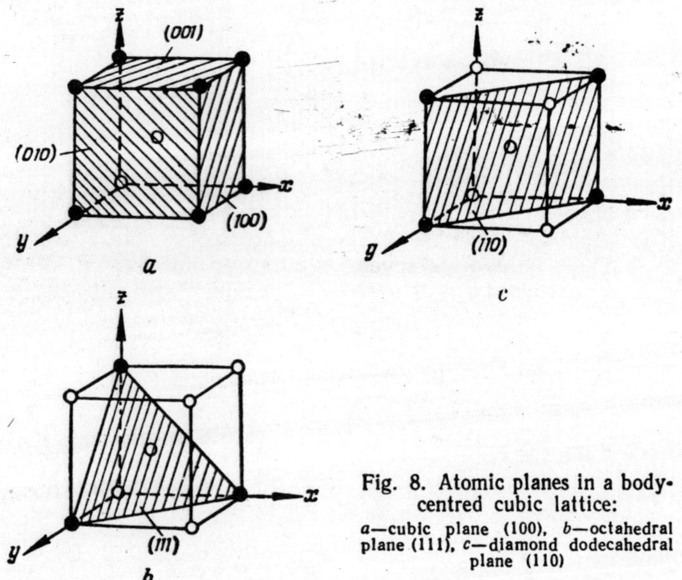

Fig. 8. Atomic planes in a body-centred cubic lattice:
a—cubic plane (100), *b*—octahedral plane (111), *c*—diamond dodecahedral plane (110)

intercepts of a given plane are the distances from the origin of the coordinate axes at which the plane intersects each axis.

The units of length along these axes are equal to the lengths of the edges of the unit cell.

As a first example in designating such planes we may establish the indices of the planes in a cube. It is evident from Fig. 8 that each plane of a cube intersects only one axis and the intercepts will equal (1, ∞, ∞), (∞, 1, ∞), and (∞, ∞, 1).* The reciprocals of the intercepts will respectively equal 1, 0, 0; 0, 1, 0; and 0, 0, 1. The indices of planes are usually enclosed in parentheses and are not separated by commas, as (100), (010), and (001).

In addition to the planes of the cube (Fig. 8, *a*), a cubic lattice has octahedral planes (111) (Fig. 8, *b*) and diamond dodecahedral planes (110) (Fig. 8, *c*).

It should be understood that the indices refer not only to one plane, but to whole sets of parallel planes.

It is not difficult to see that the density in the arrangement of atoms may vary in different planes. In a body-centred cubic lattice, for example, the plane (100) has but one atom ($^1/_4 \times 4$). The diamond dodecahedral plane (110) of the same lattice has two atoms; one from the atoms at the corners ($^1/_4 \times 4$) and one in the centre of the cube. The plane with most closely packed atoms of a face-centred cubic lattice will be the octahedral plane (111); in a body-centred cubic lattice, it will be plane (110).

1-4. The Structures of Actual Metal Crystals

Wide application of the electron microscope and X-ray analysis for investigations of the metal structure made it possible to establish the fact that no actual single crystal (i.e., monocrystal) or even any grain of a polycrystalline metal has a perfectly regular arrangement of atoms throughout its volume.

Each actual crystal consists of separate blocks, often called blocks of mosaic structure, from several microns to hundredths of a micron in size, each tipped slightly out of alignment, through an angle of several minutes, with its neighbours. A diagram illustrating mosaic structure is given in Fig. 9, *a* while Fig. 9, *b* shows a real structure as observed with the aid of an electron microscope. The reasons for the occurrence of mosaic structures have not yet been completely explained but they are undoubtedly associated with a special type of solidification (see Chapter 2). A study of mosaic (block) structure is of great importance since the size of the blocks considerably affects the properties of the metal.

Under ordinary conditions, a metal consists of a great number of randomly oriented grains (crystallites), usually of irregular form (see Fig. 16). On the boundaries of the grains, the atoms have a much less regular arrangement than in the grain body, due to the influence of atomic force fields in the neighbouring grains (Fig. 10). It should be noted also that in engineering metals, all sorts of impurities are con-

* A plane parallel to an axis does not intersect it at all and the intercept is listed as ∞; i. e., it is infinitely great.

centrated on the grain boundaries and they increase misorientation in atomic arrangement.

A certain degree of misorientation (or distortion) probably exists on the boundaries of mosaic blocks as well. The atoms in the boundary layers will possess higher energy due to distorted packing (Fig. 10). In exactly the same manner, atoms located on the surface of the crystal possess increased energy due to uncompensated forces of atomic interaction.

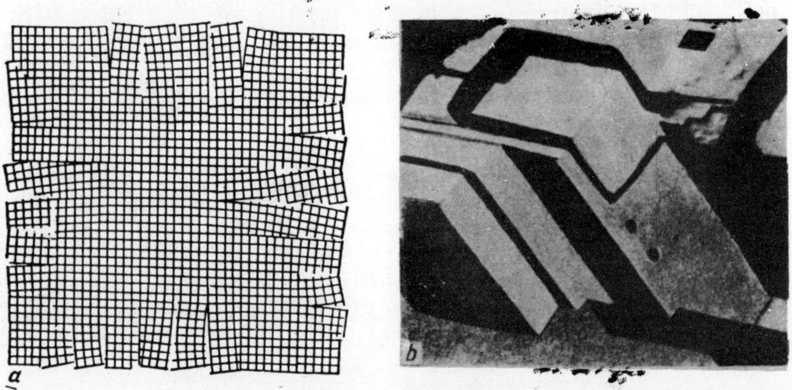

Fig. 9. Mosaic structure:
a—schematic diagram, *b*—structure of aluminium, ×700

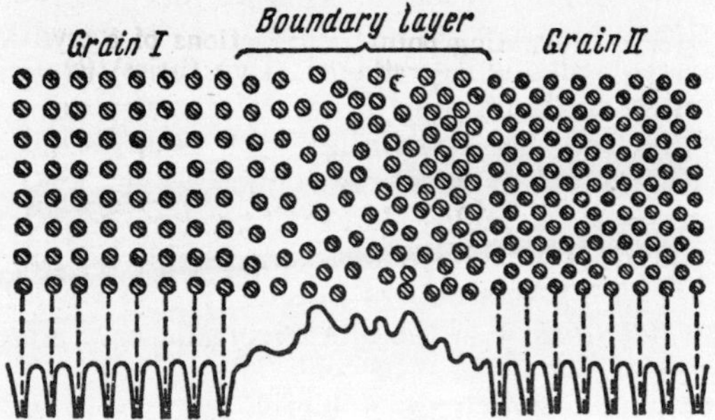

Fig. 10. Model illustrating atomic arrangement in the boundary layer between two crystals and the variation in the potential energy of the atoms (after K. P. Bunin)

This surplus energy, known as surface energy, is of prime importance in phase transformations of metals and alloys.

It is incorrect to consider each atom as being firmly bound to a definite point in the crystal lattice. Diffusion takes place in crystallic bodies, including self-diffusion, i.e., the rearrangement of atoms of a single material within its crystal lattice. The process of diffusion in a crystallic body is explained by the fact that certain atoms, whose kinetic energy substantially exceeds the average value due to the effect of thermal vibrations, may leave their regular positions at the points of the crystal lattice (Fig. 11, *a*) and jump into an interstitial position. An atom which has left its lattice site is said to be dislocated. This movement from a lattice point to an interstitial position will create a vacant lattice point (or vacancy) which may be occupied by any other atom (Fig. 11, *b*).

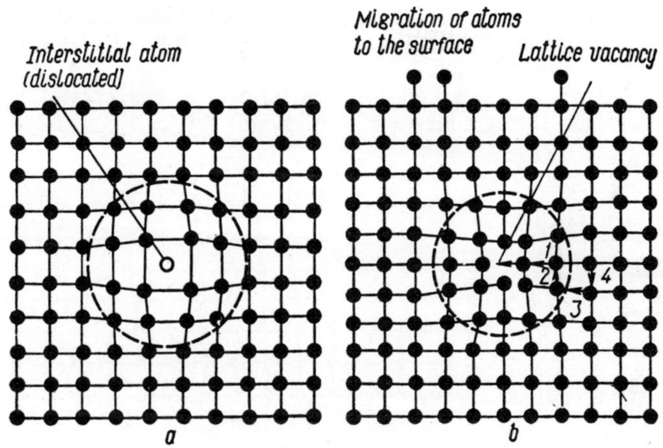

Fig. 11. Diagrams illustrating point imperfections of a crystal (*a* and *b*) and diffusive migration of atoms (holes) (*b*)

The higher the temperature, the more dislocated atoms and vacancies there will be. However, even at temperatures approaching the melting point their number will not exceed 2 per cent of the total amount.

Atoms surrounding a dislocated atom are naturally spread apart somewhat, since the diameter of an atom is always larger than the interstitial spaces. Atoms surrounding a vacant lattice point, on the other hand, are located somewhat closer.

Vacant lattice points in a crystal do not remain stationary. A vacancy may migrate in the process of thermal vibrations since, sooner or later, one of the atoms surrounding the vacancy will jump into it (for example *1* in Fig. 11,*b*). This frees another lattice point which will be taken up by *2*, etc. Thus, the vacancy continuously migrates in the crystal.

Movement of atoms to interstitial spaces, formation of vacancies and their migration, and distortion of the lattice around dislocated atoms and vacancies are reasons for departures from geometrically perfect structures of crystals.

Distortions of the crystal lattice may also occur as a result of the incorporation of foreign atoms of the impurities which, as a rule, are present in even the "purest" metal.

For example, monocrystals of germanium can now be obtained with a purity of 99.99999999 per cent (one atom of an impurity per ten billion atoms), but even in this case one gram-atom of germanium will contain 10^{12} foreign atoms. The lattice around the impurities is distorted (Fig. 11, *a*); its structure is impaired and, consequently, higher energy is peculiar to the atoms as a whole in the parts of the lattice surrounding foreign atoms.

So-called dislocations can also be formed in the crystal. These likewise impair the regularity of the crystal lattice.

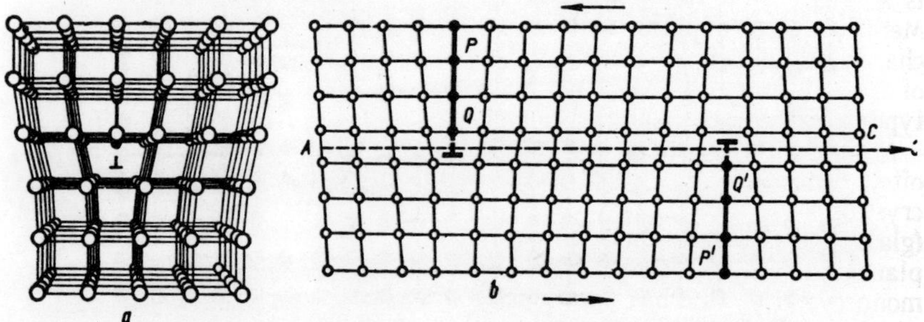

Fig. 12. Regions of positive (⊥) and negative (⊤) dislocations

Shown in Fig. 12 are so-called edge dislocations. An edge dislocation is a localised distortion of the crystal lattice due to the presence of an "extra" atomic plane (*PQ* or *P′Q′*), perpendicular to the plane of the drawing.

This extra plane (see Fig. 12, *b*) may be either above (*PQ*) or below (*P′Q′*) the slip plane (shown by the dash line *AC*). In the former case, the dislocation (*PQ*) is said to be positive and is denoted by the symbol ⊥; in the latter case, it (*P′Q′*) is said to be negative and is denoted by the symbol ⊤.

The edge of an extra plane is a line of the edge dislocation whose length may reach many thousands of interatomic distances. The crystal lattice is severely distorted in a region extending over several atomic distances from a dislocation. Distortion rapidly disappears beyond this region.

Dislocations have been observed directly by means of an electron microscope.

Dislocations are formed in the process of solidification of metals and mainly in their deformation.

The dislocation density * in a metal prior to deformation ranges from 10^4 to 10^6 per cm^{-2}; after deformation it may reach 10^{12} per cm^{-2}.

New dislocations occur in the process of plastic deformation. In this case, the increased number of dislocations is created by the so-called Frank-Read sources (or Frank-Read dislocation generators). A description of this mechanism may be found in special literature ** and is not within the scope of this book.

1-5. Properties of Metals

Certain typical properties characterise metals. These properties are associated with the internal structure of metal crystals.

The first scientific definition of the word "metal", formulated by the great Russian scientist, Mikhail Vasilyevich Lomonosov—"A metal is a bright solid that can be forged"—still holds true in our times. Metallic lustre and the ability for plastic deformation are, in fact, characteristic features of metals. The most important feature, however, of typical metals is their high electrical conductivity of the electron type, which decreases as the temperature is raised.

Many properties of metal crystals (physical, mechanical, and others) often depend on the direction along which they are measured. All crystals are anisotropic, as differing from isotropic amorphous solids (glass, plastics, and others which display the same properties in every plane and every direction). For example, the modulus of elasticity of a monocrystal of zinc has a value of 3,560 kg per sq mm in a direction parallel with its main axis and a value of 12,630 kg/sq mm in a perpendicular direction. In the same manner, the thermal coefficient of linear expansion $\left(\alpha=\frac{1}{l}\times\frac{dl}{dt}\right)$ varies from 63.9×10^{-6} to $14.1\times\times10^{-6}$ mm/mm-degree C.

Properties of a monocrystal are based on the different arrangements of the atoms in various directions. The arrangement of the atoms of an amorphous solid is more or less equally irregular in all directions and, therefore, the properties do not have a directional nature.

---

* The dislocation density is $\rho=\frac{\Sigma l}{V}$ cm^{-2}, where Σl is the total length of the dislocations in the crystal, in cm; V is the volume of the crystal, in cm^3.

** At the present time, the theory of dislocations is being extensively used to explain certain phenomena occurring in metals. In particular, it has been successfully used in the study of the processes of diffusion, plastic deformation, mechanical ageing (ageing after cold working), heat resistance, crystal growth, etc. A detailed description of the theory of dislocations may be found in the following special treatises: W. T. Read, Jr., *Dislocations in Crystals*, McGraw-Hill Book Company, Inc., New York, 1953; A. H. Cottrell, *Dislocations and Plastic Flow in Crystals*, Oxford University Press, New York, 1953; and A. J. Forty, "Direct Observations of Dislocations in Crystals", *Advances in Physics* (A quarterly supplement of the *Philosophical Magazine*, Jan. 3, 1954, No. 9).

Since actual or engineering metals are solids consisting of a great number of crystals (i.e., polycrystalline solids) with anisotropic properties, the question naturally arises as to whether the metal as a whole is anisotropic in respect to its properties.

In cases where the crystallites have the same orientation, the polycrystalline solid they compose will also be anisotropic. If the crystallites are differently oriented relative to each other, the summary properties of all the crystals will be approximately the same in all directions. This apparent isotropy will be retained only until the crystals are reoriented along a single direction by some technological process, for example, considerable cold working. After this, the metal will again be anisotropic.

Chapter 2

SOLIDIFICATION AND METAL INGOT STRUCTURE

2-1. Solidification of Metals

Transformation of a metal from the liquid to the solid state proceeds due to the conversion of matter to a more stable thermodynamic condition with less free energy F. If the transformation is accompanied by only a small change in volume, then *

$$F = H - TS$$

in which H is the total energy (enthalpy) of the system,
T is the absolute temperature, and
S is the entropy.

At a given temperature, the total energy content (H) of a system can be considered to be composed of two portions: the free energy F and the bound energy TS. Both free and bound energy may occur simultaneously in several different forms within the system, but again all of these various forms can ultimately be converted into heat. For this reason, both types of energy are usually evaluated in thermal units, their sum being referred to as the heat content of the system.

Free energy (F) is best described as usable energy in isothermal processes; in other words, under certain conditions it can be released by and removed from the system with no accompanying change of temperature in the system.

Bound energy is not usable in isothermal processes; that is, it cannot ordinarily be liberated from a system without changing the temperature of the system.

Bound energy, whose magnitude is expressed by the product TS, exists principally as sensible heat, i.e., as kinetic energy of the random motions of atoms within the system and as potential energy of their arrangements relative to each other. Fig. 13 schematically represents the variation in free energy of the liquid and solid states in relation to the temperature. At a temperature of T_s, the magnitudes of the free

* According to the Second Law of Thermodynamics, not a single process will proceed spontaneously without a decrease in the free energy of the system concerned.

energy for the two states are equal. This temperature is the equilibrium temperature of freezing (or melting) for the given material at which both phases (liquid and solid) may exist simultaneously. At temperatures above T_s, liquid metal will be more stable, having less free energy; at lower temperatures, solid metal will be more stable.

For the development of the solidification process (freezing), there must be a difference in free energy which is due to the lesser free energy of solid metal in comparison with the liquid state.

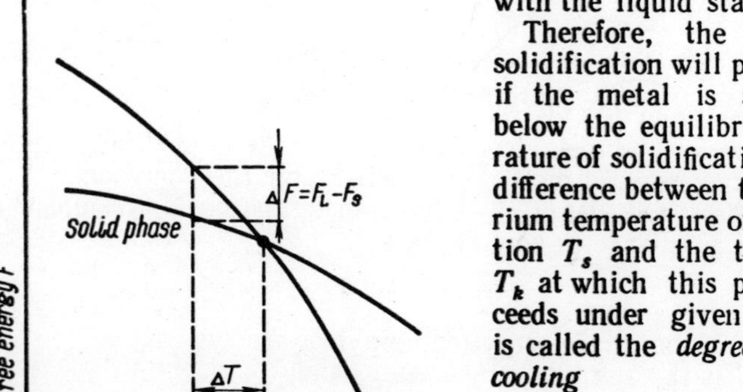

Fig. 13. Free energy vs temperature for the liquid and solid states

Therefore, the process of solidification will proceed only if the metal is supercooled below the equilibrium temperature of solidification T_s. The difference between the equilibrium temperature of solidification T_s and the temperature T_k at which this process proceeds under given conditions is called the *degree of supercooling*

$$\Delta T = T_s - T_k$$

The higher the rate of cooling, the larger the degree of supercooling will be, since T_k will be correspondingly lower.

Thermal curves, characterising the solidification process in pure metals and obtained at various cooling rates, are given in Fig. 14. At very low rates of cooling the degree of supercooling is small and the solidification process proceeds at a temperature near to the equilibrium temperature (Fig. 14, curve *1*). The horizontal section of the curves (representing a stop in temperature fall) is due to the evolution of the latent heat of fusion which compensates the heat dissipated to the surrounding atmosphere.

The degree of supercooling increases with an increase in the rate of cooling (see Fig. 14, curves *2*, *3*, *4* and *5*) and the solidification process proceeds at temperatures substantially below the equilibrium temperature of freezing.

The maximum degree of supercooling depends on the nature and purity of the metal but usually it does not exceed 30°C. The purer a liquid metal is, the higher its stability will be and the more it is subject to supercooling.

In studying the structure of steel ingots, the great Russian metallurgist Dmitri Konstantinovich Chernov first established that the

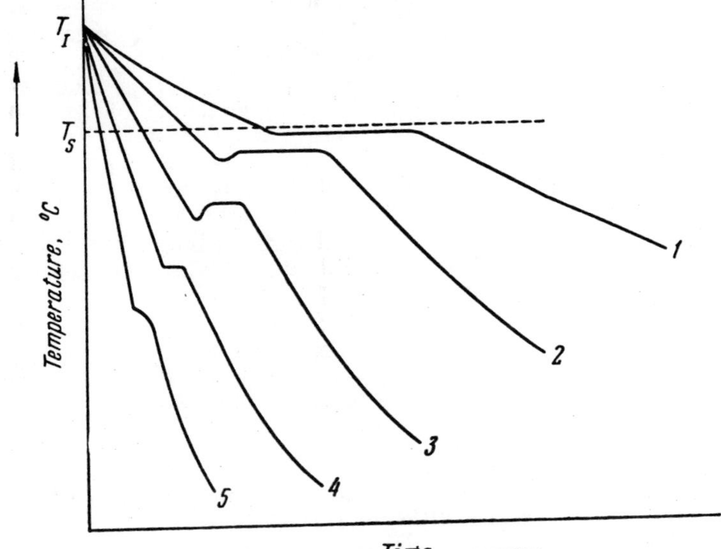

Fig. 14. Cooling curves of a pure metal

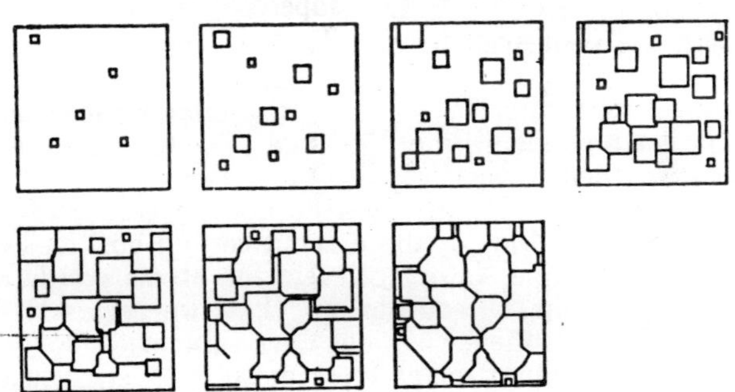

Fig. 15. Solidification of a metal (schematic) (after I. L. Mirkin)

process of solidification begins with the formation of nuclei (embryos) or centres of crystallisation and proceeds with their growth.

A schematic picture of the process of solidification is shown in Fig. 15. Upon supercooling of the liquid metal below T_s, separate crystals form, as shown in the diagram, and begin to grow. As long as the crystals grow freely, their geometry may be considered to be almost perfect, but as soon as the growing crystals mutually obstruct each other, their regular form is violated. At the places where they interfere with each other, further growth of the faces ceases; it continues only in those directions in which free liquid is still accessible. As a result of this process, the growing crystals, having a regular internal arrangement of atoms, take on an irregular shape after solidifying. They are called crystallites or grains (Fig. 16).

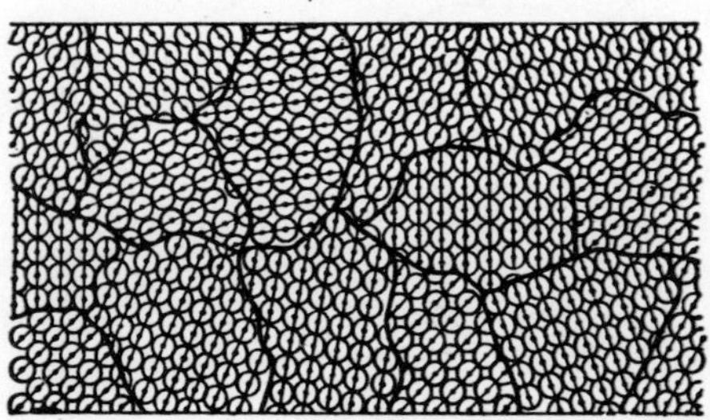

Fig. 16. Structure of a polycrystalline metal

Phenomena associated with the process of solidification are complex and varied. It is especially difficult to conceive the initial stages of the process when the first crystals or centres of crystallisation appear. Evidently, to explain the appearance of these centres it will be essential to understand the structure of the initial (matrix) phase,. i.e., the liquid metal.

It has been established at the present time that in a liquid metal, as distinct from the vaporous state where atoms are chaotically arranged, a certain "order" may be observed in atomic arrangement, notwithstanding the high mobility of the atoms, due to the forces of interaction between them.

The atoms (ions) in a liquid metal do not move about freely. As in a crystalline metal, they oscillate about certain positions of equilibrium. In a crystalline solid, however, the atoms oscillate about certain mean positions of equilibrium that are regularly located in space and maintain these positions over sufficiently long periods of time; while in a liquid metal these mean positions, themselves, are in continuous motion.

X-ray research has shown that at temperatures near to solidification, groups of atoms may come together in small volumes of the liquid metal; they have an atomic arrangement near to that in the solid metal. These groups or clusters may appear, exist for a certain time and then fall apart to appear again at some other place. Under definite conditions, however, such clusters may acquire higher stability and become centres of crystallisation or nuclei.

The formation of nuclei is facilitated by unequal distribution of energy between the atoms of a substance. At each given temperature, most of the atoms have an energy equal to a certain average value. In small volumes of substance, however, there are always a certain number of atoms possessing energy of a quantity either smaller or larger than the average value.

These random and tentative deviations in the energy of separate atoms or regions of atoms from the average value for the given temperature are called *fluctuations of energy*. Correspondingly, the temperature, which determines the thermal energy of the matter, will vary over a wide range in these regions.

As a result of the aforesaid, nuclei, capable of subsequent growth, will appear in certain small volumes of the liquid having a temperature lower than the mean value.

Numerous investigations have proved that nuclei of a great variety of sizes may appear in freezing. Not all nuclei, however, are capable of further growth. The formation of nuclei changes the free energy of the system (Fig. 17). The appearance of a nucleus and its growth as a crystal will reduce the amount of free energy per unit volume (Fig. 17, curve *1*) due to the lesser free energy of a unit volume of a solid in comparison to the liquid state. At the same time, the appearance of an interface between the nucleus and the liquid medium causes an increase in the free energy of the system (Fig. 17, curve *3*) since atoms on the interface possess higher potential energy. Therefore, a certain amount of energy is expended to form this interface between the phases. This additional energy is equal to the product of the nucleus surface area and the surface tension.

The change in total free energy due to the formation of the nucleus and the phase interface is represented by curve *2* in Fig. 17 *.

It is evident from curve *2* of Fig. 17 that the growth of a nucleus having a radius less than R_c is impossible since this is accompanied by an increase in free energy.

In nuclei with radii less than R_c, the ratio of the surface area to the volume is large. The increase in free energy, associated with the formation of the interface, is greater than the reduction in free energy due to the formation of a nucleus. This increases the free energy of the system.

If a nucleus appears with a radius exceeding R_c, it will be stable and capable of further growth since an increase in its size will reduce the free energy of the system (Fig. 17, curve *2*). The minimum size of

* The variation in free energy of a system ΔF, associated with the appearance of a solid phase nucleus, is expressed by the equation $\Delta F = -V(F_L - F_S) + S\sigma$, in which V is the volume of the nucleus, S is its surface area, σ is the coefficient of surface tension between the liquid and solid phases, and F_S and F_L are the free energies per unit volume of the solid and liquid phases, respectively.

nucleus (R_c), capable of growth at a given temperature, is called the *critical size* of the nucleus.*

Additional energy, however, equal to $^1/_3\ \sigma\ S_c$, is required for the formation of critical nuclei. This energy is provided by the system and is due to energy fluctuations.

Thermodynamic calculations show that the size of the critical nucleus depends on the degree of supercooling.

The higher the degree of supercooling, i.e., the lower the temperature at which freezing proceeds, the smaller the size of the critical nucleus will be (Fig. 18). The formation of nuclei in the liquid metal or nucleation, in accordance with the above-described mechanism, is called *spontaneous nucleation*. Nucleation is much facilitated if solid particles are present in the liquid metal (walls of the ingot mould, various inclusions, etc.) which serve as ready-made substrates for the formation of centres of crystallisation.

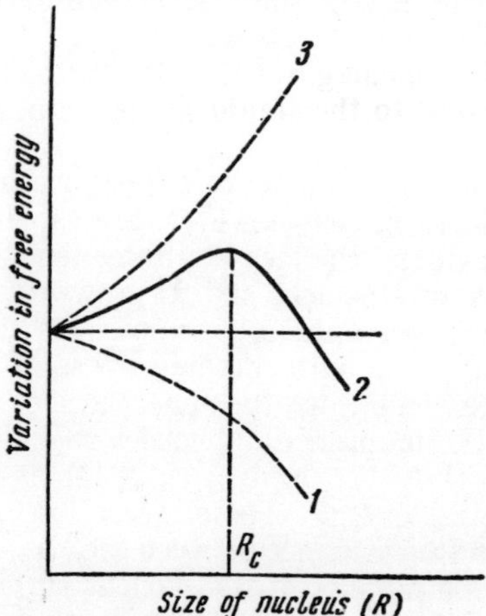

Fig. 17. Variation of the free energy of a metal upon the formation of a nucleus

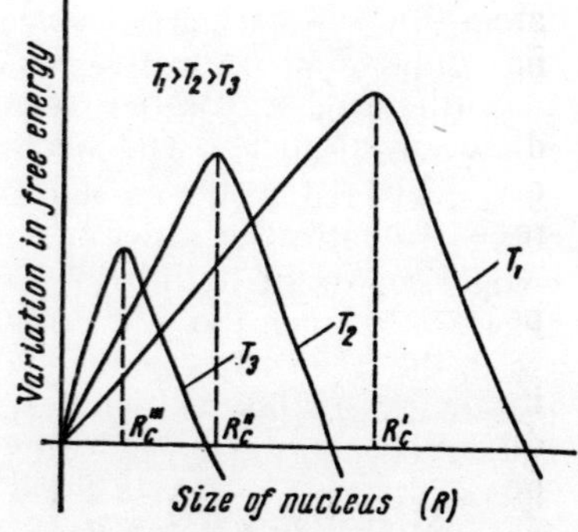

Fig. 18. Variation in the free energy of a metal versus nucleus size for various degrees of supercooling

This is explained by the fact that the work required to form a nucleus on an existing surface is substantially less than that required for spontaneous nucleation in the liquid metal. It must be noted, however, that foreign inclusions will facilitate nucleation only if the surface tension between the nuclei and the inclusion is less than that at the interface between the liquid phase and the solid crystals. It has been established that the nearer the crystallographic structure of the nuclei being formed is to that of the solid particles in the liquid metal, the less the surface tension at their interfaces.

---

* A critical nucleus (with a radius R_c) may either grow or melt again.

The formation of centres of crystallisation on such ready-made substrates complies with the principle of structural and size conformity. The nucleus forming on a ready-made substrate is oriented so that the atoms on its external face are arranged, in their configuration, in the same manner as those of the contacting face. The parameters of the lattices concerned should differ but slightly.

The growth of a formed nucleus, usually called a *three-dimensional nucleus*, proceeds by building up new single-atom layers on its surface. These are called two-dimensional nuclei and they must be of a size not less than a definite critical value. The addition of single atoms or groups of atoms, less than critical in size, to the growing nucleus is accompanied by an increase in the free energy of the system and the added atoms will be transformed back to the liquid phase.

The formation of a two-dimensional step (i.e., layer of single-atom thickness) takes place in the same way as that of three-dimensional nuclei.

With increased degrees of supercooling, the critical size of a two-dimensional nucleus and, consequently, the fluctuations of energy required for its formation, are reduced.

The rate of solidification depends on:

1) the rate of nucleation, i.e., upon the number of three-dimensional nuclei appearing per unit time in a unit volume $\left(\frac{1}{mm^3 sec}\right)$ and 2) the linear rate of crystal growth, i.e., upon the rate of increase in the linear dimensions of crystals per unit time (mm per sec).

The linear rate of crystal growth (*RG*) and the rate of nucleation (*RN*) both depend on the degree of supercooling (Fig. 19).

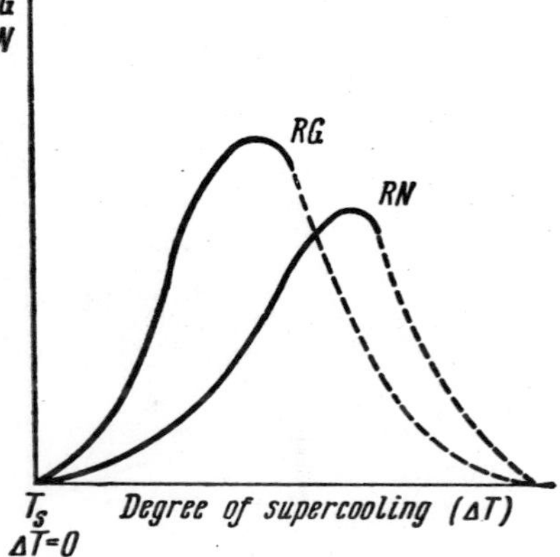

Fig. 19. Effect of the degree of supercooling on the rates of nucleation (*RN*) and crystal growth (*RG*)

It is evident from Fig. 19 that at the equilibrium temperature (T_s) both *RG* and *RN* equal zero and, therefore, no freezing takes place. Upon an increase in the degree of supercooling both *RG* and *RN* increase and reach a maximum at a definite rate of supercooling and then drop again. The maximum for *RG* is always reached at less supercooling than that for *RN*.

Increased rate of solidification, due to an increased degree of supercooling, may be explained by the reduction in the critical size of the nucleus and, consequently, by less work required for its formation *. At high degrees of supercooling, however, the mobility of the atoms is reduced by the increase in viscosity of the liquid phase. This impedes nucleation and reduces RG and RN (see Fig. 19).

Only the rising sections of RG and RN curves (solid lines in Fig. 19) have a practical significance for metals. Metals solidify sooner than degrees of supercooling are obtained which cause reduction of RG and RN.

The larger the number of nuclei and the slower their growth, the smaller the crystals will be that grow out of each nucleus (i.e., the smaller the metal grains).

The number of grains (n) (and, consequently, their size) may be computed from the following equation as a function of RG and RN:

$$n = k\sqrt{\frac{RN}{RG}}$$

Since RN increases faster than RG for higher degrees of supercooling (Fig. 19), the higher the degree of supercooling, the smaller the grains will be.

As a very high degree of supercooling cannot be achieved in liquid metal, smaller grain size is obtained in actual practice, at the present time, not by increasing the rate of cooling, but by introducing dispersion particles (inoculation) into the metal. These particles either facilitate nucleation or create conditions for slow crystal growth. This principle is the basis for the widely used industrial method of reducing grain size, known as modification.

2-2. Metal Ingot Structure

The crystals which form in the process of solidification of a metal may have many different structures (dendritic, lamellar, needle-type or acicular, etc.) depending on the rate of cooling, and the type and amount of admixtures or impurities in the melt.

Perfect crystals of proper external shape can be obtained only if crystallisation develops under conditions when the degree of supercooling is very slight and the metal has a very high purity.

In the great majority of cases, branched or tree-like crystals are obtained, which are called dendrites (Fig. 20).

The nucleus develops to form a dendritic crystal chiefly along the directions of maximum linear rate of growth. As a result, the long branches are formed first. They are the so-called axes of the first order (primary axes k of the dendrite) and they branch out in various directions from the initial nucleus.

---

* The work A_n required to form a nucleus is equal to one third of the sum of the products of the area S of each face on the surface of the nucleus and the surface tension σ on that face.

While the primary dendrites grow in length, branches of the second order (*m*) evolve from their edges and grow in a perpendicular direction. Axes of the third order (*n*) evolve and grow on the second-order axes and so forth. A stage is eventually reached at which the spaces between the branches are filled with solidifying metal (Fig. 20). Upon further freezing of the metal and development of the dendritic crystal all of the liquid metal in the spaces is solidified.

After the complete solidification of the metal, dendritic crystals can sometimes be observed directly on the surface of an ingot, in a characteristic pattern, or on the surface of a pipe or shrinkage cavity, i.e., in regions with an insufficient supply of liquid metal. More often, dendritic structure is revealed only after special etching of microsections (Fig. 21). Etching reveals not only the structure of the crystallites but the boundaries between them as well.

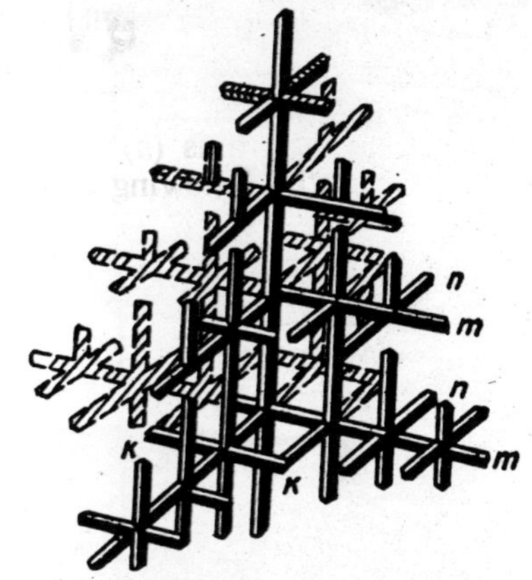

Fig. 20. Dendritic growth (after D. K. Chernov)

The structure of a steel ingot is shown schematically in Fig. 22. It is evident from Figs. 21 and 22, that three structural zones may be distinguished in the cast metal. Solidification of the liquid metal begins at the surface of the mould and proceeds, at first, chiefly in the thin layer of highly supercooled liquid adjacent to this surface. This leads to the formation of a zone of fine crystallites (due to the high degree of supercooling). A second zone of elongated columnar crystallites (zone of transcrystallisation) follows immediately the first zone. These crystallites grow in the direction of heat removal, i.e., normal to the mould walls.

The subsequent growth of the columnar crystallites from the mould walls proceeds by advancement of first-order branches into the molten metal and the evolution of higher-order branches as described previously (Fig. 20).

In cases when the metal is highly overheated and then rapidly cooled, the zone of columnar crystallites may extend through the whole volume of the ingot (Fig. 23).

Slow cooling of metal which is not overheated (as in castings of heavy cross-section) will create conditions favourable for crystal nucleation in the central part of the ingot. This will form a third structural zone in the central part of the ingot consisting of equiaxed randomly-oriented crystallites (Figs. 21 and 22) of a size depending

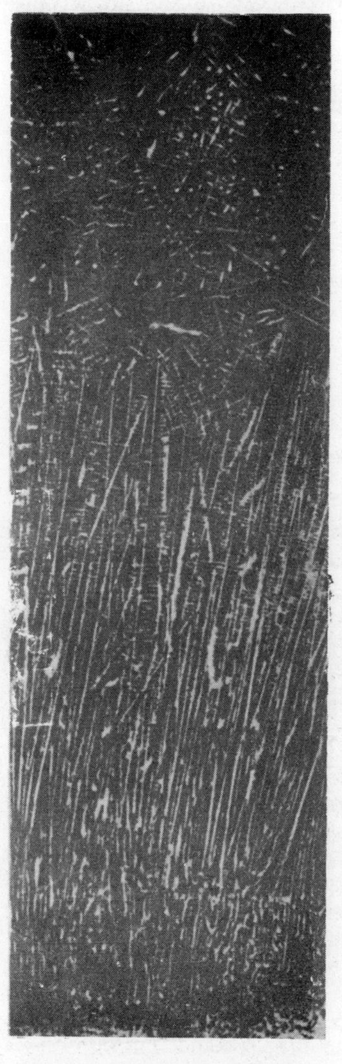

Fig. 21. Macrostructure of cast steel containing 0.8 per cent C

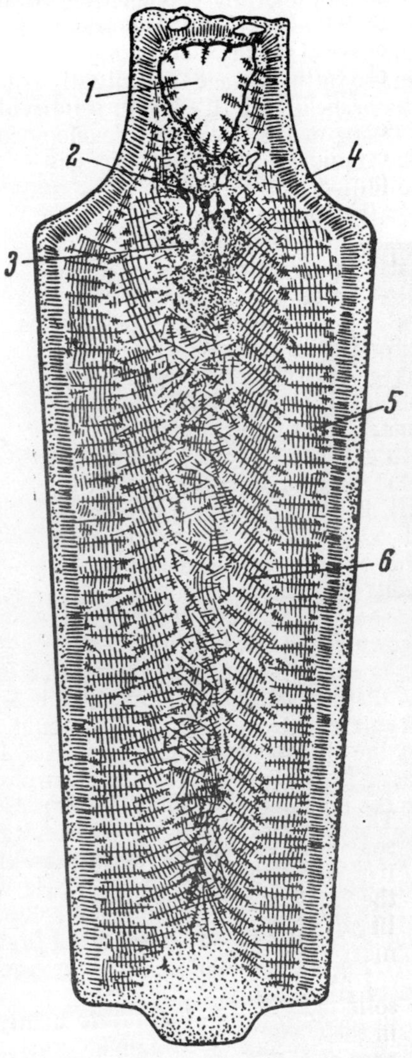

Fig. 22. Structure of a steel ingot (after N. A. Minkevich):

1—pipe, 2—shrinkage cavities, 3—shrinkage porosity, 4—thin layer of fine crystallites, 5 — zone of columnar crystallites, 6—zone of large unoriented crystallites on the degree of overheating of the metal, rate of cooling, presence of admixtures, etc. The degree to which the various structural zones develop depends on the size of the ingot, degree of overheating, rate of cooling, and other factors.

The zone of columnar crystallites possesses high density (few gas cavities, blowholes, etc.). The metal at the junctions of the columnar crystallites has a low strength, however, and in subsequent working (forging, rolling, etc.) fractures may occur at these places in the ingot.

Fig. 23. Microstructure in the cross-section of a chromium-nickel stainless steel ingot (after V. M. Doronin)

The development of columnar crystallites is undesirable, for this reason, in metals of low ductility, including steel. On the other hand, it is expedient to extend the zone of columnar crystallites throughout the ingot for plastic metals, such as copper and its alloys. High ductility of an alloy excludes the fracture of the ingot in subsequent mechanical working.

Due to the fact that liquid metal has a higher unit volume than solid metal and that solidification does not take place simultaneously in the whole volume, a cavity, called a pipe *, is formed in the upper part of the ingot which freezes last. The pipe is usually surrounded by the most contaminated portion of the metal. After solidification,

* In rimming steel, the increase in volume due to the formation of gas cavities compensates for the difference in the unit volumes of the liquid and solid metal and, therefore, no shrinkage cavities are produced.

micro- and macroporosity and blowholes are characteristic of this zone which is called the zone of shrinkage friability in the ingot.

Gases evolve from the metal in the process of freezing since their solubility in the liquid metal is considerably higher than in the solid one.

If cooling is sufficiently slow, surplus gases have time to leave the metal. They do not evolve at more rapid cooling and remain in the ingot (or casting) forming so-called gas cavities or holes which may substantially decrease the mechanical properties of the casting or finished product.

Chapter 3

PLASTIC DEFORMATION AND RECRYSTALLISATION IN METALS

3-1. Plastic Deformation

Deformation is the change in dimensions or form of matter under the action of applied forces. Deformation is caused either by the mechanical action of external forces or by various physical and physico-chemical processes (for example, change in volume of separate crystallites in phase transformations or as a result of a temperature gradient).

The process of deformation comprises the following consecutive stages: 1) elastic deformation, 2) plastic deformation, and 3) fracture.

Elastic deformation is often defined as deformation which completely disappears as soon as the action of external forces ceases. Elastic deformation does not cause any noticeable changes in the structure of metals. The application of a load elastically displaces atoms to a slight extent, relative to each other, or twists blocks of the crystal. Tensile loads applied to a monocrystal tend to increase the interatomic spacing while compressive loads tend to reduce it. Due to the forces of repulsion or attraction, the atoms return to their equilibrium positions as soon as the load is removed so that the initial form and size of the crystals are restored.

It should be noted, however, that the elastic deformation of polycrystalline metals, especially at higher temperatures, causes so-called viscous flow and after the removal of the external forces, deformation does not disappear completely even in the case of relatively small loads.

In viscous flow, the grains composing the metal are not deformed; only certain crystallites are displaced in reference to others. At high temperatures and low stresses, viscous flow becomes much more important, but permanent changes in the dimensions and shape of a metal specimen occur only after long periods of stress application. This phenomenon is called *creep*.

At stresses exceeding the elastic limit (the maximum stress that can be applied without producing a measurable permanent deformation or set after removing the stress) *plastic deformation* is observed. Plastic deformation is associated with displacements of the atoms within the grains and causes permanent changes in shape of the specimen. Plastic deformation in crystals occurs in two ways: by slip or by twinning.

Slip is due to the action of shear stresses; block-like sections of crystals are laterally displaced relative to each other on definite crystallographic planes which are typical for each type of crystal lattice. The critical resistance to shear is less along these planes than the acting shear stresses. Numerous studies of the slip process have shown that it occurs firstly along planes and directions with the most closely packed atomic arrangement. In metals having a body-centred cubic lattice (α-Fe, Cr, W, Mo, etc.) the possible slip planes are the diagonal planes (101), in a face-centred cubic lattice (Cu, Al, Ni, Au, etc.) they will be the octahedral planes (111), while in a hexagonal system (Mg, Zn, etc.), they will be the basal planes (0001).

Slip planes and directions for different crystallic systems are shown in Fig. 24. The more possible slip planes and directions a metal has, the more plastic it is.

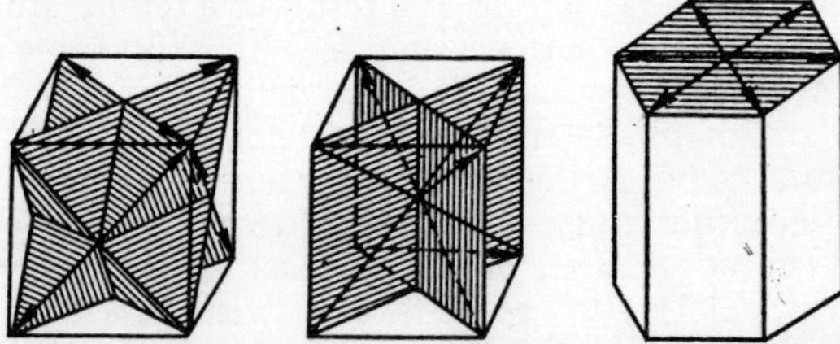

Fig. 24. Crystallographic possible slip planes (shaded) and slip directions (arrows)

Slip planes and directions are subject to changes in accordance with the temperature, degree of preliminary deformation, and alloy composition. For example, in aluminium, subjected to deformation at low temperatures, slip will occur along planes (111); at high temperatures—along planes (100).

Fig. 25 schematically illustrates the slip process in a monocrystal under a shear load. As a result of plastic deformation, the monocrystal is divided into layers or slip blocks which are displaced in reference to each other and are separated by thin layers in which a considerable displacement of atoms has taken place. These intermediate layers with strongly distorted lattices are called *slip planes.*

The slip process is not to be understood as a simultaneous displacement of the parts of a crystal. Such slip would require stresses considerably higher than those actually present in the process of plastic deformation.

Slip propagates gradually along the cross-section of a crystal and, at each moment, only the atoms of a comparatively small portion of the cross-section are being displaced.

According to the latest views on plastic deformation, the sources of slip are groups of atoms, displaced relative to each other, i.e., dislocations. Slip originates at a definite point, where a dislocation exists, and proceeds by movement of the dislocation in the given plane of the crystal due to shear stresses.

The schematic diagram of slip deformation, given in Fig. 26, shows the changes in the arrangement of the atoms during the progress of dislocation movement. At first only the upper part of the crystallite

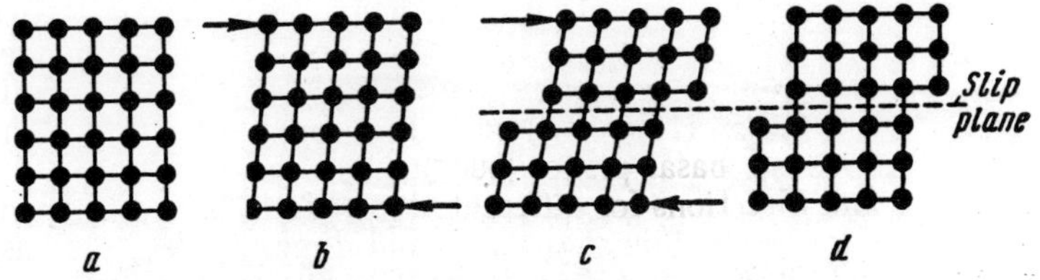

Fig. 25. Elastic and permanent deformation of a metal under the shear stresses: *a*—original crystal unstressed, *b*—elastic strain produced by shearing force below the elastic limit, *c*—increased elastic strain plus permanent strain by slip, resulting from load above the elastic limit, *d*—shearing load removed, only permanent strain remains

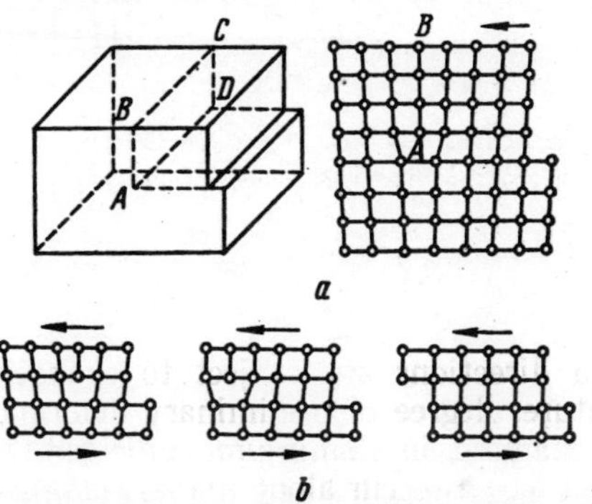

Fig. 26. Progress of a dislocation across a crystal

is displaced to the left relative to the lower part. Plane *ABCD*, in this case, will be an extra plane of atoms in the upper part of the crystallite. The edge *AD* of this plane, as mentioned earlier, is called an edge dislocation (Fig. 26, *a*).

As is evident from Fig. 26, the edge dislocation is perpendicular to the direction of slip and should be regarded as the frontal boundary of that part of the slip plane along which local displacement has occurred.

Fig. 26, *b* shows the changes that take place in the arrangement of the atoms during the movement of the dislocation. Progress of the dislocation through the whole crystallite to the opposite crystal surface causes a slip of one interatomic distance between the parts (Fig. 26, *b*).

Dislocations may be either: 1) positive (⊥)—in which the dislocation originates as a result of shear stress and the extra plane is in the upper part of the crystallite, or 2) negative (⊤)—where the extra plane is in the lower part.

In addition to edge dislocations, there are also screw dislocations which occur when the slip plane covers only a part and not the whole of the crystallite cross-section. A screw dislocation differs from the

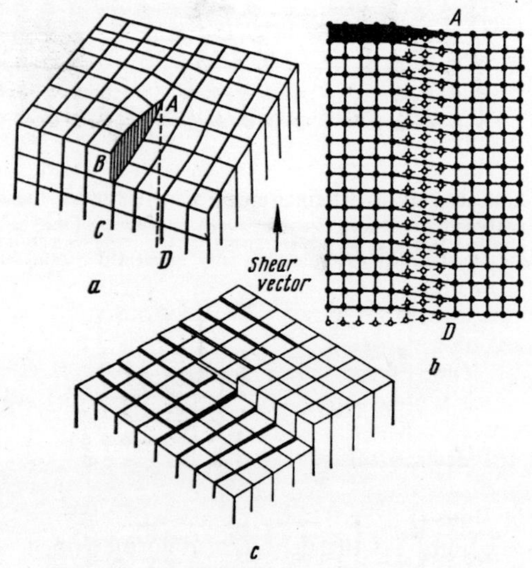

Fig. 27. Formation of a screw dislocation

edge type in that it is parallel and not perpendicular to the shear vector. Fig. 27 schematically illustrates the formation of a screw dislocation (after W. T. Read). Let us assume that a part of the crystallite is displaced in the direction indicated by the arrow (Fig. 27). Dislocation *AD* appears as a result of incomplete slip. It is a screw dislocation, in this case, as is quite evident from Fig. 27, *b* where the arrangement of the atoms around the dislocation is shown *. Upon the formation of a screw dislocation, the crystal should be regarded as consisting of a single atom plane wound into a helical surface and not of parallel atom planes.

A combination of edge and screw dislocations will give a composite dislocation. The mechanism for dislocation movement in the crystal, described above, explains why the process of deformation may proceed at comparatively low stresses at which simultaneous movement of all the atoms in a given slip plane would be impossible.

* The plane of the illustration is parallel to the slip plane.

Plastic deformation in such metals as gold, silver, nickel, cadmium, and certain others may also occur by twinning. Twinning is a plastic deformation which results in a change in orientation of one part of a crystal to a position symmetrical to the other part in reference to a separating plane called the twinning plane. As in slip, only very small parts of the crystal participate simultaneously in the twinning process.

Plastic deformation of a polycrystalline metal proceeds similarly to deformation of a monocrystal. It must be noted, however, that in this case intercrystalline deformation is imposed on the process of plastic deformation inside the individual crystals.

Intercrystalline deformation in polycrystalline metals is characterised by rotation or slip of the crystallites (grains).

Plastic deformation of a polycrystalline metal is never uniform. This is due to the fact that the separate crystals deform in various degrees because of their dissimilar orientation. This random orientation of the grains affects the slip process in the metal as a whole. Upon the application of an external load, those grains deform first which have possible slip planes most favourably positioned in relation to the applied force. At the same time, the deformation of these crystals is impeded by their neighbouring crystals in which the possible slip planes are otherwise positioned. Therefore, the crystallites will be in a complexly stressed state which hinders the development of slip throughout the whole volume. Considerable structural changes result from deformation in a polycrystalline metal.

The first sign of permanent deformation in a metal is the appearance of slip bands or lines (Fig. 37). Slip bands are observed upon even the slightest deformations. As the degree of deformation is increased, the slip blocks in each separate crystallite turn to conform with the direction in which the force acts and the metal grains are elongated in the direction of deformation to form a so-called fibrous or banded structure (Fig. 28). Preferred crystallographic orientation of the grains, revealed by X-ray analysis, arises after considerable deformation. This regular crystallographic orientation of the crystals in relation to the external forces of deformation is called *texture*.

The formation of a texture is accompanied by anisotropy of mechanical and physical properties. This is undesirable in most cases but is sometimes used to obtain optimum properties in a definite direction (see page 388).

Plastic deformation strongly affects all physico-chemical properties of a metal.

An increase in the degree of deformation *, as is evident from Fig. 29, increases those properties which characterise the resistance

* The degree of deformation is determined by the formula: $f=\frac{F_0-F_1}{F_0}100$ per cent, in which F_0 is the cross-section area of the test piece before deformation and F_1 is the same, but after deformation.

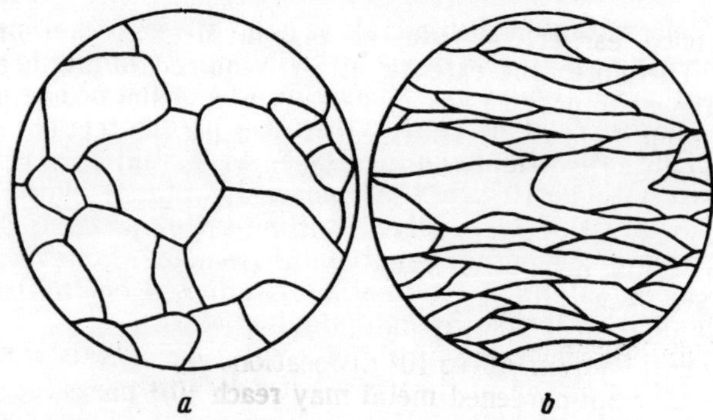

Fig. 28. Effect of plastic deformation on the microstructure of a metal, × 300 (after S. S. Steinberg):
a—structure before deformation, *b*—structure after deformation

of the metal to plastic deformation (ultimate strength σ_u, and hardness R_B). The capability for plastic deformation (relative elongation δ, per cent, and relative reduction of area ψ, per cent), on the other hand, is reduced. This phenomenon is called *strain* or *work hardening*.

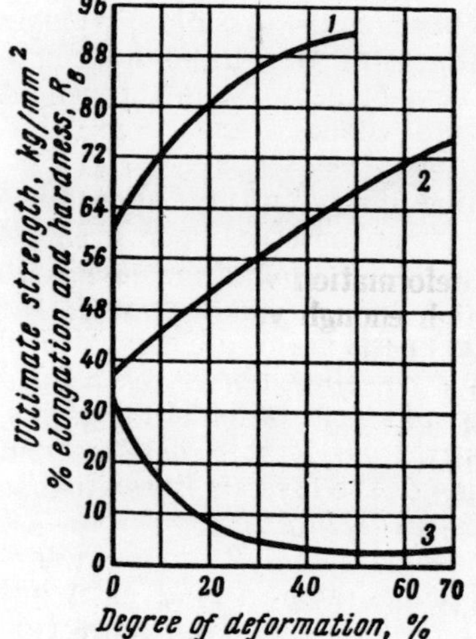

Fig. 29. Effect of cold working on the mechanical properties of low-carbon steel:
1—hardness, *2*—tensile strength, *3*—elongation

Cold plastic deformation decreases the specific weight and the electrochemical potential but increases the chemical reactivity of a metal. Deformed metal possesses less corrosion resistance than undeformed metal. Cold deformation of iron substantially increases its coercive force and decreases its magnetic permeability. The residual induction is decreased with small degrees of deformation and increased with high degrees.

Changes in the properties of metals subject to cold plastic deformation (strain hardening) are associated with fragmentation of the mosaic blocks and an increase in the angles of their disorientation, the appearance of considerable deformation stresses, and an increase in the number of dislocations.

The smaller the blocks, the higher the strength of the metal will be. As a result of a high degree of deformation (about 80 per cent), the block will have a size from 200 to 400 Å.

As mentioned earlier, a field of elastic stresses surrounds a dislocation. Therefore, the external stress, required to produce slip, must be sufficient to provide for movement of a dislocation through the counteracting field of stresses created by other dislocations or by the stress fields at the boundaries of blocks, grains, particles of other phases, etc.

The dislocation density increases with the increase in the degree of deformation. This naturally leads to interference in dislocation movement and, consequently, increases the strength of the metal. Before plastic deformation (in the as-annealed condition), the dislocation density will be from 10^6 to 10^8 dislocations per cubic centimetre. The density in strain-hardened metal may reach 10^{12} per cubic centimetre *.

It may be noted here that a strength approaching the theoretical value may be obtained if no dislocations are present in the metal. In this case plastic deformation may be produced only by simultaneous displacement, relative to each other, of all the atoms located on both sides of the slip plane.

This condition has been proved experimentally by means of filament-like crystals called "whiskers". The ultimate strength of a whisker of iron, one micron (0.001 mm) in diameter and 2.5 mm long, reached 1,300 kg per sq mm and its elastic elongation was up to 5 per cent. Unfortunately, only very small crystals of metal, free of dislocations, can be obtained at the present time. However, extensive research in this direction leads us to expect the production of whiskers of considerable size in the near future.

As mentioned above, any process of deformation will end in fracture if the stresses are increased to a high enough value. Two types of fracture are distinguished: brittle and ductile.

Brittle fracture consists in destroying the interatomic bonds by normal stresses. Brittle fracture is not accompanied by noticeable plastic deformation. The resistance to brittle fracture is called cohesive strength (S_T). Brittle fracture is comparatively rare. For example, this type of fracture may be observed in zinc and its alloys or in iron and low-alloy steel at low temperatures, as well as when brittle layers segregate on the grain boundaries. It must be stressed, however, that pure brittle fracture is practically never encountered. Fracture is always preceded by plastic deformation.

Ductile fracture is due to tangential stresses which have reached a definite fracture value called the shear strength (τ_k). This type of fracture is preceded by considerable plastic deformation.

Proceeding from the fact that S_T and τ_k are constant values for a given material, Y. B. Friedman considers that brittle fracture, due

* For a more detailed explanation see Morton C. Smith, ***Principles of Physical Metallurgy***, Harper and Brothers, New York, 1956.

to normal stresses, will occur in cases where the cohesive strength is lower than the shear strength. If the tangential stresses produce considerable plastic deformation before the normal stresses reach the cohesive strength, the fracture will be of the ductile type. Thus, fracture due to normal stresses by separation or cleavage may be either brittle or ductile.

Fracture due to tangential stresses, resulting from plastic deformation, will, of course, always be ductile. Most frequently, metal fracture does not occur as pure separation or shear but as a complex combination of these two types of fracture.

3-2. Recovery and Recrystallisation

Plastic deformation, which distorts the crystal lattice and breaks up the blocks of initial equiaxed grains to produce a fibrous structure or thin plates, increases the free energy level of a metal.

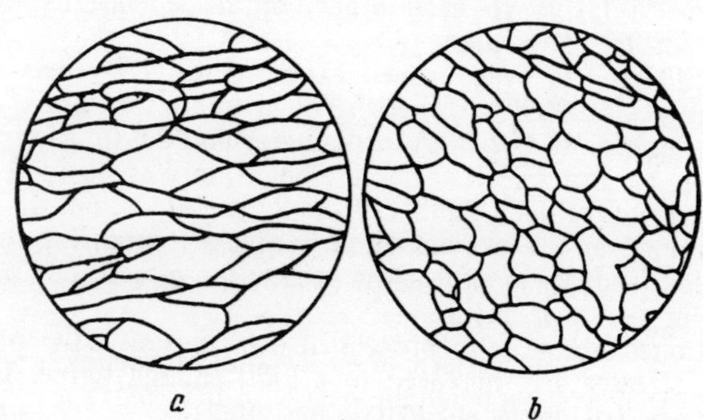

Fig. 30. Effect of recrystallisation on the microstructure of a metal (schematic), × 300 (after S. S. Steinberg):
a—after deformation, *b*—after recrystallisation

Deformed metal, in comparison with its undeformed state, is in a nonequilibrium, thermodynamically unstable state. Therefore, spontaneous processes occur in strain-hardened metal, even at room temperatures, which bring it into a more stable condition. The speed of these processes may be increased by raising the temperature.

If strain-hardened metal is heated to comparatively low temperatures, the elastic distortions of the crystal lattice are reduced due to the increase in amplitude of the thermal oscillation of the atoms. This heating will slightly lower the strength of the strain-hardened metal but the elastic limit and ductility will increase, though they will not reach the values possessed by the initial material (before strain hardening). No changes in microstructure are observed in this period. This partial restoration of the original properties, produced by reducing the distortions of the crystal lattice without noticeable changes in microstructure, is called *recovery*.

A further rise in temperature increases the mobility of the atoms. When a certain temperature (a definite value for each metal) is reached, equiaxed grains are produced (Fig. 30) and the metal is completely softened (Fig. 31).

The formation of new equiaxed grains in the heating process, instead of the oriented fibrous structure of the deformed metal, is called *recrystallisation*. The temperature required for the beginning of

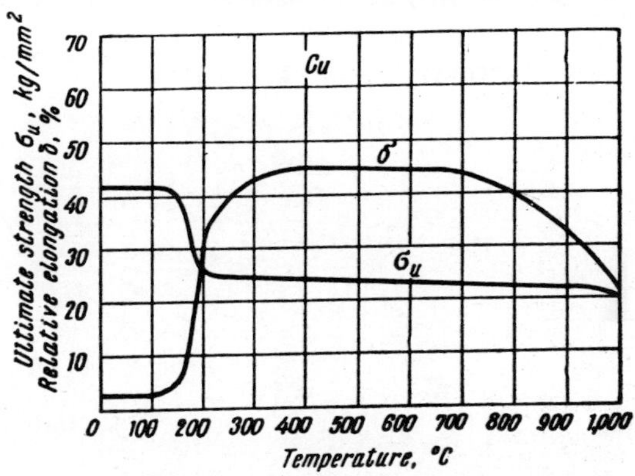

Fig. 31. Effect of temperature on the mechanical properties of strain-hardened copper

recrystallisation is characteristic of each metal but depends on a number of factors and firstly, upon the degree of deformation. The higher the degree of deformation, the lower the recrystallisation temperature will be.

According to the researches of A. A. Bochvar, the minimum recrystallisation temperature of commercially pure metals, subject to high degrees of deformation, will be

$$T_r = 0.4T_m$$

in which T_r is the recrystallisation temperature and
T_m is the melting point in the absolute temperature scale.

When the minimum recrystallisation temperature (recrystallisation threshold) is reached, a sharp drop in the resistance to deformation (σ_u) is observed, in conjunction with an increase in ductility of the metal (δ, per cent) (Fig. 31). The recrystallisation process usually proceeds in a certain temperature interval. First more favourable portions of the metal are recrystallised and then the process extends to the full volume. The presence of impurities impedes recrystallisation. The higher the temperature and degree of deformation, the more rapidly recrystallisation will proceed. Recrystallisation, in essence, consists in having the atoms of the deformed metal overcome the bonds of the distorted lattice, the formation of nuclei of equiaxed grains, and subsequent growth of these grains due to transfer of atoms from deformed to undeformed crystallites.

The nuclei of new recrystallised grains are most probably formed at places where highest distortions of the lattice are concentrated.

As a result of recrystallisation, the metal will consist of new equiaxed grains which are in a thermodynamically more stable condition. The size of the recrystallised grains depends chiefly on the degree of

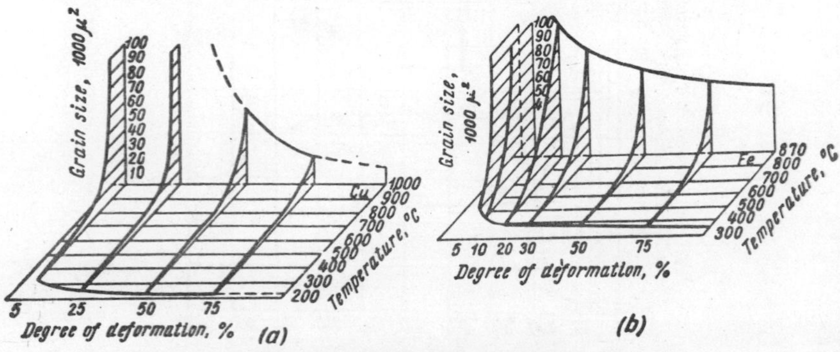

Fig. 32. Recrystallisation diagrams:
a—copper, *b*—iron

previous deformation of the metal and the temperature to which it is heated.

Relations between the size of recrystallised grains, degree of deformation, and the temperature are usually presented in the form of recrystallisation diagrams (Fig. 32). The higher the degree of previous deformation and the lower the recrystallisation temperature, the smaller the grains will be.

At low degrees of deformation (3 to 15 per cent), the grain size after recrystallisation may become very large (Fig. 32). This is called the *critical degree of deformation*.

The formation of very large grains in this process sharply reduces the ductility of the metal. Therefore, the critical degree of deformation should be avoided.

In the process of recrystallisation of such metals as iron, copper, and aluminium, the new crystals sometimes grow in an oriented arrangement and the so-called *recrystallisation texture* is obtained. This usually differs from the deformation texture but, in some cases, they may coincide (as in the recrystallisation of very pure aluminium). Recrystallisation texture leads to anisotropy of the properties. This lowers the fabrication properties of metals and, in particular, their formability.

Fig. 33. Macrostructure of deformed metal

When a previously worked metal has completely recrystallised, in the process of subsequent heating, certain recrystallised grains will grow at the expense of others due to migration of the grain boundaries. This process of growth of new recrystallised grains is called *secondary recrystallisation*.

Secondary recrystallisation is due to the tendency of the system to reduce its free energy.

As mentioned above, the energy of atoms inside the grain is less than that of atoms on its surfaces. It is natural then that grain growth, which decreases the total surface area of the grains, will also reduce the number of atoms with higher energy and generally reduce the energy in the substance. Heat treatment, consisting in heating metal above the recrystallisation temperature to soften the strain-hardened metal, is called *recrystallisation annealing*.

Deformation may be either cold or hot in accordance with its temperature as compared to the recrystallisation temperature. Cold deformation or cold working of a metal is a process accomplished at a temperature below that required for recrystallisation; cold working is accompanied by strain hardening.

Hot deformation or hot working, on the other hand, takes place at a temperature above the recrystallisation temperature. In this case, hardening due to plastic deformation is completely eliminated by recovery and recrystallisation. This is true, however, only if the rate of recrystallisation is higher than the rate of deformation.

An ingot is usually subject to hot plastic deformation. In the process of hot working, the dendritic structure of the ingot is destroyed and dendrites are elongated in the direction of deformation to form a fibrous structure (Fig. 33).

The mechanical properties of a metal are anisotropic after hot plastic deformation. They are higher as a rule along the fibres than across them.

Chapter 4

METHODS FOR STUDYING METAL STRUCTURE *

4-1. Macrostructure (Macrography)

Macrostructure is the structure of metals and alloys as seen by the naked eye or by low-power magnification.

Macrography reveals the size, form, and arrangement of crystallites (dendrites) in cast metals (Fig. 21), fibres in deformed metal (Fig. 33), shrinkage porosity and gas cavities (Table 1), cracks appearing during certain fabrication processes, chemical nonhomogeneity in the distribution of certain constituents appearing in alloys upon their solidification from the liquid state (Fig. 34), etc.

Macrostructure may be studied either directly on the surface of the work (a casting, for example) or on a fracture, or, more frequently, on specimens or samples cut out of large billets (ingots, forgings, etc.) or articles. The surface of the specimen is ground and then etched by special reagents. The action of etching reagents, or etchants, as they are called, is based on their ability to colour and dissolve various constituents in a different manner and to widen microcavities, cracks and similar defects.

To reveal the dendritic structure of cast steel, for example, a 10 or 20 per cent aqueous solution of ammonium persulfate $[(NH_4)_2S_2O_8]$ is employed. Keshian reagent (25 parts H_2O+10 parts H_2SO_4+ 65 parts HCl) is used to reveal fibres in deformed metal, as well as cracks, cavities, porosity, and chemical nonhomogeneity.

The Baumann method is extensively used to reveal the degree of nonhomogeneity in the distribution of sulphur (or phosphorus) in steel. This method consists in applying ordinary silver bromide photographic paper, soaked in a 5 per cent solution of sulphuric acid, to the surface of the sample. At the sulphur-rich areas, the paper will darken due to the formation of silver sulphide Ag_2S.

---

* Only a brief description of the methods used to reveal the structure of metals and alloys and the transformations which occur in them is within the scope of this book. A more detailed description may be found in special literature, such as: G. I. Pogodin-Alexeyev, Y. A. Geller, and A. G. Rakhshtadt, *Engineering Physical Metallurgy*, Oborongiz, Moscow, 1956.

Table 1

Principal Defects in Cast and Deformed Steel (after M. I. Vinograd)

Defect	General description	Cause	Revealed by
1. Pipe	Metal has not filled the cavity in the upper axial section of the ingot. The surface of the cavity is covered with oxides and adjacent regions are contaminated with inclusions	Volume decrease (shrinkage) during solidification in conjunction with insufficient feeding down of further metal to fill the head or insufficient heat retention by a "hot top"	1. Fracture 2. Macrography 3. X-ray examination 4. Ultrasonic flaw detection
2. Central porosity	Fine pores, sometimes with sulphide or oxide inclusions. Usually observed in articles made of the central or lower part of the ingot	Formed by insufficient filling of the central part of the ingot by metal during solidification	1. Fracture 2. Macrography
3. General porosity	Inclusions or gas holes, located along the whole cross-section of the article or at the end of the zone of columnar crystals. Have the form of dots, threadlike streaks, or hair cracks	—	1. Macrography 2. Magnetic crack detection
4. Gas holes	Channels or cavities not filled with metal. Usually at the surface, at the end of the zone of columnar crystals, or randomly located over the cross-section of the article	Formed during solidification of metal saturated with gases (H_2, N_2, and others)	1. Fracture 2. Macrography 3. Stepped turning
5. Nonmetallic inclusions and internal hair lines	Presence of impurities in the metal, such as particles of refractories, products of deoxidation of steel, etc. (alumina, silicates, sulphides, nitrides, and oxides). Certain nonmetallic inclusions of comparatively large size (larger segregations) stretch in the direction of working and produce internal hair lines	—	1. Micrography 2. Macrography

A special case of macrography is the study of fractures in metals. This reveals the size and arrangement of the grains, discontinuity flaws, etc. The character of the fracture obtained in cast aluminium

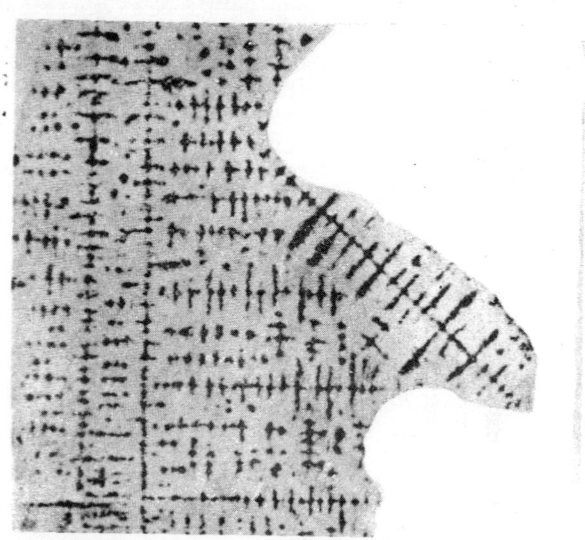

Fig. 34. Macrostructure of cast steel revealed by etching in a reagent composed of HCl, $FeCl_3$, and $CuCl_2$. Nonuniform distribution of phosphorus and carbon is observed, ×6

alloys enables the amount of slag inclusions, coarse porosity, and other defects to be estimated as well.

4-2. Microstructure (Micrography)

Micrography is the study of the structures of metals and their alloys under a microscope at magnifications from ×75 to ×1,500. The observed structure is called the microstructure.

The aims of metallographic study include the determination of the size and shape of the crystallites which constitute an alloy, revealment of structures characteristic of certain types of mechanical working operations, discovery of microdefects (nonmetallic inclusions, microcracks, etc.), and, in some cases, the determination of the chemical content of alloys (for example, annealed carbon steels). Microstructure is an indication of the quality of heat treatment, of mechanical properties, etc.

Micrographic investigations are conducted on specimens called *microsections.* These are prepared by cutting out a small specimen, for example, a 10 mm cube or a washer 10 to 15 mm in diameter and 15 mm high. One face of the specimen is smoothed with a file or emery wheel and then ground with special emery paper arranged on thick plate glass or secured to a rapidly rotating wheel. The specimen is ground first by the coarsest available emery paper and successively by finer and finer grades until the finest available grade

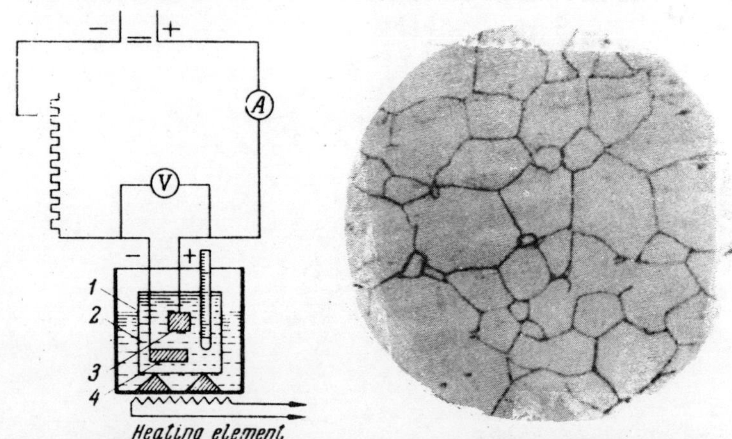

Fig. 35. Electropolishing outfit (shown schematically):
1—water bath, 2—vessel with electrolyte, 3—microsection (anode), 4—cathode of stainless steel

Fig. 36. Microstructure of a solid solution (Mg-Al)

is reached. The direction of grinding is periodically changed through 90°. After grinding, the specimen is polished by a rotating disc, covered with broadcloth, felt, or velvet, to which a very finely ground abrasive material (alumina, chromium oxide, etc.) is continuously applied.

Electropolishing or electrolytic polishing (Fig. 35) is often employed for preparing microsections. Here the microsection *3* serves as the anode and plate *4* of stainless steel is the cathode. A direct current with a density from 2 to 6 A per sq cm is passed through the electrolyte. Localised dissolving of the surface irregularities takes place and a mirror surface is obtained on the microsection. Aqueous solutions of acids, mixed with solvents that facilitate dissolution of the anode products, are usually employed as electrolytes.

The following electrolyte is most often used for electropolishing steel: orthophosphoric acid—83 millilitres, chromic anhydride—11 grammes, sulphuric acid—15 millilitres, and water 5 millilitres.

Good results are obtained for copper alloys with an electrolyte containing 980 grammes of orthophosphoric acid to 1 litre of water.

Electropolishing reduces the time required to prepare a microsection and completely eliminates deformation in the surface layers due to grinding. Such deformation may distort the actual character of the metal structure.

The finished microsection is first examined before etching to reveal nonmetallic inclusions, such as graphitic inclusions in cast iron.

Microstructure is revealed by *etching* the microsection with a *special reagent* called an *etchant*. Steel and cast iron are usually etched with

Fig. 37. Microstructure of iron. Slip bands are seen within the grains

a 4 to 5 per cent alcoholic solution of nitric or of picric acid. A 0.5 per cent aqueous solution of hydrofluoric acid is extensively used for aluminium alloys and an 8 per cent ammonia solution of $CuCl_2$ is suitable for copper alloys. Many other etchants are employed as well. Electrolytic etching has come into favour recently and is being widely used. It is based on the different rates of electrolytic dissolution of the various structural constituents of an alloy.

If a pure metal or alloy, consisting of homogeneous grains, is etched, the grain boundaries are revealed as a thin dark network (Figs. 36 and 37). In some cases, grains of the same composition are differently etched because the different grains intersect the surface of the microsection along different crystallographic planes with varying chemical activity. As a result, certain grains are almost unattacked and remain bright while others are heavily etched and appear darker, i.e., they scatter light to a greater extent.

Deep etching reveals the so-called etch figures (Fig. 38) which indicate the crystallographic orientation of the grains.

The various constituent structures of a nonhomogeneous metal are etched with varying degree of intensity. In the photograph of the microstructure of steel, shown in Fig. 39, the latter is composed in one case (*a*) of lamellae and in the other (*b*) of globules of Fe_3C. By the action of the etchant on the microsection surface, iron (ferrite) is etched to a greater extent than iron carbide Fe_3C (cementite) and a relief appears on the surface. Upon observation under a microscope with oblique illumination, the projecting boundaries of the iron carbide Fe_3C produce a shadow effect (Fig. 40) and appear bright while the crystals of ferrite on the boundaries with the carbides are dark.

Metallurgical microscopes are employed to investigate the structure of metals. Since the objects are opaque, they are observed in reflected light.

A schematic diagram of the optical system of a metallurgical microscope is shown in Fig. 41. Optical microscopes enable only those details of the structure to be observed, that are larger than 0.15 to 0.2 microns (1,500 to 2,000 Å). This corresponds to a useful magnification from ×1,000 to ×1,500.

At the present time, the electron microscope * (Fig. 42) is also used to study metals and alloys. An effective magnification of ×100,000 may be obtained (magnifications from ×7,000 to ×25,000 are usually employed). The use of an electron beam with very short wave length (0.04 to 0.12×10^{-8} cm) enables much smaller particles to be observed than in optical microscopes. Details of the object may be distinguished which are from 30 to 50 Å in size. This is about 100 times smaller than can be seen in an optical microscope. Electromagnetic or electrostatic fields are used as "lenses" to focus the electron beam. The source of the electron beam is the so-called electron gun consisting of a tungsten filament which emits electrons when heated to a high temperature. The electrons are accelerated in a strong electric field and are directed to the object. The actual surface of the metal is not usually observed in the electron microscope. A quartz or collodion replica, made from the etched surface, reproduces all details in its relief that depend on the actual structure of the metal.

The electron microscope has considerably extended the field of application of micrography. It enables the structure to be determined in cases when the high dispersity of the phases present in the metal makes observation through an optical microscope useless.

Vacuum metallography is being employed in the last years. It involves the study of microsections heated in a vacuum.

---

* The electron microscope and methods of electron microscopy are described in detail in A. I. Gardin, *Electron Microscopy of Steel*, Metallurgizdat, Moscow, 1954.

Fig. 38. Microstructure of iron, ×100

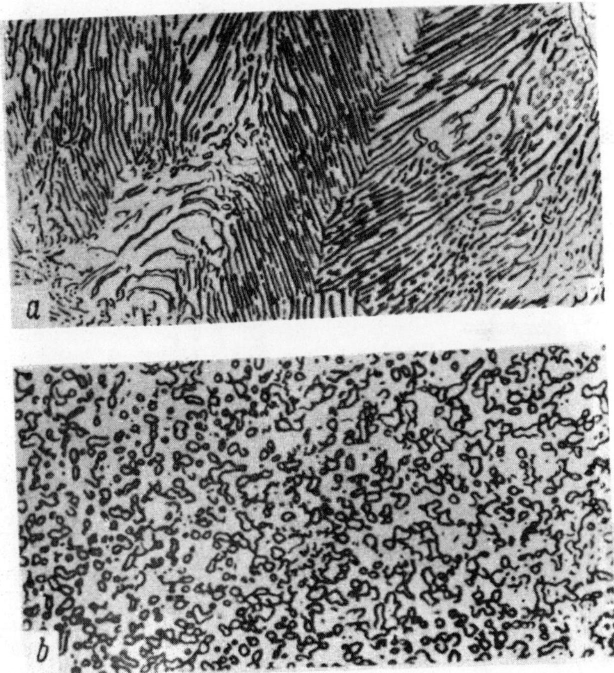

Fig. 39. Microstructure of steel containing 0.8 per cent C, ×400:
a—Fe_3C precipitated as lamellae (lamellar pearlite), *b*—as grains (globular pearlite)

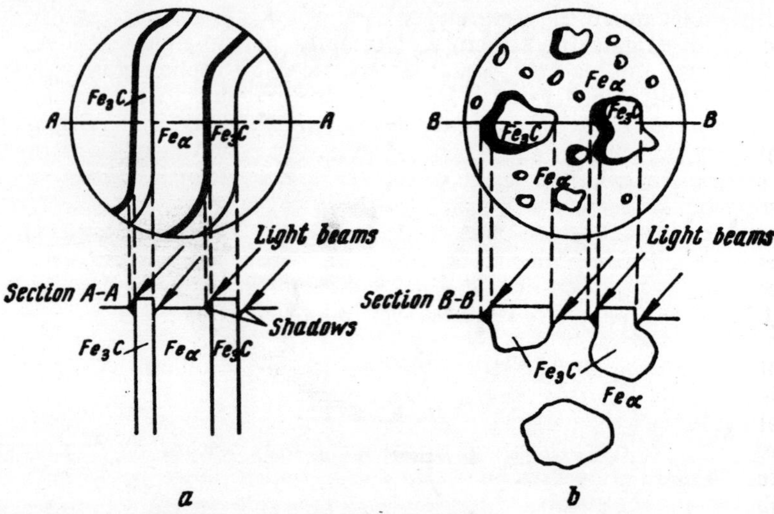

Fig. 40. Diagrams explaining features in the structure of pearlite as seen under a microscope:

a—lamellar pearlite, *b*—granular pearlite (after F. I. Maslennikov)

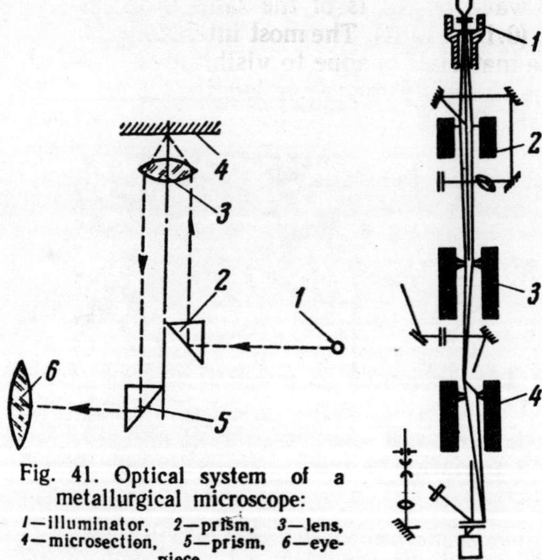

Fig. 41. Optical system of a metallurgical microscope:

1—illuminator, *2*—prism, *3*—lens, *4*—microsection, *5*—prism, *6*—eyepiece

Fig. 42. Diagram of the electron microscope:

1—electron gun, *2*—collimating magnetic coil (lens), *3*—objective magnetic coil (lens), *4*—ocular magnetic coil (lens)

Ultraviolet microscopy, developed in the U.S.S.R., is successfully used to investigate the structure of complex multiple-phase alloys. In this case, in place of visual observation, microphotography is applied in ultraviolet rays* . The latter are invisible to the human eye and have a wave length about 1.5 times shorter than visible light. For ultraviolet microscopy, either unetched microsections are used or microsections coated with oxide films. A special ultraviolet microscope is used. It has a quartz glass optical system, transparent to ultraviolet light. The source of light is a mercury-vapour lamp which intensively produces ultraviolet light. The principle of ultraviolet microscopy is based on the fact that the reflection factor of the different phases and regions of the metal varies for the ultraviolet rays falling on the specimen.

Magnetic microstructural analysis has been introduced recently. It is based on the principle that ferromagnetic structural constituents of alloys attract the particles of a magnetic colloid while nonmagnetic constituents do not. Therefore, the distribution of ferromagnetic constituents in an alloy may be observed under a microscope by the distribution of the magnetic particles which form a so-called pattern.

4-3. X-Ray Analysis

X-ray analysis is extensively employed to study the crystalline structure of metals (X-ray crystallography) and to reveal internal defects (X-ray flaw detection).

X-rays ** are a form of electromagnetic radiation but have very short wave lengths. Their wave length is of the same order as the lattice constants of metals (0.1 to 10 Å). The most interesting feature is their ability to penetrate materials opaque to visible light. X-rays are described as "soft" if they have a larger wave length and are more easily absorbed by materials and as "hard" if they have short wave lengths and a high penetration. X-rays are usually obtained by slowing down a rapid stream of electrons on the surface of the anti-cathode in a special X-ray tube.

An X-ray tube is a high-vacuum glass bulb into which two electrodes have been fused. One is the anode (or, more properly, anti-cathode) and the other is the cathode. The source of electron emission is a heated tungsten filament. The electrons, due to the high potential maintained between the anti-cathode and the cathode, pass at a high velocity to the former and interact with the atoms of its material. This interaction excites the electrons in the atoms of the anti-cathode material. Conversion of these electrons back to a lower energy level emanates the electromagnetic radiation called X-rays. Tubes for X-ray

* Photography is also done in visible rays to obtain a grey-white image.

** Discovered in 1895 by W. Roentgen.

flaw detection operate at voltages from 200 to 300 kV (X-ray tubes for 2,000 kV are also made). Tubes for structural analysis operate at voltages up to 50 kV. A tungsten anti-cathode is used in tubes for radiography having a rating up to 3 kW. Tubes for structural analysis, rated up to 1 kW, have anti-cathodes of chromium, iron, cobalt, nickel, and other metals.

X-ray crystallography is widely employed to ascertain the relative arrangement of atoms in metals, to determine the phase composition of alloys, to study recovery and recrystallisation processes, to define internal stresses (distortion) in the lattice, grain size, etc. This method of analysis is based on an interference pattern of rays diffracted by each atom of the material separately. This complex phenomenon of the interference of X-rays, diffracted by the atoms of a crystalline solid, proceeds in the same manner as if the rays were reflected from parallel atom planes of the crystal. The condition for the production of a reflected X-ray beam is specified by the Wolfe-Bragg equation (Fig. 43):

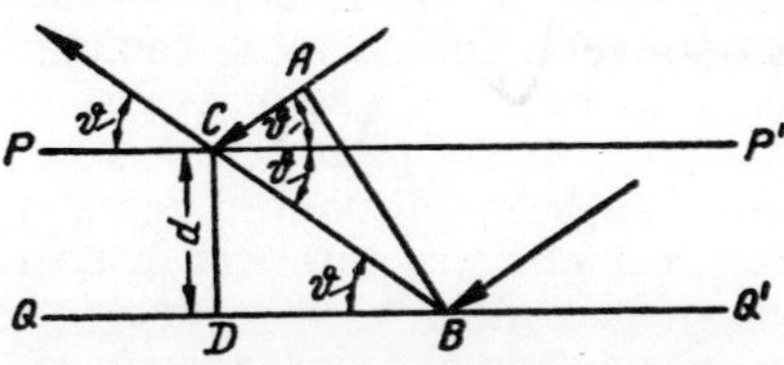

Fig. 43. Diagram illustrating the Wolfe-Bragg equation: *P-P'* and *Q-Q'* are atomic planes

$$2d \times \sin \vartheta = n\lambda$$

in which d is the distance between the parallel atom planes (interplanar spacing);
ϑ is the angle of incidence of the beam on the atom plane;
λ is the wave length of the X-rays;
n is an integral number (so-called reflection order).

The Debye and Scherrer method of structural analysis of polycrystalline samples is most often employed in physical metallurgy. It is also called the powder method. The principle involved to obtain a structural radiograph of a polycrystalline specimen consists in having a fine pencil of monochromatic X-rays strike the specimen *P* (Fig. 44, *a*). In the crystals of the specimen, which are randomly oriented in reference to the X-ray pencil, there are always a considerable number of atomic planes of such orientation as to reflect the X-ray pencil, since the angle ϑ, at which the beam strikes them, conforms to the Wolfe-Bragg equation.

At a definite angle ϑ, the beam reflected from any of the crystals will evidently pass along the elements of a cone whose axis is a prolongation of the pencil. The apex angle of the cone will equal 4 ϑ (Fig. 44, *a*). Reflections of the first order will be obtained for planes for which $d > \frac{\lambda}{2}$. Planes with a larger interplanar spacing d produce several cones of reflected radiation of varying order.

If the number of reflecting crystals is very large (practically, when their size does not exceed 1 to 3 microns) the cones will be continuous.

A flat photoplate film set at a certain distance behind the specimen, perpendicular to the direction of the X-ray pencil, will show a series of concentric rings which represent intersections of the cones by a plane. Usually radiographs of polycrystalline samples are obtained not on a photoplate, but on a comparatively narrow strip in a cylindrical holder surrounding the specimen. The radiograph, then, consists of a series of arcs (Fig. 44, *b*), symmetrical in pairs.

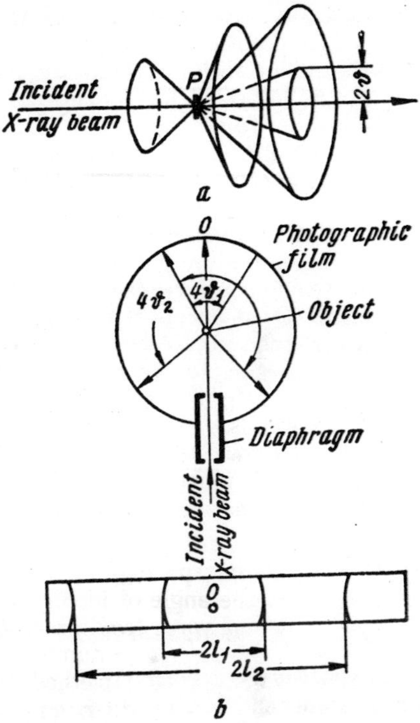

To describe the crystal lattice it is necessary to determine the size and form of the unit cell, the number of atoms it contains, and the coordinates of each atom. For this purpose, the angles of reflection of the planes and their corresponding indices are determined by the distance between the symmetrical lines on the radiograph. The intensities of the Debye lines are also determined *.

Fig. 44. Method of obtaining a radiograph of a polycrystalline specimen:
a—formation of diffracted radiation cones, *b*—principle of the Debye-Scherrer camera and a radiograph (diffraction pattern) obtained by this method

A detailed analysis of a radiograph is not often required in practice. Sometimes it will be sufficient to draw the theoretical diffraction pattern of the metal being investigated and to compare it with the radiograph obtained in a Debye camera. Each ring (pair of arcs) on the radiograph is a result of diffraction from planes with definite indices (*hkl*) in different, suitably oriented, crystals. If the structure of a certain crystalline substance is known, the positions of the rings (lines) on the radiograph may be computed, as well as their relative intensity.

* Methods used in measuring and analysing radiographs are given in special literature; see, for example, Y. S. Umansky, A. K. Trapeznikov, and A. I. Kitaigorodsky, *Radiography*, Mashgiz, Moscow, 1951.

To determine the content of various phases in an alloy, it is sufficient to make a radiograph and to determine whether it has symmetrical lines spaced at a distance equal to that computed beforehand for the most intensive lines of the phase being determined. The number of lines on a radiograph, their form, position, and uniformity of darkening enable conclusions to be made concerning the changes that have taken place in a metal or alloy. Thus, the appearance of a new phase will produce additional lines on the radiograph, an increase in grain size will cause a lack of continuity or gaps in the lines, formation of a texture will darken the Debye lines to the maximum extent, etc.

In addition to X-ray crystallography, *electronographic analysis* is also widely used in laboratory practice. The phenomenon of diffraction is produced by the passage of a pencil of electrons through a crystalline solid in the same manner as for a pencil of X-rays. Electron diffraction can be expediently applied as a method of investigation in the same cases that X-ray structural analysis is used.

Metal structure is revealed by electron diffraction in a special instrument called an electronograph.

A narrow pencil of accelerated electrons, obtained in the electronograph, is directed to the metal specimen and the resulting diffraction pattern is recorded on a photographic plate or on film.

Electronographs of latest designs enable diffraction patterns to be observed at different temperatures. This provides for a direct study of phase transformations in metals and alloys.

In principle, the electronograph resembles the electron microscope (Fig. 42) which is frequently used to obtain electron diffraction patterns.

RADIOGRAPHY (X-RAY FLAW DETECTION)

Radiography is extensively employed to inspect cast, welded, or forged components for the purpose of detecting internal flaws (blowholes, porosity, lack of weld penetration, etc.). A schematic diagram of an X-ray flaw detection unit is shown in Fig. 45.

X-rays, passing through a solid, are partially absorbed by the material. The absorption will vary along the cross-section, if internal defects are present, and therefore a photographic film, placed so as to intercept the emergent beam, will show lines or areas of varying intensity. The negative will be darkened more at places corresponding to the location of the defects, due to less absorption of radiation.*

Up-to-date X-ray units provide radiation with a penetrating power that enables steel specimens up to 100 mm thick, copper and its alloys up to 60 mm thick, and aluminium and its alloys

* Regions more penetrable to X-ray radiation will be darker, areas of high absorption will be lighter.

up to 400 mm to be tested. The γ-rays of radioactive isotopes give better results for thicker sections. The atom nuclei of such isotopes are of unstable structure and spontaneously emit much energy in the form of α-particles (helium nuclei), β-particles (rapid electrons), and γ-rays. These γ-rays resemble X-rays but have

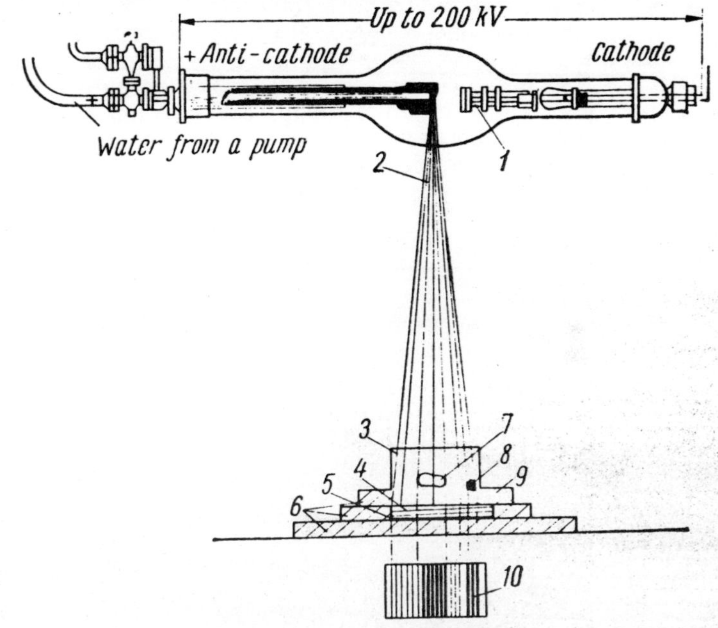

Fig. 45. X-ray flaw detection unit:

1—X-ray tube, *2*—direction of beam, *3*—specimen, *4*—cassette, *5*—X-ray film, *6*—lead shield, *7*—blowhole, *8*—inclusion (of higher density), *9*—fluorescent screen, *10*—X-ray film or screen

even shorter wave lengths (from 10^{-10} to 10^{-11} cm). This makes them more penetrative than X-rays. The source of γ-radiation is usually the radioactive isotope of cobalt, enclosed in a special ampule.

4-4. Physical Methods for Studying and Inspection of Metals and Alloys

*Thermal analysis.** Phase transformations in metals and alloys are accompanied either by the evolution or absorption of heat. Thermal analysis consists in determining the temperature at which heat is absorbed or evolved. This can be done by simply measuring the

* The development of thermal analysis has been based, to a great extent, on the work of Academician N. S. Kurnakov, the outstanding Soviet chemist.

temperature of the metal (or alloy), being heated or cooled, at equal intervals of time. This temperature-time relationship is plotted as a heating or cooling curve (Figs. 70 and 81). Broken lines or horizontal steps on the otherwise smooth curves indicate the temperatures at which heat is evolved (in cooling) or absorbed (in heating) due to transformations in the alloy. Temperatures at which changes occur in an alloy are called the *critical points.*

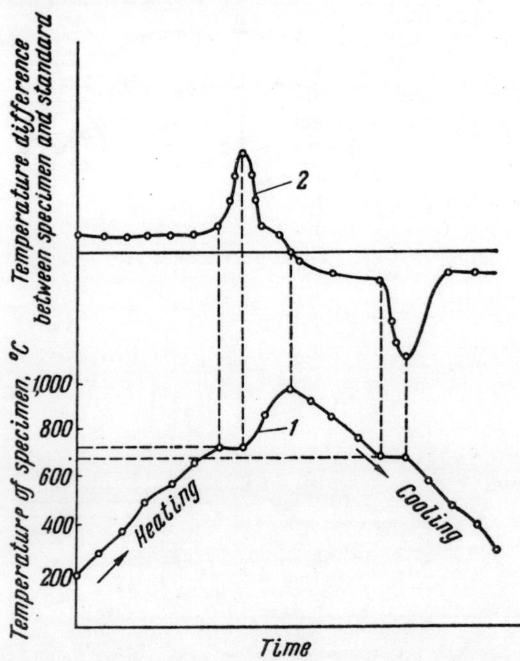

Fig. 46. Heating and cooling curves for a metal: Curve *1* is plotted in "temperature-time" coordinates while curve *2* is plotted in "temperature difference (between test-piece and standard)-time" coordinates

This type of thermal analysis is applicable only if a considerable amount of heat is evolved. Differential thermal analysis is employed to record transformations accompanied by very small thermal effects. In this procedure, the difference in temperature is determined between the specimen being investigated and another material (standard), which does not undergo transformation in the range considered. Both are heated or cooled simultaneously, under identical conditions. The results of a differential thermal analysis are plotted as temperature difference-time curves (Fig. 46). At the moment of transformation, the temperature difference increases sharply and does not depend on the heating or cooling conditions.

Dilatometric analysis * is also used to determine the temperature of phase transformations (critical points). For this purpose, changes in length are noted in heating or cooling the specimen, or under isothermal conditions (constant temperature). Changes in the length of the specimen indicate the volumetric changes that accompany a transformation in the metal (Fig. 47, *a*). This method is applied for determining the critical points in steel since its transformations are always accompanied by substantial changes in volume (due to the different specific volumes of α- and γ-iron). The differential method of dilatometric analysis is used to obtain higher accuracy. In this case, the difference in the expansion between the specimen and the standard is measured in reference to the temperature. The temperature is determined by a dilatometric pyrometer, i.e., by the expansion of a standard material having a known relation of expansion to temperature.

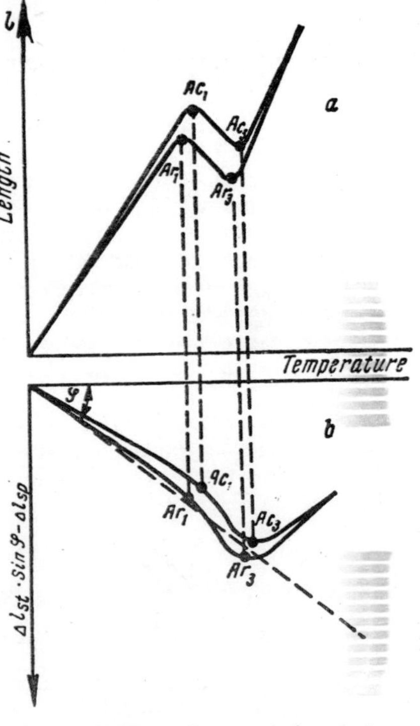

Fig. 47. Dilatometric curves for steel containing 0.8 per cent C:

a—simple record in "extension-temperature" coordinates, *b*—differential record in "extension difference (between test-piece and standard)-temperature" coordinates

A dilatometric curve for an 0.8 per cent C steel is illustrated in Fig. 47, *b*. When the temperature reaches a critical point in heating a specimen, the changes in its length as differing from the standard will depend, not so much upon thermal expansion (in heating) or contraction (in cooling), as upon the transformations in the metal or alloy (for example, allotropic transformations). At this point an abrupt expansion or contraction of the specimen occurs. This produces a sharp break in the hitherto smooth differential curve which turns sharply upward or downward.

* Dilato is the Latin for "I expand".

Special instruments, called dilatometers, are used in dilatometric analysis. They may also be employed to determine the coefficient of thermal expansion in a given temperature interval.

Electrical research methods. The physical parameter determining the electrical properties of metals is their specific resistance or electrical resistivity ρ, expressed in ohm-mm^2/m.

The electrical resistivity of metals and alloys not subject to phase transformations in heating, increases upon an increase in temperature according to the linear equation

$$\rho_{t2}=\rho_{t1}\left[1+\alpha\left(t_2-t_1\right)\right]$$

in which α is the temperature coefficient of resistivity. The resistivity of metals and alloys depends on their specific physical nature and may change substantially in accordance with their composition and structure (this was first discovered by N. S. Kurnakov).

The resistivity of alloys differs considerably from that of pure metals (Figs. 85, 89, and 94). Its variations, depending on the concentration of the elements comprising the alloy, are essential indications of its nature and structure.

Electrical resistivity of alloys is ordinarily measured by the duplex Thomson bridge or by potentiometric methods (which are more accurate).

Magnetic methods of research and inspection. The chief characteristics of the magnetic properties of materials are the permeability μ, which is defined as the ratio of the induction B (in gs) to the magnetising force H (in oersteds):

$$\mu=\frac{B}{H}$$

and the magnetic susceptibility K

$$K=\frac{I}{H}$$

in which I is the intensity of magnetisation. The susceptibility K is related to the permeability μ by the equation $\mu=1+4\pi K$.

Depending on the character of their magnetic properties, materials are classified as:

1. *Diamagnetic materials*, having a small negative susceptibility, independent of the temperature and magnetising force. Such materials include zinc, gold, mercury, etc.

2. *Paramagnetic materials*, for which $K>0$. These include lithium, sodium, rubidium, and other metals.

3. *Ferromagnetic materials*, such as iron, nickel, cobalt, and their alloys. These metals have a positive magnetic susceptibility which greatly exceeds that of paramagnetic materials.

The permeability of ferromagnetic materials may reach high values (up to 3.5×10^6) and depends not only on the temperature but also upon the magnetising force

Magnetic materials are characterised by the initial permeability μ_0 and the maximum permeability μ_{max}. The relationship between the degree of magnetisation and the magnetising force or field is illustrated by magnetisation curves and hysteresis loops (Fig. 48).

The magnetic induction *B* first increases with an increase of the magnetising force *H* and then remains practically constant.

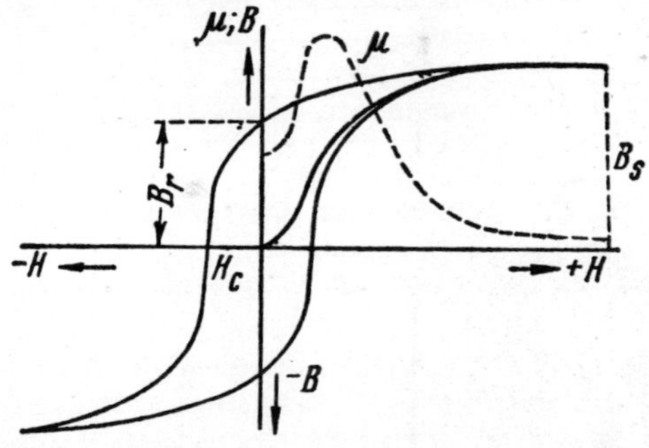

Fig. 48. Magnetisation curve and hysteresis loops

The phenomenon of hysteresis is typical of ferromagnetic materials. When the magnetising force is reduced to zero, a ferromagnetic material, magnetised to saturation, will retain residual induction B_r (Fig. 48).

The reverse magnetising force required to reduce B_r to zero is called the coercive force H_c (Fig. 48).

The magnetisation of ferromagnetic materials decreases as the temperature increases. Above a definite temperature for each given metal (Curie temperature), ferromagnetic properties disappear and the metal becomes paramagnetic.

When a ferromagnetic material is placed in a magnetic field, a small change in linear dimensions and volume results. This phenomenon is called *magnetostriction*.

The saturation intensities of magnetisation and magnetostriction are structure-insensitive properties. They are determined by the composition, atomic structure, and the quantitative ratios of the phases of which the alloy is comprised. Coercive force (H_c), permeability (μ) or susceptibility (K), and residual induction (B_r) are structural-sensitive properties. These properties depend to a great extent upon the dispersity and shape of the crystallites, their orientation and relative arrangement, as well as distortion of the crystal lattice, i.e., upon stresses, etc. (see Sec. 13-12).

A study of the changes that occur in magnetic properties when composition and treatment are varied enables one to understand the transformations taking place in ferromagnetic metals and their alloys. Magnetic research methods are extensively employed, for example, in studying allotropic transformations in iron and its alloys. They are particularly useful to study the transformations of supercooled austenite (Sec. 8-3) and for determining the amount of retained austenite (Sec. 8-5). They are also used to study the ordering of solid solutions (page 108), the ageing process (Sec. 8-7), and others.

The magnetic method is also used for nondestructive testing of the composition and structure and, therefore, the properties of many articles after heat treatment. The testing methods are based on the measurement of a definite magnetic characteristic such as: magnetic saturation, residual induction, coercive force, and magnetic permeability.

A magnetic testing method will satisfy all requirements of shop inspection for those cases in which a definite relationship has been established between the magnetic properties and the structure of steels.

Magnetic testing is distinguished for its high sensitivity, rapidity, simplicity, and the possibility of being fully automated.

Magnetic inspection (magnetic flaw detection) is most extensively used to reveal surface defects on machine components (quenching, grinding, and fatigue cracks, hair cracks, lack of weld penetration, etc.). The method is based on the principle that when a ferromagnetic component is magnetised, stray fields are produced at all discontinuities in the surface (at cracks). If the component is immersed in an oil (or gasoline, alcohol, etc.) bath containing suspended particles of magnetic powder (magnetic suspension), the particles will be attracted by the dispersion field and will collect on the edges of the flaw so as to reveal it.

The radioactive isotope method (radioactive tracers) has been widely used in physical metallurgy in recent years. The application of radioactive tracers ("tagged" atoms) became possible after the discovery of artificial radioactivity by Frédérick Joliot-Curie and Irene Curie in 1934.

The atomic nucleus of a radioactive material has a complex and unstable structure. Therefore, such materials continuously and spontaneously emit considerable energy in the form of radiation that cannot be controlled. As mentioned above, this radiation includes: 1) α-rays which are positively charged helium nuclei and are only slightly deflected by a magnetic field. When α-rays pass through a gas, it is ionised. This enables the passage of the rays to be observed by their traces or, as they are called, "tracks". The velocity of α-particles is about 17,000 km per hr. 2) β-rays which are negatively charged electrons and are strongly deflected by magnetic fields. β-rays possess less ionising capacity than α-radiation. The velocity of β-particles approaches that of light. 3) γ-rays which are electromagnetic waves of the X-ray type but with considerably shorter wave length. γ-rays are undeflected by a magnetic field, they are not charged, and do not ionise gases.

Radioactive or unstable isotopes are the ordinary source of radiation. Radioactive isotopes of most of the elements may be obtained artificially as by-products of nuclear reactions.

More than 1,000 radioactive isotopes have been obtained artificially at the present time. Radioactivity is associated with processes which take place in the atomic nucleus. The number of atoms that decays in a unit of time is proportional to the initial number of radioactive atoms. The same fraction of the radioactive atoms, strictly defined for each element, decays in each time unit. This fraction is known as the *radioactive decay constant*. The decay constant determines the possibility for decay or disintegration in a unit of time. The larger this constant, the more unstable the atomic nucleus is.

Another value is usually employed to characterise the rate of radioactive decay. It is called the half-life period and is equal to the time required for the initial number of radioactive atoms to be reduced by one half. The half-life of various elements ranges from 10^{-6} second to several milliard years. The half-lives of certain more frequently used isotopes are: Fe^{59}—57 days, S^{35}—88 days, C^{14}—from 10^3 to 10^5 years, B^{12}— 0.022 second, etc.

The application of tagged atoms or tracers for studying metals and alloys is based on mixing radioactive and nonradioactive substances. The introduced radioactive substance cannot be chemically separated from the nonradioactive substance and they act in exactly the same way in all processes that occur in the metal. Due to their radioactivity, however, movements of these tracers can be observed.

Special counters are used to register radioactivity (radioactive radiation). They are based on the ionisation of the substance in the detector or tube of the counter by radioactive particles. The counter enables each elementary act of decay of a radioactive substance to be registered. The autoradiography method may also be used for the same purpose. Here, a specimen containing the radioactive element is placed on a special photographic plate (photographic emulsion). The location of the blackened areas, that appear after the plate is developed, corresponds to the distribution of radioactive substance in the specimen. The degree of blackening or density is proportional to the amount of the isotope in the given volume. These blackened areas or spots are due to the action of radioactive emission on the photographic emulsion. The photograph obtained by this method is called a micro-autoradiogramme (Fig. 49).

Due to its high sensitivity and accuracy, the radioactive tracer method is finding more and more extensive application. In physical metallurgy, this method is successfully used for studying the process of diffusion (self diffusion), the distribution of impurities and of elements specially introduced into an alloy, and also for phase

5 *

analysis. Radioactive tracers are of exceptional value for studying the mechanism of friction and wear in metals *, as well as the behaviour of metals in forging and press-working.**

Ultrasonic inspection *** is being employed to a certain extent, lately, for revealing all sorts of internal flaws (blowholes, porosity, certain cracks, etc.).

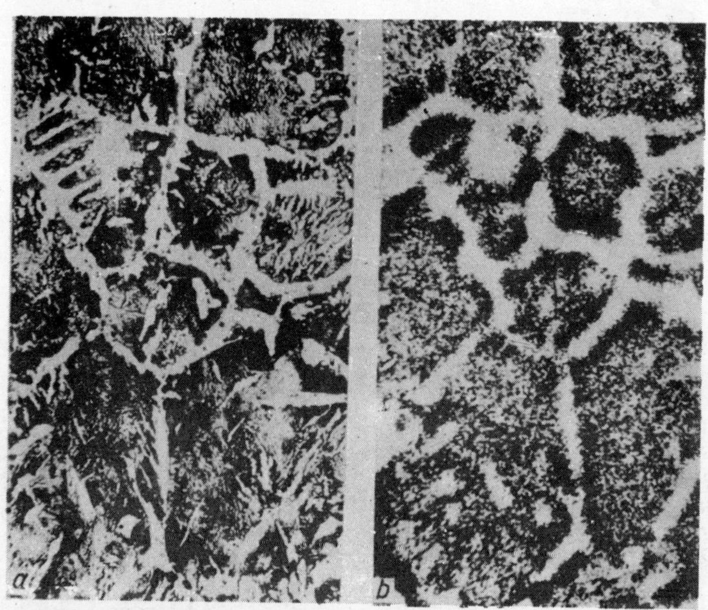

Fig. 49. Microstructure (*a*) and micro-autoradiogramme (*b*) of the same specimen

Ultrasonics are elastic waves of frequencies far beyond the audio range. Frequencies of the order of 10^6 cycles per second are used in ultrasonic flaw detection.

The directivity of ultrasonic vibrations increases with an increase in their frequency. At a frequency in the order of one megacycle per second, the angle of divergence is so small that they may be called "ultrasonic beams".

Ultrasonic inspection is based on the property of ultrasonic vibra-

---

* E. P. Nadeinskaya, *Research on Wear in Cutting Tools Conducted with Radioactive Isotopes*, Mashgiz, Moscow, 1955.

** S. I. Gubkin, S. A. Dovpar, *Proceedings of the Academy of Sciences of the U.S.S.R.*, Vol. 91, No. 5, 1953.

*** For more detail see *Up-to-Date Methods of Testing Engineering Materials*, Mashgiz, Moscow, 1956.

tions for passing through metals over large distances, as directed from the surface of various defects. A flaw, deep within the metal, may be detected and its shape may be determined by registering the reflected waves or the energy of the waves that have passed through the metal.

General methods employed in ultrasonic flaw detection are the following:

1. *The pulse method.* This more extensively used method is designed for detecting discontinuities in the metal (blowholes, porosity, etc.), located at depths from several millimetres to several metres, and for establishing coarse-grained zones in blanks and finished articles.

The reflection of ultrasonic waves from the boundaries of flaws is used to reveal blowholes, porosity, and cracks. Coarse-grained zones are detected by the scattering of the waves by large crystallites.

In the operation of a pulse-type ultrasonic flaw detector, the emitter * transmits short pulses of elastic vibrations (with a duration from 0.5 to 10 microseconds) separated by pauses from 1 to 5 milliseconds long. Upon the presence of an obstacle (flaw in this case), which may be considered as a discontinuity in the acoustic properties of the medium, the echo-signal is reflected from the boundary of the flaw to the receiver and is registered by an indicating device. The distance to the defect may be determined by the time that elapses between the transmission of the pulse and the reception of the echo-signal. If the zone being inspected has a coarse-grained structure, the ultrasonic vibrations will be scattered at the boundaries of the crystals. The echo-signal will be considerably weakened and may even disappear completely. The pulse method is used when the inspected item is accessible from one side only.

2. *The resonance method* is based on the generation of a standing wave in the metal due to interference. In this method, the variations in the operation of the emitter are observed at the moment that standing waves appear. The resonance method is chiefly employed for testing items up to 15 mm thick. It is also used to measure the thickness of metal, zones of corrosion, etc.

3. *The shadow method* is based on the "sound shadow" which appears beyond a defect located in the path of ultrasonic vibrations. This method has found comparatively few applications due to the cumbersome apparatus required and to a number of inconveniences in operation. This method, however, may be further developed to obtain an image of the defect on a screen.

In conclusion, it must be noted that ultrasonics are being employed more and more extensively in the processing of metals to speed up diffusion processes, to change the conditions of solidification (and consequently, the structure of metals) in the casting of metals, for cleaning metal items, etc.

---

* A piezo-electric transducer excited by radio impulses of a special generator.

Chapter 5

THE MECHANICAL PROPERTIES OF METALS

5-1. Mechanical Testing of Metals

In the course of operation or use, all articles and structures are subject to the action of external forces which create stresses that inevitably cause deformation. To keep these stresses, and, consequently, deformations within permissible limits, it is necessary to select suitable materials for the components of various designs and to apply the most effective heat treatment. A comprehensive knowledge of the chief characteristics of semi-finished metal products and finished metal articles (strength, ductility, toughness, etc.) is essential for this purpose. For this reason, the specifications of metals, used in the manufacture of various products and structures, are based on the results of mechanical tests.

Mechanical tests are those in which specially prepared specimens (test pieces) of standard form and size (and less frequently, the articles and structures) are tested on special machines to obtain the strength, ductility, and toughness characteristics of the metal. Mechanical tests may be conducted under various loading conditions. Tests may be: 1) *static*, when the load is increased slowly and gradually and the metal is loaded by tension, compression, torsion, or bending; 2) *dynamic*, when the load increases rapidly as in an impact, and 3) *repeated* or *fatigue* (both static and impact type), when the load repeatedly varies in the course of the test either in value or both in value and in direction.

5-2. Tension Tests

Tension tests are of the static type, i.e., they are tests in which the load is increased comparatively slowly, from zero to a certain final value. Standard specimens (GOST 2055-43 *), used for tension tests, are shown in Fig. 50. The specimens may be of round (*a*) or flat (*b*) cross-section. The ends of the specimen are secured in the grips of the testing machine. The mechanical properties in tension are determined on the gauge length l_0 of the specimen. Such testing machines as the universal lever-pendulum type (Model P5), the Gaga-

---

* Here, and throughout the book, GOST (Russian—ГОСТ) is the abbreviation for the U.S.S.R. State Standard.

rin press, model ИМ-4Р (designed by the TsNITMASh Institute), and others * are extensively used for tension tests.

All tension testing machines have two chief parts:

1. unit for applying a load to the specimen (with a hydraulic or mechanical drive), and

2. unit for measuring the applied load (lever with adjustable weight, pendulum-type lever, pressure gauge and hydraulic capsule). Testing machines are often equipped with recording devices which

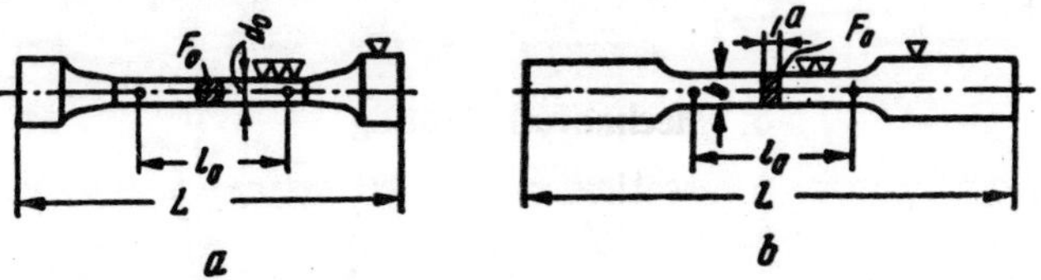

Fig. 50. Tension-test specimens. The gauge length l_0 for a long specimen is taken as $l_0=11.3\sqrt{F_0}$ and for a short specimen $l_0=5.65\sqrt{F_0}$, in which F_0 is the cross-sectional area in sq mm. The standard dimensions of the long specimen should be: $l_0=200$ mm, $d_0=20$ mm, $F_0=314$ mm^2

automatically record changes in length of the specimen in accordance with the applied load (stress-strain diagram). The most accurate tension stress-strain diagrams may be obtained on the Gagarin press and the model ИМ-4Р testing machine.

Fig. 51, *a* illustrates a tension load-elongation diagram for low-carbon steel. It is evident from the diagram that at first the elongation is proportional to the load. This law of proportionality remains valid up to a certain load P_p which is called the load at the limit of proportionality.

The stress, calculated on the basis of the initial cross-sectional area of the specimen

$$\sigma_p=\frac{P_p}{f_0} \text{ kg per sq mm}$$

is called the *limit of proportionality* or *proportional limit.*

The conventional proportional limit σ_p is the stress at which the deviation from a linear relationship between the stress and deformation has reached a value for which the tangent of the angle between the curve and the stress axis has increased by 10 per cent.

Stresses not exceeding the proportional limit result practically in elastic deformation only, i.e., strain that disappears when the load is removed. For this reason the proportional limit is frequently identified with the elastic limit, so that $\sigma_e=\sigma_p$. This is actually not so but accurate enough for practical purposes. Usually the elastic

* A detailed description of these and other machines is given in D. O. Slavin and D. M. Shapiro, *Mechanical Testing of Metals*, Metallurgizdat, Moscow, 1950.

limit σ_e is defined as the stress at which the permanent set is equal to 0.005 per cent of the initial gauge length (l_0) of the specimen.

Thus, σ_e and σ_p characterise the resistance of the metal to small plastic deformations.

The proportional and elastic limits are essential characteristics of a metal. Any structure or machine component must be designed so that its working stresses do not exceed σ_e and σ_p.

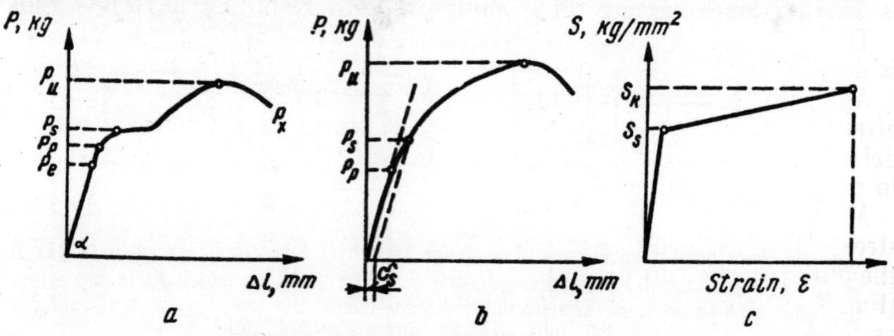

Fig. 51. Tension-test diagrams

There is a line relation in the region of elastic deformation between stress and strain for metals and alloys. It conforms to the law of proportionality (Hooke's Law):

$$\sigma = E\delta, \text{ and}$$

$$E = \frac{\sigma}{\delta} = \tan\alpha$$

The coefficient of proportionality E called the *modulus of elasticity* (Young's modulus) characterises the rigidity of a material, i.e., its resistance to elastic deformation in tension.

The modulus of normal elasticity E depends only to a comparatively small extent on the structure, heat treatment, and composition of an alloy. It is determined, mainly, by the type of crystal lattice. The moduli of elasticity for certain important engineering metals are as follows:

Metal	Al	Cu	Fe	Mg	Zn	W
E, kg/mm²	7,200	12,100	21,000	4,500	10,300	36,000

Iron, which has a high modulus of elasticity, is the most important of engineering materials.

At an increase of load beyond P_p, the relation between load and elongation will depart from a straight line. The straight line will become a curve and, at a certain load, a horizontal step will sometimes appear on the diagram. This horizontal step indicates that the metal is elongated (yields) without an increase in load. The load P_s is called the load at the yield point and the corresponding stress

$$\sigma_s = \frac{P_s}{f_0} \text{ kg/mm}^2$$

is called the *yield point.*

Thus, the yield point is the minimum stress at which the specimen is deformed (yields) without a noticeable increase in the load.

Appreciable plastic deformation, i.e., deformation which remains after the load is removed, occurs above the proportional limit. Elastic deformation causes only elastic distortion of the crystal lattice while in plastic deformation, slip of some magnitude takes place.

Most metals do not have a clear-cut yield step ("sharp-kneed" stress-strain diagram).* In this case, the curve passes smoothly from the elastic section to the section corresponding to plastic deformation (Fig. 51, *b*). For such diagrams the conventional yield point is defined as the stress at which the specimen receives a permanent elongation equal to 0.2 per cent of the initial gauge length denoted as $\sigma_{0.2}$. The yield point is easily determined and, for this reason, is extensively employed as an indication of strength in engineering materials. At the present time, it has often supplanted such characteristics as σ_e and σ_p, since the latter require precise instruments for their determination.

A further increase in load will cause more marked deformation in the whole volume of the metal. The maximum load which the specimen can withstand without failure is called the load at the ultimate strength and corresponds to the stress

$$\sigma_u = \frac{P_u}{f_0} \text{ kg/mm}^2$$

which is called the *ultimate strength* or *tensile strength.*

The ultimate strength σ_u is the stress corresponding to the maximum load before failure of the specimen. Beginning with the stress σ_u, the deformation is concentrated in one section of ductile metals. A localised reduction of cross-sectional area or "necking down" appears on the specimen, the load falls, and at a certain moment, failure occurs. Thus, for ductile metals, the ultimate strength characterises the resistance of the metal to considerable plastic deformation.

In a tension test, in addition to strength indices such as σ_e, σ_p, σ_s, and σ_u, *ductility indices* are also determined. They indicate the ability of the metal to deform in static tension tests without failure.

* The reasons for the occurrence of a yield step are not suffiently clear as yet. It is most frequently explained by the segregation on the grain boundaries of brittle components that impede grain deformation.

Ductility indices comprise:

Relative elongation δ, which is the ratio of the extension in length of the specimen after fracture to its initial gauge length, expressed in per cent:

$$\delta = \frac{l_1 - l_0}{l_0} 100 \text{ per cent}$$

in which l_1 is the gauge length of the specimen after fracture.

Relative reduction of area ψ, which is the ratio of the reduction of cross-sectional area in the fractured specimen to the initial cross-sectional area, expressed in per cent:

$$\psi = \frac{f_0 - f_1}{f_0} 100 \text{ per cent}$$

In determining σ_p, σ_s, and σ_u, the corresponding loads P_p, P_s, and P_u are divided by the initial cross-sectional area f_0 of the specimen.

The tension test diagrams shown in Fig. 51, *a* and *b* do not take into consideration the considerable reduction of area of ductile metals at stresses above the yield point and particularly upon necking down of the specimen. A more accurate relationship between specimen deformation and actual stresses is obtained from *a true stress-strain diagram*. This diagram is plotted in the coordinates true stresses (S)-true elongation (ε). The true elongation ε is the length increment of the specimen at each instant referred to the actual length at that instant, instead of to the initial length:

$$\varepsilon = \ln\left(\frac{1}{1-\psi}\right) = \ln \frac{f_0}{f_k}$$

in which: ψ is the reduction of area of the "neck",
f_0 is the initial cross-sectional area, in sq mm,
f_k is the cross-sectional area in the process of deformation, in sq mm.

A true stress-strain diagram is shown in Fig. 51, *c*. On this diagram the yield strength is denoted as S_s and the failure stress as S_k. The values of σ_u and S_k are practically the same for all brittle metals which fracture without considerable plastic deformation. S_k is always larger than σ_u for ductile metals. It is evident from Fig. 51, *c* that the true stress-strain diagram has two straight line sections corresponding to the ranges of elastic and plastic behaviour. Since the plastic deformation is much more important than the elastic deformation, the range of elastic behaviour is frequently considered as coinciding with the Y-axis; such a diagram is drawn as in Fig. 52.

One of the most important values that characterise the properties of a metal in the range of plastic behaviour is the modulus of strain hardening D. This modulus is equal to the tangent of the angle that the straight line, in the plastic range of a true stress-strain curve, makes with the X-axis (Fig. 52). It indicates the capability of the metal to be strain hardened (hardened by plastic deformation). The steeper the straight line in the plastic range of the curve, the higher the modulus of strain hardening, i.e., the higher the stress must be to obtain the required amount of plastic deformation. The larger D is, the higher the cold hardenability of the metal and the more it is subject to cold hardening in the process of plastic deformation.

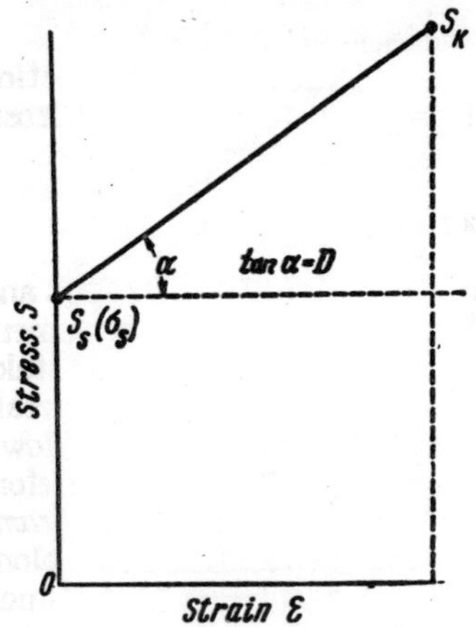

Fig. 52. True stress-strain diagram

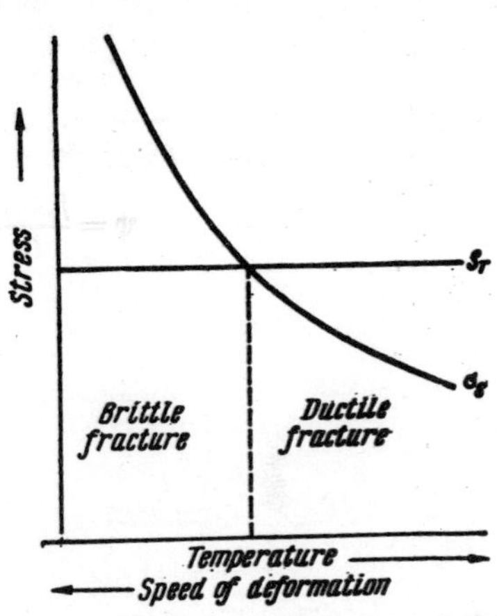

Fig. 53. Diagram showing the relation of the brittle and ductile state of solids to the temperature and rate of deformation (after A. F. Joffe)

The ductility of a material depends upon the main values characterising its mechanical properties:

$$e = \frac{S_k - \sigma_s}{D}$$

Therefore, the larger σ_s and D are, the less the ductility will be. An increase in S_k increases the ductility as well. The amount of work to carry a material to failure, or the toughness (see impact tests, Sec. 5-4), is equal to the area under a complete stress-strain curve. It may be expressed as a function of S_k, σ_s and D:

$$a_k = \frac{S_k^2 - \sigma_s^2}{2D}$$

Hence, comparatively small variations in the values of S_k and σ_s, which are in the second power in the equation, may substantially affect the toughness though they do not have a noticeable effect on the ductility.

Mechanical properties depend on the test conditions. It has been established that lowering the temperature or increasing the rate of deformation considerably increases the resistance to plastic deformation (σ_s, σ_s, as well as D) (Fig. 53), though they do not change the cohesive strength (see Sec. 3-1). Thus, at low temperatures (or high rates of deformation), metals and alloys, which are ductile at normal room temperatures (or low rates of deformation), may fail with a brittle fracture because the cohesive strength S_T is reached at stresses below the yield point. The higher the cohesive strength S_T, the lower the temperature at which a metal or alloy will pass over from a ductile to a brittle state, i.e., the less it is susceptible to brittle fracture.

In tension tests, normal tension stresses are the more important since they are twice as large as the tangential stresses. Therefore in testing high-strength alloys having low ductility, compression and torsion tests are being more frequently employed. Here, the effect of normal stresses is considerably less than that of tangential stresses.

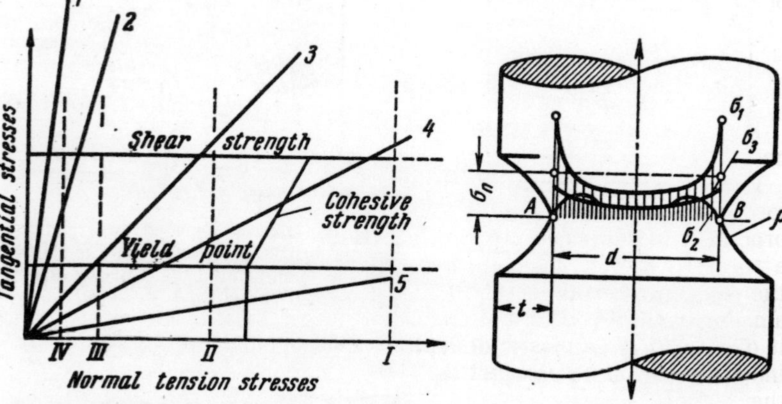

Fig. 54. Mechanically-stressed state diagram showing the effect of the stressed state on the mechanical properties:

1, 2—penetration, scratching, 3—torsion, 4—tension, 5—tension of a notched specimen, I—soft metals, II—tool steels, III—cemented carbides, cast iron, IV—glass, silicates (after Y. B. Friendman)

Fig. 55. Stress distribution in a tension test of a notched specimen:

σ_1—axial stress, σ_2—radial stress, and σ_3—transverse stress, directed along the tangent to the profile of the cross-section at A-B

The diagram of the mechanically stressed state (Fig. 54) shows the qualitative effect of the method of loading on the type of fracture (separation or shear) obtained.

If $\tau_{max} > S_{max}$, i.e., if the tangential stresses are large in comparison to the normal stresses, the stressed state will be of the "mild" type and the metal will fracture by shear which will be preceded by considerable plastic deformation. If, on the other hand, $S_{max} > \tau_{max}$, then the metal will be in the "severe" stressed state and will fail by separation (brittle fracture) without noticeable plastic deformation.

Notches, i.e., sharp changes in cross-section, have a large effect on the mechanical properties of metals. Notching in a specimen is similar, in effect, to actual machine components of intricate shape, with through holes, thread, etc., or to internal defects in a metal in the form of various discontinuities (nonmetallic inclusions, graphite in cast iron, cracks, etc.). As is evident from Fig. 55, a notch will cause nonuniform distribution of stresses and thus lead to stress concentration. Nonuniform distribution of stresses, though they often increase the resistance to plastic deformation (σ_s, σ_u), will always contribute to lowering the ductility of a metal (δ, ψ). The effect of a notch must be regarded among the factors that contribute to the transition of a metal to the brittle state. A notch reduces the ultimate strength of high-strength and brittle materials. The sharper the notch, the more its effect will be on the mechanical properties of a metal. Notch sensitivity, which is defined as the change in properties of a notched specimen, is one of the most important characteristics of materials and is determined by special tests.*

5-3. Hardness Tests

Hardness is the resistance of a metal to the penetration of another harder body which does not receive a permanent set.

Hardness tests consist in measuring the resistance to plastic deformation of layers of metal near the surface of the specimen. In the process of hardness determination when the metal is indented by a special tip (steel ball or diamond cone), the tip first overcomes the resistance of the metal to elastic deformation and then, a small amount of plastic deformation. Upon deeper indentation of the tip, it overcomes large plastic deformation. Therefore, hardness is not an independent characteristic of the mechanical properties. It determines the same properties as in other testing methods, for example, tension tests, but under different loading conditions.

This fact enables a relation to be established between the hardness and ultimate tensile strength of ductile metals. For metals of low ductility, in which the tensile strength is characterised by the cohesive strength and not by the resistance to considerable plastic deformation, the relation between hardness and ultimate tensile strength is not sufficiently reliable.

It is necessary to distinguish: 1) macrohardness, which is the hardness of a material determined by its resistance to plastic deformation in a volume so large that the differences in actual hardness of its component microvolumes are of no influence, and 2) microhardness, which is the hardness of materials in microscopically small volumes (hardness of the separate structural components).

---

* For more detail see N. A. Shaposhnikov, *Mechanical Testing of Metals*, Mashgiz, Moscow, 1954.

Hardness measurements for determining the properties of a machine component have found extensive application in the quality control of metals and metal products in all branches of industry due to the rapidity and simplicity of the tests and their nondestructive character.

The most widely employed hardness testing methods are:

Ball-indentation tests (Brinell principle). This method consists in pressing a hardened steel ball under a constant load P into a specially prepared flat surface on the test specimen (Fig. 56). After re-

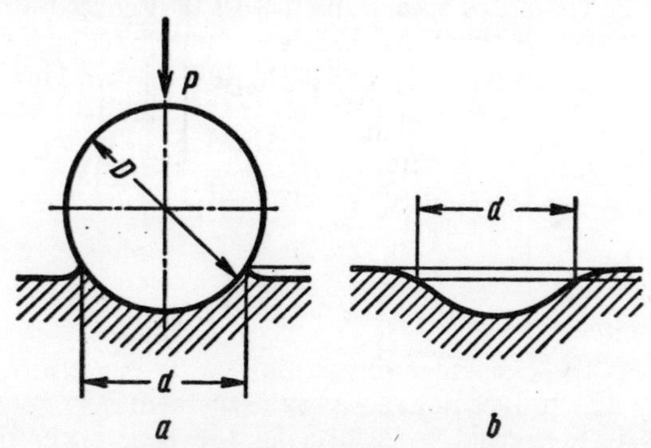

Fig. 56. Brinell hardness test:
a—indentation having encircling ridge, b—indentation having encircling depression

moving the load, an indentation remains on the surface of the test specimen.

If the area of the spherical surface in the indentation is denoted as F sq mm, the Brinell number *Bhn* will be

$$Bhn=\frac{P}{F}\ \text{kg/mm}^2$$

Thus the Brinell number *Bhn* is the ratio of the load, applied to the ball during the test, to the area of the indentation produced by pressing the ball into the test specimen.

If the indented area F is expressed by the ball diameter D and the diameter of the indentation d, then the Brinell hardness number may be determined from the formula

$$Bhn=\frac{2P}{\pi D\left(D-\sqrt{D^2-d^2}\right)}\ \text{kg/mm}^2$$

The ball diameter and applied load are constant and are selected to suit the composition of the metal, its hardness, and the thickness of the test specimen (Table 2). The diameter of the indentation is measured with a special magnifying glass containing a scale graduated in tenths of a millimetre.

Table 2

Brinell Test Conditions (according to GOST 10241-40)

Material	*Bhn*	Thickness of test specimen, mm	Ratio of load *P* to ball diameter *D*	*D*, mm	*P*, kg	Time of load application, sec
Steel, cast iron	up to 450	over 6		10	3,000	
		from 6 to 3	$P=30D^2$	5	750	10 to 30*
		less than 3		2.5	187.5	
Copper and its alloys, magnesium alloys, etc.	31.8 to 130	over 6		10	1,000	
		from 6 to 3	$P=10D^2$	5	250	30
		less than 3		2.5	62.5	
Aluminium, babbitts	8 to 35	over 5		10	250	
		from 6 to 3	$P=2.5D^2$	5	62.5	60
		less than 3		2.5	15.6	

* For hardnesses up to *Bhn* 140, the time of load application is 30 sec; for harder materials — 10 sec.

In practice, the Brinell numbers corresponding to a given observed diameter of indentation *d* are taken from tables to simplify calculations.

The following relationship exists between the Brinell hardness number and the ultimate strength of metals which form a neck in tension tests:

$$\sigma_u = k\ Bhn$$

in which $k=0.36$ for rolled steel and $k=0.3$ to 0.4 for steel castings.

It is not advisable to apply Brinell tests to materials having a hardness which exceeds *Bhn* 450 since the ball may be easily deformed and this will introduce errors into the test results. The hardness of the ball should be at least 1.7 times higher than the test specimen to prevent permanent set in the ball.

Determining hardness by the depth of indentation of a diamond cone or small steel ball (Rockwell principle). This test, like the Brinell test, is based on the indentation of a hard tip or indenter under a static load into the test specimen. In the Rockwell test the hardness is determined not by the diameter but by the depth of the indentation or impression.

The indenter or "penetrator" in a Rockwell test may be either a conical-shaped diamond (called a "brale") with a 120° apex angle or a hardened steel ball 1.5875 mm ($^1/_{16}$") in diameter. The brale is used for testing materials with a high hardness and the steel ball for softer materials. The brale and the ball are indented by two consecutive loads: the preliminary load P_0 (equal to 10 kg) and the additional load *P* which equals 90 kg for the ball (Scale *B*) and 140 kg for the brale (Scale *C*).* A brale with an applied load of 50 kg (Scale *A*) is employed for very hard materials and thin test specimens. The Rockwell superficial-hardness tester is applied for testing thin layers near the surface or thin specimens (nitrided steel, safety-razor blades, etc.). In this machine, the preliminary load is 3 kg and the total loads may be 15, 30, or 45 kg.

After loading, the additional load is removed. The Rockwell hardness number is the difference in depths of the indentations made by applying the total and preliminary loads, measured after removing the additional load (Fig. 57).

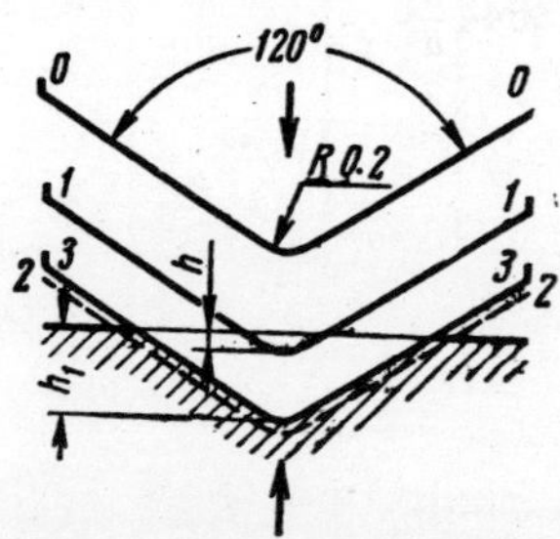

Fig. 57. Hardness determination with a diamond cone (brale) in a Rockwell test: *1-1*—penetration of cone under preliminary load P_0, *2-2*—penetration of cone under total load (P_0+P), and *3-3*—penetration of cone after removing additional load P (under preliminary load P_0)

The Rockwell hardness number is read directly on the dial of the instrument having a scale graduated in hardness units which are determined by the formula

$$R_c = \frac{k-(h_1-h)}{c}$$

in which k is a constant equal to 0.2 for the brale and 0.26 for the steel ball;

h_1 is the depth of penetration, in mm, of the brale (or ball) after removing the additional load (Fig. 57);

h is the depth of penetration, in mm, of the brale (or ball) under the preliminary load P_0 (Fig. 57);

c is the value of the scale divisions (0.002 mm).

Rockwell tests are widely applied in industry due to the rapidity and simplicity with which they may be performed, high accuracy achieved, and due to the small size of the impressions produced. The Rockwell hardness number may be converted into the Brinell number using special tables or charts.

Determining hardness by indentation of a diamond pyramid (Vickers principle **). This testing method is used for determining hardness of

* The total load is the sum of the preliminary and additional loads.
** According to GOST 2999-45.

specimens of small cross-section or of their external layers on case-hardened, nitrided, etc., specimens having a very high hardness. The test consists in forcing a square-based diamond pyramid (with an angle of 136° between opposite faces) into the ground or even polished surface to be tested. The hardness number is defined as the ratio of the load in kilograms to the surface area of the indentation in square millimetres. The latter is calculated by measuring the length of the diagonals of the indentations. The hardness number is determined from the formula

$$VHN = \frac{2P \sin \frac{\alpha}{2}}{d^2} = 1.8544 \frac{P}{d^2}$$

in which P is the applied load (5, 10, 20, 30, 50, 100, or 120 kg);
α is the angle between opposite faces of the pyramid ($\alpha = 136°$);
d is the arithmetical mean, in mm, of the two diagonals measured after removing the load.

The load is selected in accordance with the size and hardness of the specimen. The thinner the specimen, the less load required. On the other hand, the higher the load, the more accurate the results will be. As a rule, the Vickers hardness number is determined from special tables in accordance with the measured value of d (length of the diagonal). Vickers and Brinell hardness numbers are expressed in the same units and coincide for hardnesses up to about 400. At higher hardnesses, the Vickers number is larger than the Brinell number.

Microhardness. At the present time microhardness tests are widely used to determine the hardness of exceedingly thin layers, very small specimens, and even separate structural components of alloys. This method is based on combining a hardness tester, using a diamond pyramid indenter at low loads, with a metallurgical microscope. Loads from 1 to 200 g are applied. The microhardness number is determined from the formula

$$H = 1.8544 \frac{P}{d^2}$$

in which P is the load, in g,
d is the length of the diagonal, in microns.

Specimens for microhardness tests must be prepared in the same manner as microsections. It is advisable to use electropolishing as this removes the thin layer of cold-hardened metal, that is inevitable in grinding the surface and which may affect the results of the test. The results obtained in measuring the microhardness of the various structural components in a nitrided layer are shown in Fig. 58 (see also page 265).

Fig. 58. Determining microhardness of various phases in a nitrided layer, × 400

5-4. Impact Tests

Static tension tests of unnotched specimens do not always reveal the susceptibility of a metal to brittle fracture. This important factor is determined in impact bending (or simply impact) tests.

Of all types of impact tests, the notched-bar tests are most extensively used (Fig. 59, *a*). Tests are conducted on a pendulum-type impact testing machine (Fig. 59, *b*). The specimen is placed on its supports or anvils so that the blow of the striker is opposite the notch. The pendulum of weight G is then raised to the height h_1, from where it is released to fracture the specimen and rise again to the height h_2. The energy required to fracture the specimen will equal

$$A = G(h_1 - h_2) \text{ kg-m}.$$

The heights of the pendulum before and after the blow may be expressed by the angles through which it is raised. In this case, the energy will be

$$A = Gl(\cos\alpha_2 - \cos\alpha_1) \text{ kg-m.}$$

The initial angle through which the pendulum is raised is a constant value and the upward swing after fracturing the specimen is

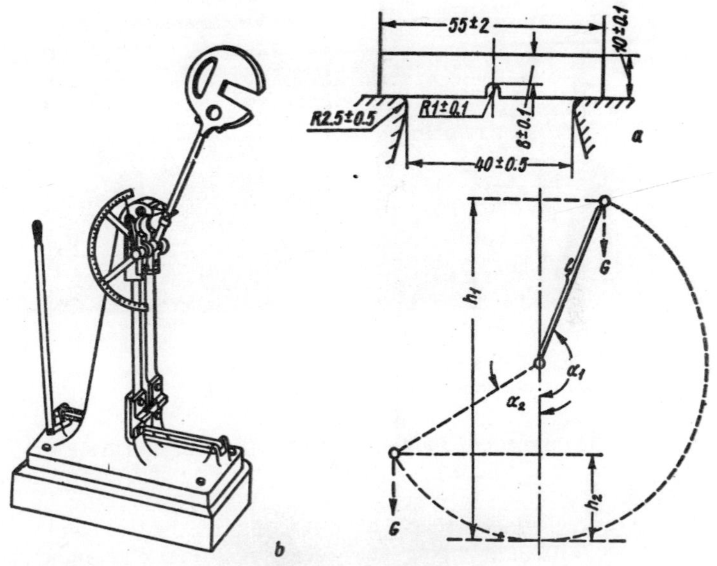

Fig. 59. A notched-bar test:
a—notched-bar test specimen, *b*—pendulum-type impact testing machine

measured on a special circular scale on the machine. When the energy *A*, required to rupture the specimen, is known, the impact strength may be determined:

$$a_k = \frac{A}{f} \text{ kg-m/cm}^2$$

in which f is the cross-sectional area of the specimen, in sq cm, at the fracture.

Impact strength is a complex characteristic which takes into account both toughness and strength of a material.

The main purpose of notched-bar tests is to study the simultaneous effect of stress concentration and high-velocity load application. Impact tests are of the severest type and facilitate brittle fracture. Impact strength values cannot as yet be used for design calculations but these tests are, as a rule, provided for in the specifications of carbon and alloy steels.

In impact tests, the fracture may be either brittle or ductile. Brittle fracture is performed by separation and is not accompanied by noticeable plastic deformation. The fracture, in this case, will have a bright granular or crystalline appearance (Fig. 60, *a*). In ductile fracture, the surface will have a dull-grey fibrous appearance (Fig. 60, *b*) and

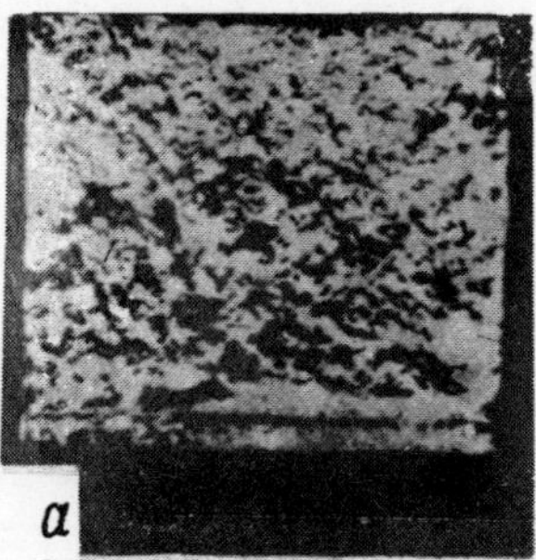

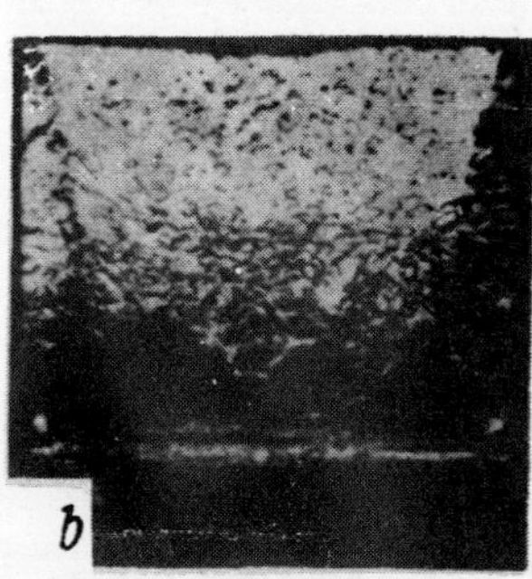

Fig. 60. Fracture in impact test of steel, × 2: *a*—brittle fracture, *b*—ductile fracture

rupture will be preceded by considerable plastic deformation. Frequently a compound fracture, with both crystalline and fibrous areas, is observed.

The type of fracture depends to a great extent on the test conditions. In accordance with the temperature, state of stress, and velocity of load application, many metals and alloys, as mentioned earlier (see page 76), may be transformed from the ductile to the brittle state.

Quality tests in industrial plants are usually conducted at normal room temperature, under the assumption that the test conditions are not less severe than the operating conditions.

It must be noted, however, that many metals, including iron and steel and also zinc, are cold-brittle or cold short. The tendency of a metal for transition from ductile to brittle behaviour upon temperature changes is determined by the "serial" testing method in which impact tests are conducted on a series of specimens at several temperatures, each lower than the preceding one.

Curves indicating the variation of impact strength with the test temperature are shown in Fig. 61. The temperature (t_{cs}) at which the transition from ductile to brittle fracture takes place in a metal is called the *cold-shortness threshold*. Most often, a gradual transition from ductile to brittle behaviour is observed on the impact strength vs temperature curve (Fig. 61, *b*), i.e., a critical temperature range of brittleness is noted. In this case, N. N. Davidenkov suggests that a temperature corresponding to 0.4 $a_{k_{max}}$(Fig. 61, *b*) is accepted as the transition temperature (t_{cs}) or cold-shortness threshold. The higher the transition temperature, the more susceptible a metal is to cold shortness.

Impact strength depends on the structure of a metal. Coarse grain structure and precipitation of brittle layers at the grain boundaries do not appreciably alter the mechanical properties in static tension but

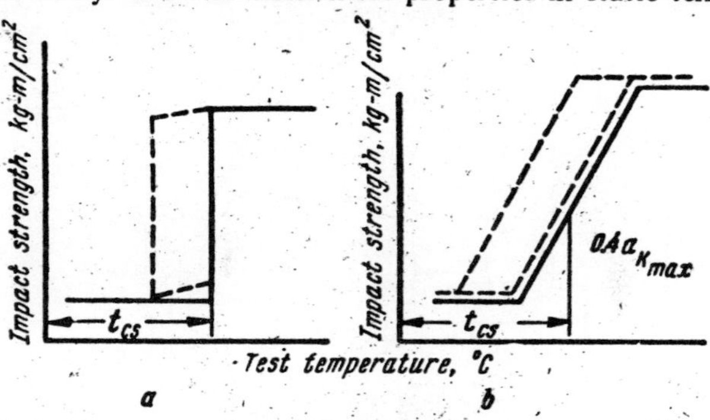

Fig. 61. Critical temperature ranges of brittleness for steel, revealed by impact tests (after N. A. Shaposhnikov):

a—sharp transition, *b*—gradual transition, t_{cs}—critical transition temperature (cold-shortness threshold)

they substantially reduce the impact strength and raise the cold-shortness threshold.

The shape of the notch and the size of the specimen greatly affect the impact strength. Therefore, to obtain comparable results, tests must always be conducted under the same conditions and on the same type of specimen.

In conclusion, it may be noted that impact tests are most frequently conducted for cold-short metals and alloys, steel, in particular, in which the relationship between the cohesive and shear strength varies sharply upon changes in structure and test conditions.

Impact tests on many steels enable the cohesive strength to be determined indirectly when this is impossible for static loading at low temperatures.

Impact tests are of little value for most wrought nonferrous metals (aluminium, copper, and many of their alloys) since there is no transition to brittle behaviour at any temperature. In some cases, due to their high ductility, no fracture whatsoever is observed in specimens of this group of metals subject to impact notched-bar tests. There is also no necessity of conducting impact tests for many cast alloys (cast iron, cast aluminium and magnesium alloys) which show a brittle fracture in ordinary static tension tests.

5-5. Fatigue Tests

Most machine components and structural assemblages are subject to a variation, in both value and sign, in the applied loads (Fig. 62).

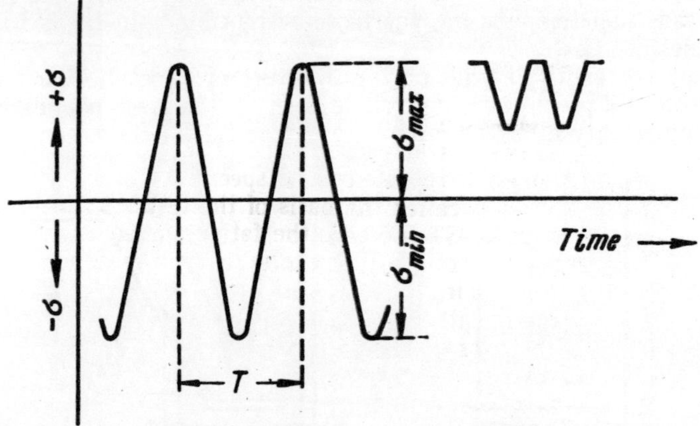

Fig. 62. Examples of cyclic loading

It has been established that metals subject to reversing stresses [compression (—σ) and tension (+σ)], repeated at definite intervals of time (Fig. 62), will fail at stresses which are not only lower than the ultimate strength σ_u, but even lower than the yield point σ_s.

Fig. 63. A fatigue failure

Failure of the specimen will also occur after repeated variations only in the value of the stress even if its direction does not change. The failure of a metal under repeated pulsating or reversing stresses is called *fatigue*. A typical fracture, containing two distinct zones (Fig. 63), is obtained as a result of fatigue failure. The first zone, called the fatigue zone, has a smooth surface and is gradually formed. First, a microcrack is formed in the weakest region of the cross-section. This gradually develops into a macrocrack. After the crack has propagated to occupy a considerable part of the cross-section, brittle fracture occurs in the remaining part. The fracture in this second zone is granular or crystalline in appearance as is typical for brittle fracture.

The exact nature of fatigue phenomena has not as yet been sufficiently explained but, probably, the chief reason lies in the nonuniform distribution of elastic and plastic deformation in micro- (or macro-) volumes of the metal.

In selecting a metal for manufacturing a machine component it is necessary to take into consideration its resistance to fatigue failure, i.e., its ability to withstand a large number of repeated varying stresses.

The resistance of a metal to fatigue failure is characterised by its fatigue limit (endurance limit). The fatigue or endurance limit σ_w is defined as the maximum stress which a specimen can endure without failure when this stress is repeated for a specified number of cycles. This number of cycles is called the basis of the test.

In accordance with GOST 2860-45, the fatigue or endurance limit is determined on the basis of 5 million cycles for steel specimens and on a basis of 20 million cycles for light cast alloys. Many experiments have proved that if a metal withstands the above-indicated number of cycles without failure, it can withstand this same stress for a considerably larger number of cycles.

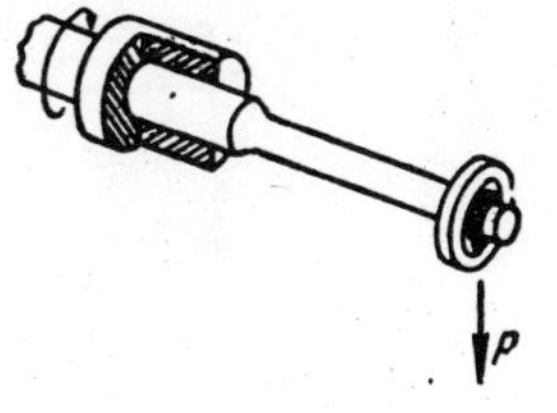

Fig. 64. Principle of a fatigue test

The fatigue limit is usually determined by subjecting a rotating specimen to completely reversed flexural loads which develop reversing stresses (tension-compression) in a symmetrical cycle (called an alternating cycle). A schematic diagram of a fatigue test, with flexural cantilever loading of a rotating specimen, is shown in Fig. 64.

At least six specimens must be tested to determine the fatigue limit. The first specimen is tested at a stress (σ_1 kg/mm²) equal to 0.6 σ_u for steel and the number of cycles *N*, at which failure occurs, is determined. The stresses σ_2, σ_3, etc., for the second and subsequent specimens are increased or reduced by 2 or 4 kg/mm², depending on the number of cycles which caused the failure of the first specimen.

The results of the tests are plotted on a diagram having the coordinates stress (σ kg/mm²) vs number of cycles (*N*) in either rectilinear or logarithmic scales. Such a curve is shown in Fig. 65. The horizontal section is a straight line and is the maximum stress at which failure will not take place after an infinite number of loading cycles. This value is the fatigue limit, which depends to a large extent on the size of the specimen, stress concentrators (fillets, grooves, etc.), surface finish, corrosion, and upon other factors. An increase in the size of the specimen will reduce the fatigue limit. The presence of stress concentrators, such as notches (sharp changes in cross-section, grooves, etc.), on the specimen will sharply reduce the fatigue limit. The finer the surface finish of the specimen (or machine component), the higher the fatigue limit will be. Corrosion substantially reduces the fatigue limit while surface hardening of machine components (carburising, nitriding, case hardening, etc.) raises the fatigue limit and considerably reduces the harmful effect of stress concentrators (see page 76).

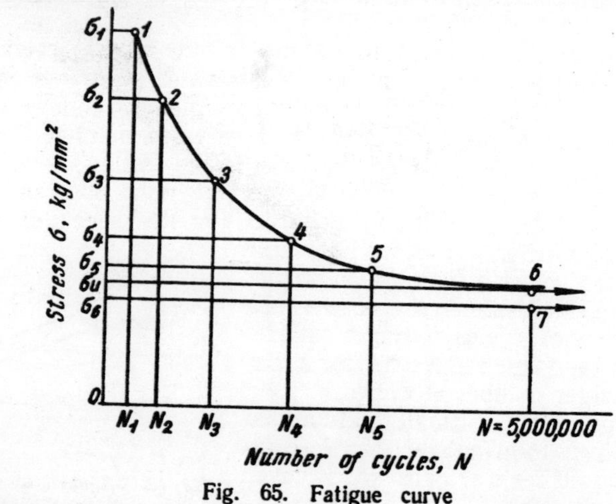

Fig. 65. Fatigue curve

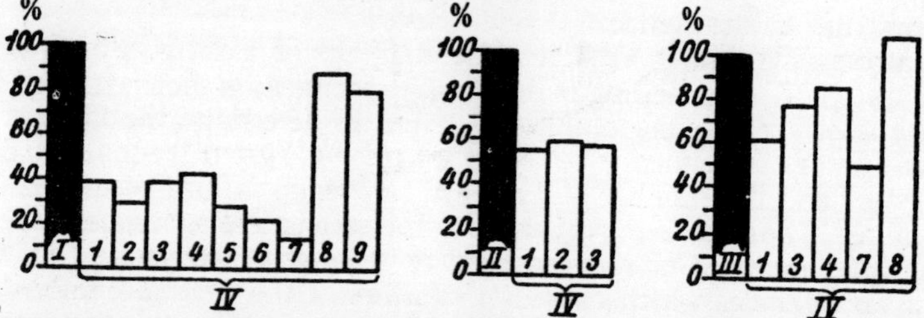
Fig. 66. Relationship between the fatigue limits for a metal in small (*I*), large (*II*), and notched (*III*) specimens, taken as 100 per cent, and the fatigue limits of various machine components (*IV*):

1—crankshaft of an aviation engine, *2*—crankshaft of a diesel engine, *3*—railroad car axle, *4*—plain shaft with a press-fitted hub, *5*—plain shaft with a keyed hub, *6*—toothed wheel, *7*—threaded bolt, *8*—riveted joint of two plates, and *9*—welded joint of two plates

It must be noted, in conclusion, that fatigue or endurance tests do not provide sufficient information concerning the strength of machine components since they do not fully reproduce the forces and treatment that the metal is subject to in manufacturing the component and in subsequent operation in a machine.

This principle is well illustrated in Fig. 66. Therefore, only tests conducted on the components themselves may properly determine their fatigue limits. In the last years, fatigue tests are conducted on the machine components or on specimens of a shape and size near to these components. These are called fullsize fatigue tests.

5-6. Tests at Elevated Temperatures

Many components of modern machinery operate under high-temperature conditions. An increase in temperature has a large effect on all mechanical properties. Research has shown that elevated temperatures reduce the modulus of elasticity (by decreasing interatomic binding), yield point, ultimate strength, and especially the modulus of strain hardening in the process of plastic deformation (see page 74).

If a metal is heated above the recrystallisation temperature, the yield point and ultimate strength may be reduced several times while the modulus of strain hardening may be reduced tens and hundreds of times.

At elevated temperatures, the mechanical properties of metals are a result of the simultaneous processes of strain hardening, due to plastic deformation, and the softening effect of recovery and recrystallisation.

Continuous plastic deformation takes place in metals and alloys, subject to constant loads at elevated temperatures, and it may lead to failure.

The slow continuous plastic deformation of metals under a constant load (especially at high temperatures) is called *creep*.

Special tests are conducted to determine the strength of metals at elevated temperatures. The following chief types of high temperature tests are most frequently conducted:

Short-time tension tests are similar to tension tests at room temperatures except that the specimen is placed in a furnace. The aim of these tests is to determine the strength and ductility at a specified temperature.

Creep tests. The purpose of these tests is to determine the creep limit, i.e., the maximum stress that may be applied for a long period of time at a given temperature if the rate of creep (over this given period of time) is not to exceed a specified value.

A creep testing installation is illustrated in Fig. 67. The specimen is placed in an electric furnace where it is heated to the given temperature and is constantly subject to a load applied by a lever and weights. The strain in the specimen is measured by an optical extensometer with an accuracy of 0.001 mm. Four or five specimens are tested at each temperature under different loads and elongation vs time curves (creep curves) are plotted for each specimen.

The section *Oa* on creep curves (Fig. 68) corresponds to elastic and plastic deformation due to the momentary application of the load. In the following section *ab*, the metal is elongated at a nonuniform (decreasing) rate and in section *bc* the rate of creep is uniform.

The creep limit may be characterised, for example, by the stress which produces a total elongation of 1 per cent in 300 hrs. On the creep curves, this will correspond to the stress σ_3. If the specifications

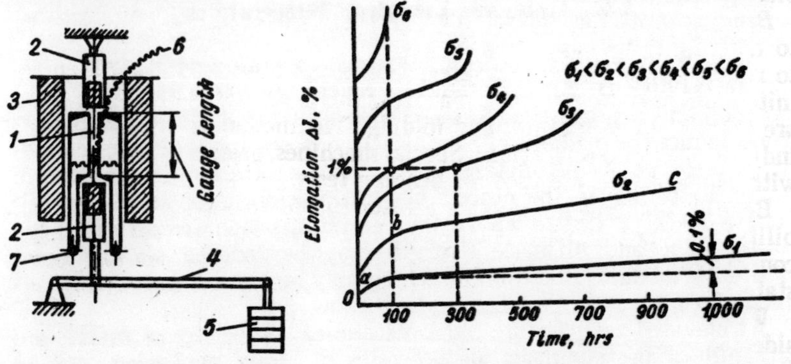

Fig. 67. Installation for creep testing:
1—specimen, 2—grips, 3—furnace, 4—lever, 5—weight, 6—thermocouple, 7—mirror-type instrument for strain measurement

Fig. 68. Creep curves

require that the creep limit be determined for an elongation not exceeding 1 per cent in 100 hrs, the stress value will be σ_4 (Fig. 68).

Only the creep rate for a steady process is taken into consideration for machine components that are designed for continuous operation at elevated temperatures. Limiting conditions are specified for such components. In practice, the stresses are most frequently determined for a total elongation of 1 per cent in 1,000, 10,000, and 100,000 hrs. These correspond to average creep rates of 10^{-3}, 10^{-4}, and 10^{-5} per cent per hour, respectively.

Creep-rupture tests are creep tests which are carried on to the failure of the specimen. In these tests, the time to fracture the specimen, at a given temperature and stress, is measured. Often permanent elongation and reduction of cross-sectional area are also measured.

5-7. Fabrication Tests

Simple tests are often used to check the suitability of metals for particular fabrication techniques. Their purpose is to determine whether a metal can withstand the deformation it will be subject to during operation or during cold or hot working procedures employed in manufacturing.

The results of fabrication tests are usually evaluated by the appearance of the metal after the test; the absence of tears, cracks, lamination, or fracture shows that the metal has passed the test.

Cupping tests are used to check the suitability of thin sheet metal for cold pressing, drawing, and similar forming operations requiring plastic deformation.

Bending tests (on sheets, bars, various sections, etc.) also serve to determine the capability of a metal for plastic deformation and to reveal flaws. These tests are classified as bending through a definite angle, bending about a mandrel until both ends of the specimen are parallel to each other, and folding. The method required may be indicated in the specifications. Special machines, presses or simply vises with rounded jaws are used for bending tests.

Bending-and-unbending tests are conducted to ascertain the capability of a metal (wire, strip, or sheet material) for withstanding considerable plastic deformation. Tests are conducted in the cold state only.

Wire twisting tests are also of the cold type. The specimen is considered to have passed the test if, after twisting (or untwisting), no lamination, cracks, or other defects are revealed.

Fabrication tests on pipe include: 1) expanding tests, 2) reducing tests, 3) flanging tests, and 4) flattening tests.

Chapter 6

BINARY ALLOYS. CONSTITUTION AND EQUILIBRIUM DIAGRAMS

6-1. General Principles of Phase Transformations in Alloys

Most combinations of metals have the property of unlimited mutual solubility in the liquid state.

Certain metals, for example, lead and iron, are almost completely insoluble in each other and separate according to their specific gravity in the liquid state. Complete insolubility is rarely encountered. Limited solubility in the liquid state is much more frequent. In such a case, a homogeneous liquid solution will be obtained if the amount of metal *B* added to metal *A* does not exceed its maximum solubility in metal *A* at the given temperature. If, on the other hand, the amount of added metal *B* exceeds its maximum solubility in *A*, the liquid will separate into two layers. These layers will comprise saturated solutions of *B* in *A* and of *A* in *B*. With few exceptions (Fe-Cu, Ni-Ag, Cu-Cr, and others), limited solubility in the liquid state is characteristic of metals whose melting points and atomic volumes sharply differ (Cu-Pb, Zn-Pb, Cr-Sn, etc.).

When a metal is alloyed with another metal or with a nonmetal, chemical interaction occurs between the atoms of the elements forming the solution. The valence electrons of the alloying elements, weakly bound to their nuclei, do not belong to the separate atoms but are shared by the whole aggregate, as in the solid phase, comprising a single system for the whole liquid solution. The distribution in a liquid solution of atoms (ions) of elements constituting an alloy depends on the relation of the interacting forces between like and unlike atoms.

The components of an alloy may either be dissolved one in the other in the solid state to form solid solutions or chemical compounds or, after solidifying, they may produce mechanical mixtures of the two components *A* and *B*, solid solutions, etc. In the formation of a solid solution, atoms of the dissolved component (the solute) will substitute for a part of the atoms in the lattice of the solvent or they will be accommodated in the interstices (see Sec. 6-4). In the majority of cases, the components of an alloy have limited mutual solubility in the solid state. Many metals, however, are mutually soluble to an unlimited extent.

If components are present in an amount exceeding the solubility limit, the surplus metal forms an independent phase comprising either a saturated solid solution or a chemical compound or, finally, crystals of one of the components. A chemical compound differs from a solid solution in that it often has its own crystal lattice and in many cases a definite quantitative relation of the components is required for its formation.

The transition of an alloy from the liquid to the solid state is associated with the conversion of the system to a state of less free energy. As in the solidification of pure metals, this transition will be initiated only if the alloy is supercooled to some extent and will then proceed by nucleation and subsequent growth of the nuclei as dendrites or as holohedral crystals. In actual conditions, the crystals, produced in the solidification of metal alloys, are usually dendritic.

When alloys solidify, the newly-formed phases differ substantially in composition from the initial liquid solution. Therefore, to form a stable nucleus or embryo, it will not be sufficient to have fluctuations of energy, as in a single-component system; there must also be fluctuations of concentration which correspond to zones not less than critical in size.

Concentration fluctuations are chance, tentative deviations of the chemical composition from its average analysis and occur in separate small volumes of the liquid (or solid) solution. The reason for the occurrence of fluctuations of concentration lies in the diffusion of atoms in a liquid (or solid) solution due to their thermal vibrations. Nuclei of the new phase may form only in those microvolumes of the liquid phase where, due to fluctuations, the concentration conforms to the composition of the solidifying phase. If then, such fluctuations in energy and concentration correspond to microvolumes of a size not less than the critical value, a stable embryo will be formed, one that is capable of subsequent growth.

The rate of crystal growth in solutions is considerably less than in pure crystals. The reason is that crystal growth (in a solution) is not only associated with the formation of two-dimensional nuclei on the crystal faces, as in pure metals, but also with atomic diffusion of the components in the liquid solution. The more the difference in concentration between the liquid solution and the crystals into which it solidifies, the slower the rate of crystal growth will be.

After an alloy has solidified, structural changes may occur that are caused by polymorphic (allotropic) transformations and by the decomposition of the solid solution due to changes in mutual solubility of the components in the solid state.

Transformations in the solid state are of a crystallisation type only and proceed by nucleation of the new phases and subsequent growth.

A distinguishing feature of solid state transformations is that the principle of structural and size conformity is usually observed in nucleation of the new phase. This means that the nucleus of the new phase is oriented so that it joins the initial phase on definite crystallographic planes which more closely match each other as to atomic arrangement and interatomic distances.

As long as the lattices on the boundary of the old and new phases match along definite crystallographic planes (Fig. 69) or, as it is

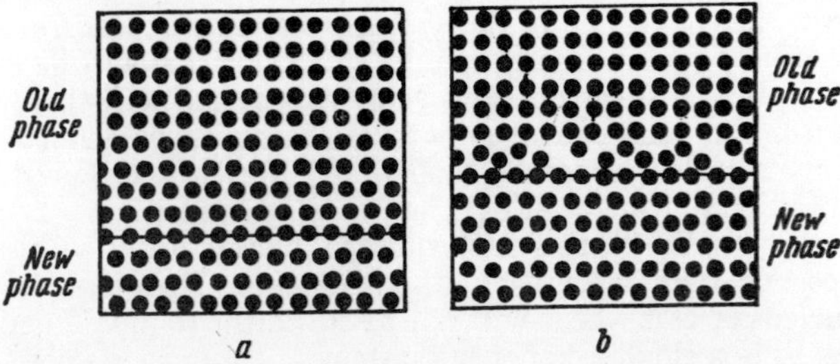

Fig. 69. Schematic diagrams of the relationship between the lattices of the old and new phases:

a—continuous transition of one lattice into the other upon crystal growth of the new phase coherent interface), *b*—lack of continuity, as well as disorder at the interface, stops coherent growth (after G. V. Kurdyumov)

called, the lattices are coherent, the growth of the new phase proceeds rapidly since the atoms pass orderly over small distances. The nucleation of the new phase, however, will create elastic energy due, for example, to the difference in specific volume of the initial and new phases. At a certain moment, the magnitude of the elastic energy may exceed the elastic limit of the medium. This will result in shear deformation and loss of coherency. Further coherent growth will be impossible. It must be noted that coherent growth may also be violated by the thermal vibrations of the atoms.

As soon as the lattices no longer match on the boundary between the phases, further growth of the nucleus proceeds at a much slower rate since the atoms pass over from one phase to the other in a disorderly manner. At low degrees of supercooling, coherent growth rapidly ceases and the transformation will have a disorderly character. Upon high degrees of supercooling, due to rapid cooling, in many cases the new phase will be subject to coherent growth. Violation of coherency will stop the growth of the new phase since the diffusion processes, required for the passage of atoms from one phase to the other, become impossible at low temperatures.

Nuclei of the new phase are easily formed on grain boundaries and around various inclusions. This is explained by the fact that grain boundaries, as well as the surface of inclusions, have a layer of atoms with increased free energy where the new phase may appear.

Therefore, the smaller the grains, the more rapidly transformations will proceed. The formation and growth of nuclei of a new phase in the solid state are hampered by the difficulty in obtaining the required fluctuations of concentration. This difficulty, in turn, is due to the low rate of diffusion and the influence of elastic energy. This explains why phase transformations proceed at a high rate only upon considerable supercooling of the initial phase.

6-2. The Phase Rule and Equilibrium Diagrams

The solidification of metal alloys is clearly demonstrated by means of equilibrium diagrams which are convenient graphic representations of changes in state due to variations in temperature and concentration. Equilibrium diagrams, also called constitutional and phase diagrams, enable the phase content of the alloy to be determined at any temperature and composition. They enable the phase transformations to be followed in heating or cooling the alloy under equilibrium conditions, i.e., when all processes in the given system are reversible. This means that changes occurring in a system as a result of processes proceeding in one direction are fully compensated by changes due to the reversal of the process in the system.

A system is here understood as the whole complex of phases of one or several components at different pressures and compositions.

The *components* are the substances (chemical elements or compounds) whose presence is necessary and sufficient to make up a system. For example, a pure metal comprises a one-component system, an alloy of two metals is a two-component (binary) system, etc.

A *phase* is a physically and chemically homogeneous portion of a system, separated from the other portions by a surface, the interface. For instance, a homogeneous liquid solution is a single-phase system; a mixture of crystals of two types, differing in composition and structure and separated by an interface, or the coexistence of the liquid alloy and its crystals comprise two-phase systems.

All changes which take place in a system consisting of several phases, in accordance with external conditions (temperature and pressure), conform to the so-called phase rule. The *phase rule* establishes the relationship between the number of degrees of freedom, the number of components, and the number of phases. It is expressed mathematically as follows *:

$$F=C+n-P$$

* For a thermodynamical derivation of the phase rule see: N. M. Wittorf, *The Science of Alloys*, Moscow, 1927; and A. A. Bochvar, *Physical Metallurgy*, Metallurgizdat, Moscow, 1956.

in which F is the number of degrees of freedom in the system (the number of variable factors);
C is the number of components in the system;
P is the number of phases in equilibrium;
n is the number of external factors (for example, temperature and pressure).

The number of degrees of freedom is the quantity of independent external or internal variable factors (temperature, pressure, and concentration) which may be altered without causing the disappearance of a phase or the formation of a new phase in the system.

In studying chemical equilibrium, temperature and pressure are regarded as external factors determining the state of the system. The effects of pressure may be neglected in applying the phase rule to metal systems leaving only one variable external factor—temperature. The equation will then be:

$$F=C+1-P$$

All internal and external factors (concentration and temperature, respectively) have definite values in a system that is in equilibrium.

Since the degrees of freedom cannot be less than zero

$$C-P+1\geqslant 0,$$

then

$$P\leqslant C+1,$$

i.e., the number of phases in a system cannot exceed the number of components plus one. Therefore, no more than three phases may be in equilibrium in a binary system, no more than four in a ternary system, etc. In cases when the maximum possible number of phases are in equilibrium, the number of degrees of freedom equals zero ($F=0$). This is called *nonvariant* equilibrium.

A system in nonvariant equilibrium may exist only under entirely definite conditions: at a constant temperature and at a definite composition of all phases involved.

A pure metal at the solidification temperature, for example, is a one-component system consisting of two phases of identical composition

$$F=1+1-2=0$$

This means that it is a nonvariant system. In this case the temperature cannot be selected or changed arbitrarily. There is only one temperature at which the system is in equilibrium. This is the melting (or solidifying) point for the given metal. If the number of phases is less than the maximum possible number by one, the number of degrees of freedom will also increase by one ($F=1$). Such a system is said to be *monovariant*.

An alloy of two metals, for example, is a two-component, two-phase system, in the general case, at the beginning of solidification. Therefore, $F=1$ in this case. A system with $F=2$ is *divariant*. Therefore, the system may be in equilibrium at different temperatures and concentrations.

A number of examples of practical application of the phase rule for studying the phase diagrams of two-component alloys ($C=2$) are given below.

For practical purposes it is essential to know the aggregate state and constitution of alloys depending on the temperature and concentration. If the state of an alloy of given concentration is known at any temperature, it is possible to predict its properties, its capability for heat treatment and plastic deformation (forging, stamping, rolling, etc.), its behaviour in casting, its suitability for various operating conditions, etc. All transformations occurring in an alloy and depending on temperature and concentration can be clearly represented by equilibrium diagrams. These diagrams are plotted with the concentration as the abscissa and the temperature as the ordinate.

Thermal analysis is employed as the basis in plotting equilibrium diagrams and especially for determining freezing points. By this method, extensively developed by N. S. Kurnakov, cooling curves are plotted first. Then the *critical points* are determined by the abrupt inflexion of the curves due to thermal effects in transformations. At the present time, all methods of physico-chemical analysis are employed in studying solid-state transformations. They include: micrography, X-ray, dilatometric and magnetic analyses. A complete equilibrium diagram may be drawn after experimentally determining the critical points of a series of alloys.

In principle, equilibrium diagrams represent quite definite types of equilibrium: complete or partial solubility, presence or absence of chemical compounds, etc., and may be drawn on the basis of theoretical considerations.

In some cases, actual equilibrium diagrams of alloys almost completely conform to these theoretical types of diagrams. More frequently, they are a combination of several types of diagram.

Equilibrium diagrams are valid only under the condition that all processes, which may occur in the given system, are of the equilibrium type.

The main types of two-component or binary equilibrium diagrams are considered in the succeeding sections.

7—1378

6-3. Equilibrium Diagram of a Binary System in Which the Components Form a Mechanical Mixture of Crystals in the Solid State and Are Completely Mutually Soluble in the Liquid State

After solidification the alloy will consist, in this case, of a mechanical mixture of the two initial components. *

Metals, which form alloys of this type, retain their crystal lattices. This has been proved by radiographs which reveal lines corresponding to both components.

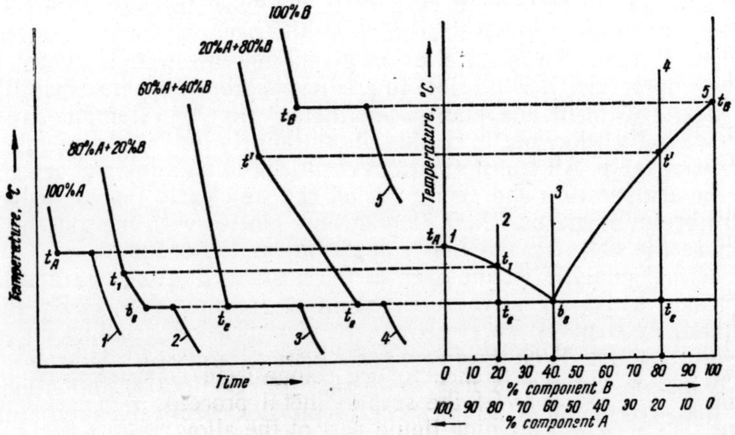

Fig. 70. Constructing an equilibrium diagram for a system of components which are completely mutually soluble in the liquid state and form a mechanical mixture upon freezing

We shall consider equilibrium diagrams of the first type using alloys composed of metals with the generalised symbols A and B as an example.

We shall first melt metal A and then cool it slowly, measuring the temperature after equal intervals of time.

As is evident in Fig. 70, the temperature of the pure metal A falls gradually to t_A at which metal A solidifies. A horizontal step appears on the curve because, according to the phase rule, the solidification of a pure metal proceeds at a constant temperature ($F=1+1-2=0$). When the metal has solidified completely the uniform drop in temperature is resumed.

* Strictly speaking, complete mutual solubility of two metals never exists. A consideration of this diagram, however, is of methodological value as it facilitates the study of the subsequent types of equilibrium diagrams.

If, now, an alloy of 80 per cent component *A* and 20 per cent component *B* is taken instead of the pure metal, the temperature will fall uniformly to t_1.

First we note the inflexion in the cooling curve at t_1 (first critical point). After this point, the rate of cooling is markedly reduced due to the evolution of the latent heat of fusion resulting from the freezing out of a definite number of component *A* crystals.

Solidification proceeds at the new rate until temperature t_e is reached. At temperature t_e, the fall in temperature ceases (the horizontal step on the curve). This corresponds to the completion of solidification of the alloy at a constant temperature (second critical point).

In cooling an alloy of 60 per cent *A* and 40 per cent *B*, it will remain liquid until it reaches temperature t_e. At t_e there is a single long temperature stop associated with solidification of the whole alloy at constant temperature.

In the solidification of alloys containing 20 per cent *A* and 80 per cent *B*, crystals of component *B* begin to freeze out of the liquid phase at temperature t'. The alloy completely solidifies at temperature t_e.

The cooling curves indicate that all alloys of the *A-B* system solidify completely at the constant temperature t_e. From this it can be assumed that the part of the liquid alloy which solidifies at the constant temperature t_e is also of constant composition. This composition corresponds to the alloy consisting of 60 per cent *A* and 40 per cent *B* which completely solidifies at temperature t_e.

In all other compositions of the alloy, the metal that solidifies first is surplus in reference to the composition containing 40 per cent *B*. Solidification of the surplus metal proceeds until the concentration of the remaining liquid part of the alloy reaches a composition of 40 per cent *B* and 60 per cent *A*.

In alloys of this composition, the components *A* and *B* solidify at the same time at the constant temperature t_e. This occurs because, according to the phase rule, there is nonvariant equilibrium when three phases (one liquid and two solid phases, *A* and *B*) exist simultaneously in a two component system, i. e., $F=2+1-3=0$.

The surplus component (*A* or *B*) freezes out of the liquid alloy in a temperature interval since the system has one degree of freedom:

$$F=2+1-2=1.$$

Alloys in which the components solidify simultaneously at a constant temperature, the lowest for the given system, are called *eutectic alloys*.

A mechanical mixture of two (or more) phases, which solidify simultaneously from the liquid alloy, is called a *eutectic*. *

* From the Greek word *eutektos* which means: easily melted.

If all the critical points that have been determined are plotted on a diagram with the compositions (concentrations) used as abscissas and the temperatures as ordinates (Fig. 70) and then if all these points of the same type (representing the same processes) are connected by smooth curves, the equilibrium diagram of the system of alloys *A* and *B* will be obtained.

The alloys begin to solidify at temperatures which form the line *acb* (Fig. 71) called the *liquidus*. As indicated above, pure metal *A*

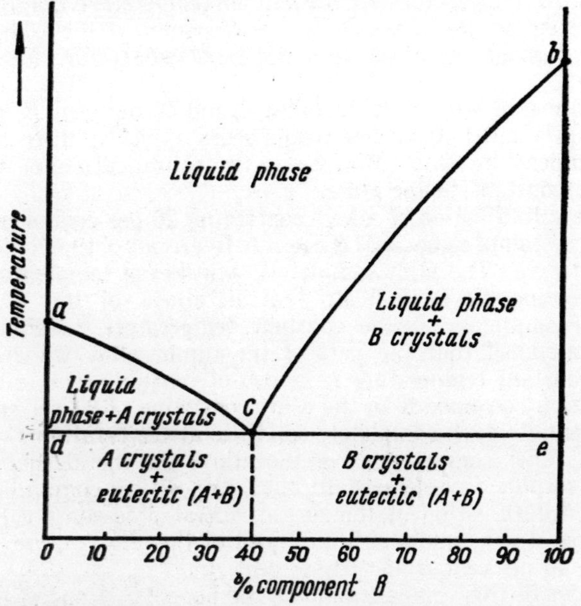

Fig. 71. Equilibrium diagram of a system of components *A* and *B* which are completely soluble in the liquid and insoluble in the solid state

begins to solidify out of the liquid alloy along the line *ac* and pure metal *B* along the line *cb*.

The complete solidification of any alloy takes place at the constant temperature t_e which, on the diagram, is represented by the horizontal line *dce* (Fig. 71). This line is called the *solidus* and, in the given case, the *eutectic temperature line*.

At point *c* both components, *A* and *B*, solidify simultaneously to form only the eutectic.

The eutectic alloy (40 per cent *B*), as is clear in the diagram, solidifies at a constant temperature t_e to produce the eutectic which consists of crystals of the components *A* and *B*. When the eutectic temperature is reached, the liquid alloy is saturated by both components at the same time. Solidification begins only after a certain degree of supercooling is reached, when the solution becomes supersaturated by both components and freezing of the eutectic becomes thermodynamically possible.

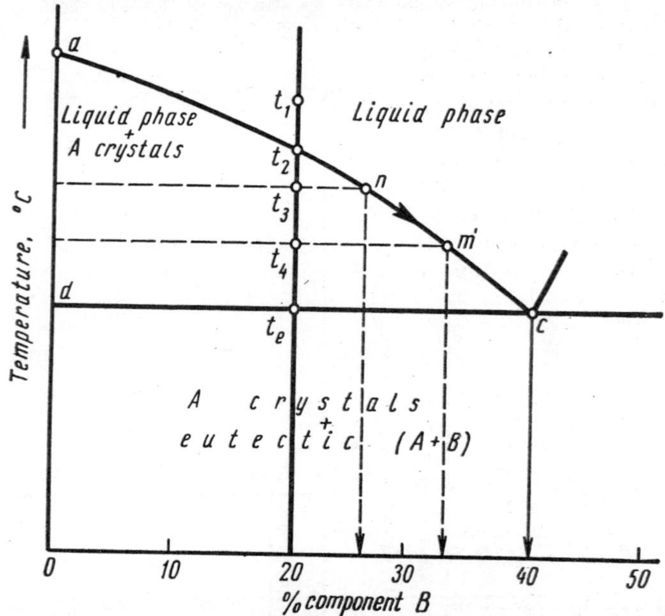

Fig 72. Portion of the equilibrium diagram for the system of components *A* and *B* (hypoeutectic alloys)

First, crystals of one component (for example, component *A*) appear and begin to grow. The liquid surrounding these crystals (*A*) becomes richer in the second component (*B*) whose crystals begin to freeze out as a result. Next, the liquid adjacent to the *B* crystals becomes richer in metal *A* whose crystals again begin to freeze out. Thus, by alternate supersaturation of the liquid by components *A* and *B*, the eutectic embryo is formed. This nucleus continues to grow since new atoms pass over from the liquid to the surface of crystals *A* and *B*.

Thus, a continuous diffusion process takes place in the liquid during eutectic solidification. In this process, atoms of component *A* continuously approach *A* crystals and atoms of component *B* approach *B* crystals.

The solid phase, first formed in the liquid, is called the primary phase. Crystals of the primary phase may have various forms. They may be dendritic, lamellar, or spheroidal (globular).

The secondary phase fills the spaces between the crystals or between the branches of the dendrites of the primary phase.

We may now consider the solidification process in certain alloys of the system formed by components *A* and *B* (Figs. 72 and 73).

An alloy containing 20 per cent *B*, begins to solidify when it reaches temperature t_2 (with a certain degree of supercooling, of course) (Fig. 72) where crystals of component *A* begin to separate from the liquid alloy. As the temperature falls, component *A* continues to solidify so that the composition of the remaining liquid phase of the alloy changes toward a higher *B* content.

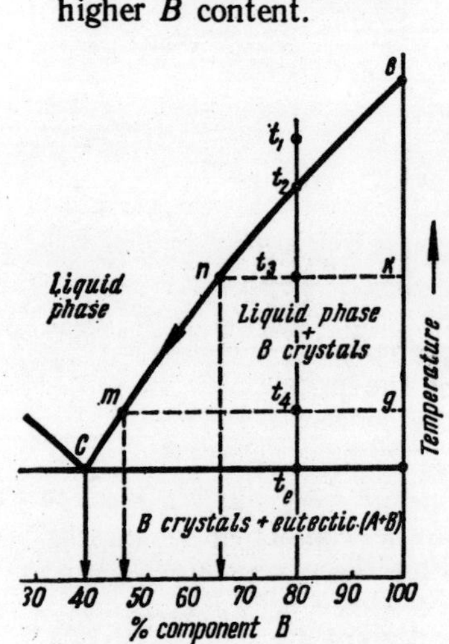

Fig. 73. Portion of the equilibrium diagram for the system of components *A* and *B* (hypereutectic alloys)

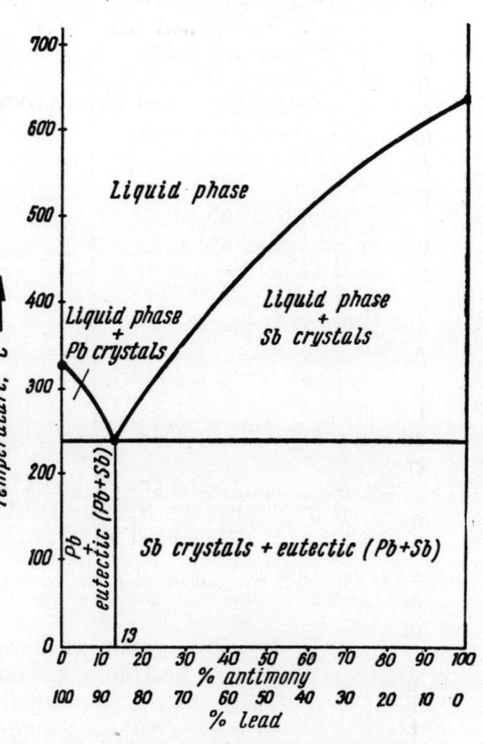

Fig. 74. Equilibrium diagram for Pb-Sb alloys

The composition of the liquid phase that is in equilibrium with the crystals of component *A* may be determined for any temperature between the liquidus and solidus (for example, t_3 and t_4) by drawing lines through the temperatures parallel with the axis of concentrations (Fig. 72). The intersections of these lines with the liquidus (*n* and *m*) indicate the compositions of the liquid phase at the given temperatures. It is evident from Fig. 72, that the composition of the liquid portion of the alloy varies along the liquidus, when the temperature is lowered, approaching the eutectic concentration of 40 per cent B (point c). At the temperature t_e, the liquid portion of the alloy solidifies at constant temperature with simultaneous freezing of components A and B which form the eutectic mixture.

After complete solidification, the alloy will contain crystals of component A, which initially separated in the temperature interval from t_2 to t_e, and the eutectic, produced upon final solidification of the alloy at temperature t_e.

Alloys containing less than 40 per cent B are located to the left of point c on the diagrams (Figs. 71 and 72) and are called *hypoeutectic alloys*. After solidifying, all hypoeutectic alloys have a structure containing the eutectic and surplus component A (Figs. 71 and 72). The nearer an alloy is in composition to the eutectic alloy (40 per cent B), the more eutectic and less surplus crystals of component A it will contain.

The solidification of an alloy containing 80 per cent B (Fig. 73) begins at temperature t_2 where surplus crystals of metal B start to precipitate from the liquid alloy. Lowering the temperature will change the composition of the remaining liquid portion of the alloy along the curve bc (liquidus) due to further separation of B crystals. At the temperature t_3, for example, the composition is determined by point n and at temperature t_4, by point m. At temperature t_e, the liquid alloy will reach the eutectic composition (40 per cent B + 60 per cent A) and will solidify to form the eutectic mixture of A and B crystals.

All alloys located on the diagram to the right of point c (containing more than 40 per cent B) are called *hypereutectic alloys*. After solidifying, they will consist of crystals of the primary component B surrounded by the eutectic. The nearer the concentration of the alloy approaches the eutectic composition, the more eutectic and less crystals of surplus component B will it contain.

The equilibrium diagram enables the quantitative ratio of the phases to be determined at any point during solidification. The relative amounts of the liquid and solid phases present at temperature t_3, for example, in a hypereutectic alloy containing 80 per cent B (Fig. 73) are determined by the lever rule. This rule states that the ratio of the amount by weight of the liquid phase to that of the solid phase is inversely proportional to the corresponding sections of the straight line nt_3k:

$$\frac{\text{liquid phase}}{\text{solid phase}} = \frac{t_3k}{t_3n}$$

Fig. 74 illustrates the actual equilibrium diagram for the system Sb-Pb.* It closely resembles the typical diagram described above.

* For the sake of simplicity, it has been assumed that Pb and Sb do not form solid solutions with each other. Actually these metals exhibit slight mutual solubility in the solid state.

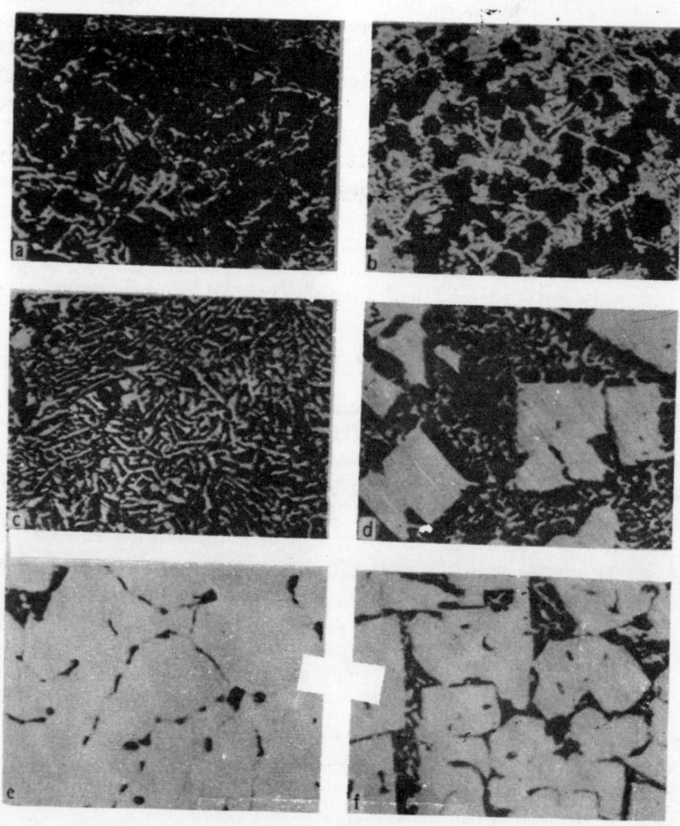

Fig. 75. Microstructure of Pb-Sb alloys, × 250:
a—hypoeutectic alloy of 96 per cent Pb and 4 per cent Sb, lead crystals (darker) and eutectic (Pb+Sb); *b*— hypoeutectic alloy of 94 per cent Pb and 6 per cent Sb, lead crystals (darker) and eutectic; *c*—eutectic alloy, 87 per cent Pb and 13 per cent Sb, eutectic (darker areas are lead and lighter areas are antimony); *d*—hypereutectic alloy of 70 per cent Pb and 30 per cent Sb, antimony crystals (lighter) surrounded by eutectic (Pb+Sb); *e*—hypereutectic alloy of 26 per cent Pb and 74 per cent Sb, antimony crystals and eutectic; *f*—hypereutectic alloy of 95 per cent Sb and 5 per cent Pb, antimony crystals and eutectic

Photomicrographs of Pb-Sb alloys, shown in Fig. 75, illustrate their structure. After solidification, these alloys form a simple mechanical mixture of the components.

A phenomenon known as gravity segregation is frequently observed in the solidification of alloys whose behaviour conforms to the first type of equilibrium diagram. This segregation (or liquation) is also evident in Pb-Sb alloys. The Pb or Sb crystals that precipitate in the process of solidification differ considerably in specific

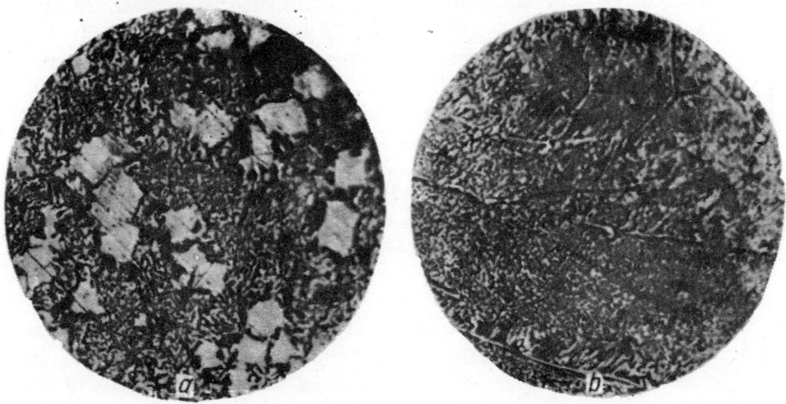

Fig. 76. Microstructure of the Pb-Sb alloy with 20 per cent Sb, × 250 (after Turkin and Rumyantsev):

a—upper part of the ingot showing segregation of antimony crystals (lighter crystals); *b*—lower part of the ingot containing almost pure eutectic (Pb+Sb)

weight from the remaining liquid portion of the alloy and, consequently, they either rise to the surface (Sb) or sink to the bottom (Pb). As a result, the ingot will not be homogeneous in composition. The photomicrographs in Fig. 76 show the structure of microsections cut out of the upper and lower parts of an ingot (Fig. 76, *a* and *b*). The alloy, in this case, contains 20 per cent Sb and 80 per cent Pb. These photomicrographs clearly show that the upper part of the ingot is rich in antimony due to segregation while the lower part is practically free of surplus antimony crystals and consists of the eutectic only. A number of measures are employed to prevent gravity segregation. They include: rapid cooling, stirring of the liquid alloy, and adding a third component which freezes first in the form of branching dendrites that prevent separation of the alloy into layers by purely mechanical means.

The properties of alloys, consisting of a mixture of two phases, depend on the properties of the phases formed in the alloy. N. S. Kurnakov (1860-1941) first showed the definite relationship between the composition, type of equilibrium diagram, and properties of an alloy. If a simple mechanical mixture is formed under conditions of low degrees of supercooling (Fig. 77), low temperatures, and different structural state of the alloys, certain of their properties will vary along a straight line.

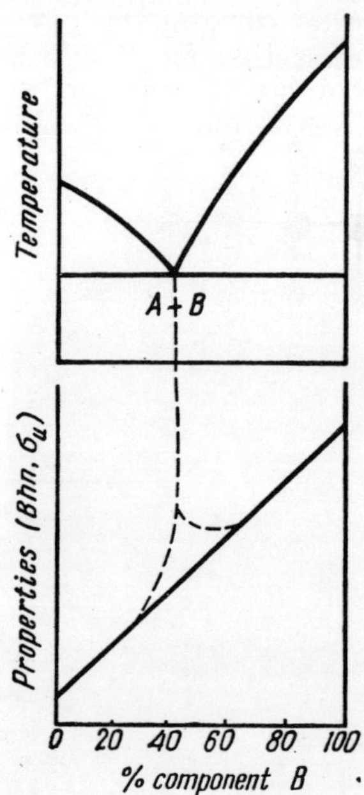

Fig. 77. Relationship between properties and composition of alloys comprising a simple mechanical mixture of the components

This means that they will be the average of the properties possessed by the initial components (or component phases). If the phases comprising the mixture form fine crystals, as in eutectic alloys, the strength and hardness are increased. In this case, the linear relationship between properties and composition does not hold true.

6-4. Equilibrium Diagram of a System Whose Components Are Mutually Soluble in Both the Liquid and Solid States

General features of solid solutions. Most metals form homogeneous liquid solutions in the liquid state. In their transformation to a solid crystalline state, the homogeneity of many alloys is retained. Consequently, their solubility is also retained. Solid phases in which the ratio of the components may vary without violating the homogeneity are called *solid solutions.*

In a solid solution, the different types of atoms of the components in the alloy form a common crystal lattice. The component, whose lattice is retained, is called the solvent.

Solid solutions are classified as *substitutional* (Fig. 78, *a*) and *interstitial* (Fig. 78, *b*). In a substitutional solid solution, a part of the atoms at the points of the solvent crystal lattice have been replaced by solute atoms (Fig. 78, *a*).

Since the sizes of the solvent and solute atoms differ, the crystal lattice is always distorted and the lattice constants are varied (Fig. 79) in a substitutional solid solution.

As a rule, if the solute atom has a larger radius than that of the solvent atom, the lattice constant will be increased and vice versa.

In interstitial solid solutions, the atoms of the solute metal occupy the vacant positions or interstices between the atoms of the solvent. Interstitial solid solutions may form only if the diameter of the solute atom is small and may be accommodated in the interstices of the solvent crystal lattice. Thus, solid solutions of this type are only found when elements with a small atom radius such as carbon, nitrogen, and hydrogen are dissolved in metal (for example, iron). An interstitial solid solution always increases the lattice constant of the solvent.

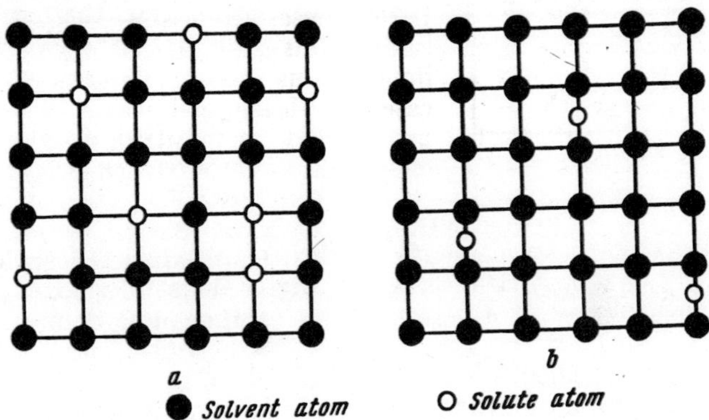

Fig. 78. Crystal lattice of a solid solution:
a—substitutional, *b*—interstitial

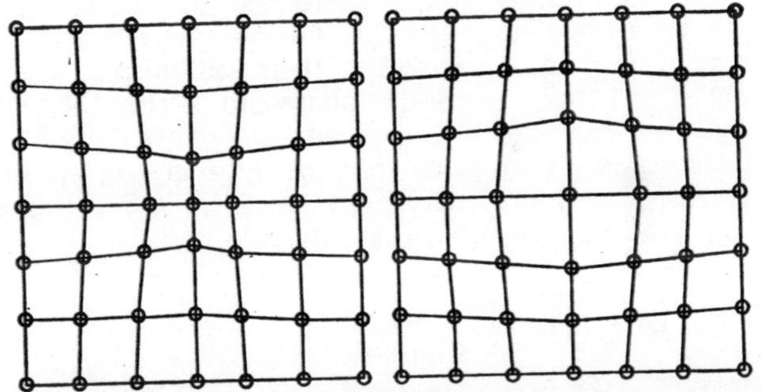

Fig. 79. Distortion of the crystal lattice in substitutional solid solutions

The metallic type of bond is found in all metal-base solid solutions without regard for the type of bond peculiar to the solute in the solid state. Therefore, nonmetallic atoms (C, N, Si, etc.) do not change the metallic bond of metallic solid solutions.

In a number of metal systems (for example Cu-Au, Fe-Al, Fe-Si, Ni-Mn, etc.) the atoms of both components occupy entirely definite positions in the solid solution lattice (Fig. 80) if the alloy has been slowly cooled or has been heated for a long time at a definite temperature. To distinguish such solid solutions from those in which the atoms of the components assume statistically random repiacement positions (Fig. 78), the former are called *ordered solid solutions*. This new order of symmetry superimposed on the solvent lattice results in a structure called a "superlattice". Ordered solid solutions usually occur when the ratios of the alloyed components in atomic per cent are equal to whole numbers: 1 : 1, 1 : 3, etc. In this case, the formula of a chemical compound, for example, Cu_3Au (Fig. 80), Fe_3Al, etc., may be assigned to the ordered solid solution.

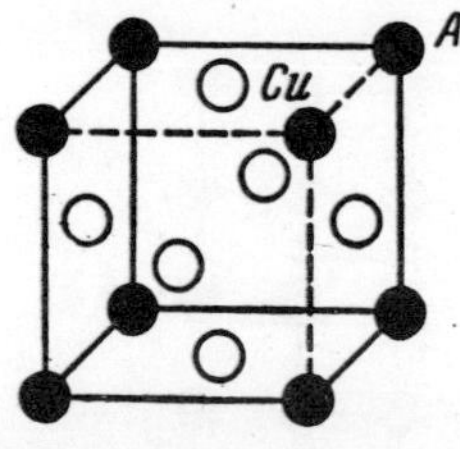

Fig. 80. Unit cell of an ordered solid solution (Cu_3Au)

The formation of ordered solid solutions is accompanied by sharp alterations in the properties of the alloy. Resistivity falls sharply, magnetic properties change, the alloy becomes very hard and brittle, etc. Ordered solid solutions may be regarded as intermediate phases between solid solutions and chemical compounds.

Ordered arrangement of the atoms of both components in the lattice and a substantial change in properties are also characteristic of chemical compounds. Ordered solid solutions differ from chemical compounds in that the solvent lattice is retained and in that a disordered solid solution is gradually formed upon heating without altering the type of lattice.

A solid solution, observed under an optical microscope, can be seen to consist of homogeneous crystalline grains (Fig. 36) as in pure metals.

Substitutional solid solutions may be limited and unlimited or continuous, i. e., they may occur only in a limited region of alloy compositions or at any composition.

a. Equilibrium Diagrams of Systems Whose Components Have Unlimited Solubility in Both the Liquid and Solid States (Fig. 81)

The chief conditions for unlimited solubility in the solid state are that the two components should have the same type of crystal lattice and that the sizes of the atoms be very similar; the difference in size for iron-, nickel-, or cobalt-base alloys must not exceed 8 per cent. A difference in size over 15 per cent prevents the formation of solid solutions due to the extreme distortion of the solvent crystal lattice.

If the atoms differ in size by more than 15 per cent, the formation of a chemical compound with a different crystal lattice or of a simple mechanical mixture of the components may be more probable from an energetical point of view.

The fact that elements have melting points near to each other and that they occupy neighbouring positions in the periodic table will facilitate the formation of a continuous series of solid solutions.

As mentioned earlier, the crystal lattice of a metal consists of positively charged ions held together by uniformly distributed electrons which are free to move from atom to atom.

The ratio of the number of valence electrons to the number of atoms is called the *electron ratio*. Each type of crystal lattice has its critical electron ratio at which it reaches its maximum stability. The critical

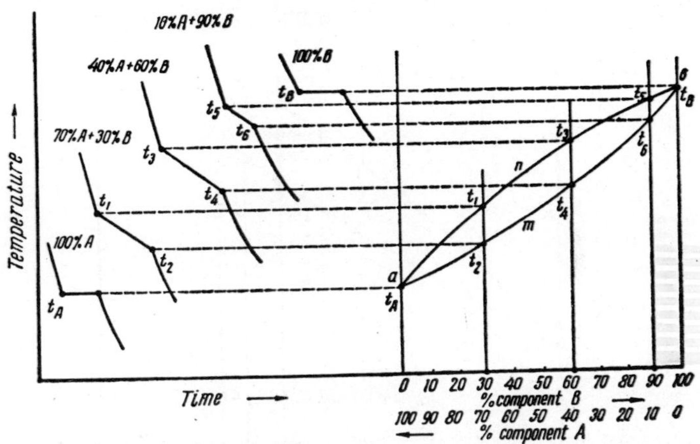

Fig. 81. Constructing an equilibrium diagram for a system of components that are completely mutually soluble in both the liquid and solid states

electron ratio, determined on the basis of theoretical calculations, for a face-centred cubic lattice is 1.36 and for a body-centred cubic lattice 1.48 electrons per atom.

A continuous series of solid solutions may be formed only when the electron ratio remains constant or does not exceed the critical value at any composition of the components.

Conformation with electron ratio rule is an important condition for the formation of a continuous series of solid solutions in many systems. It must be noted, however, that such series are not always formed even when the electron ratio is suitable. This shows that solid solution formation is a complex chemical interaction whose laws are still insufficiently clear.

Fig. 81 shows how an equilibrium diagram is drawn for components that are completely mutually soluble in both the liquid and solid states. In the complete form this diagram is illustrated in Fig. 82.

The upper line corresponds to the temperatures at which the alloys begin to solidify, the liquidus. The lower line, the solidus, indicates the completion of solidification.

In the temperature interval between the liquidus and solidus, the alloys are in a semi-solid state, i. e., they consist of crystals of a solid solution of components *A* and *B* and the liquid alloy.

Solidification begins with nucleation and subsequent growth of the nuclei. Nucleation occurs in regions of the liquid phase where,

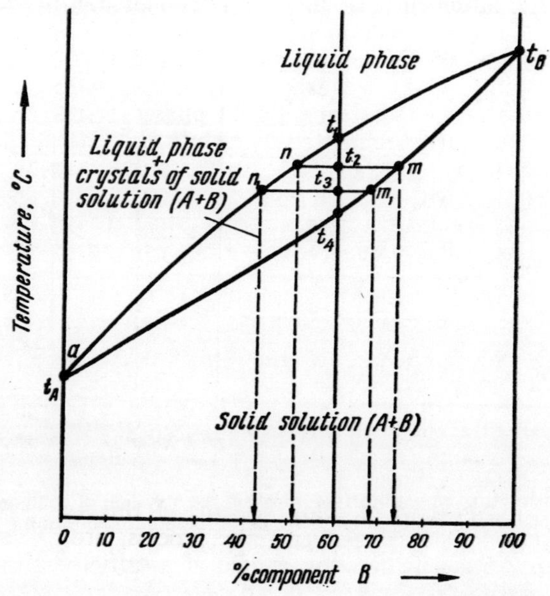

Fig. 82. Equilibrium diagram for a system of components that are completely mutually soluble in both the liquid and solid states

due to fluctuations in concentration, the chemical composition corresponds to the equilibrium composition of the solid solution at the given temperature. The critical size of the nuclei depends upon the degree of supercooling and supersaturation of the initial liquid solution. The higher the degree of supercooling and supersaturation, the smaller the critical size of the nuclei.

We shall now consider the solidification of an alloy containing 60 per cent *B* (Fig. 82). Freezing begins at temperature t_1 where the first crystals of the solid solution of the components *A* and *B* separate from the liquid alloy. The alloy is in a two-phase state below temperature t_1. As indicated by the phase rule for a two-component system in which two phases are present (liquid and solid), the number of degrees of freedom will be

$$F=2+1-2=1$$

Therefore, each temperature corresponds to definite concentrations of the phases.

The concentration of the solid solution, in equilibrium in the liquid phase, is determined by the point of intersection of a horizontal line, passing through the given temperature, with the solidus (Fig. 82). The concentration of the liquid phase is determined by the intersection of the same temperature line with the liquidus. At a temperature of t_2, for example, points m and n represent the concentrations of the solid and liquid phases, respectively. At temperature t_3, they are represented by points m_1 and n_1. Thus, in solidification, the composition of the liquid phase continuously varies along the liquidus while that of the solid phase varies along the solidus.

The ratio by weight between the liquid and solid phases at a temperature t_2, for example, may be calculated using the lever rule:

$$\frac{\text{weight of liquid phase}}{\text{weight of solid phase}}=\frac{t_2m}{t_2n}$$

An alloy containing 60 per cent B completely solidifies at temperature t_4.

Fig. 83 illustrates the equilibrium diagram for copper-nickel alloys which are an example of complete mutual solubility in both the liquid and solid phases.

In the solidification of alloy systems of this type, the solid solution crystals, separating during the process, are of variable composition. However, if the alloy is held at a sufficiently high temperature for ample time and is cooled very slowly, diffusion between the liquid part of the alloy and the crystals and also between the crystals of the solid solution, that are of varied composition, will help to establish equilibrium between the phases. Diffusion also facilitates equalisation of the composition after the alloy is completely solid. Since diffusion is a very slow process, the composition does not have enough time, in actual practice, to be fully equalised and the crystals usually vary in composition. The axes of the crystals, which start to grow at the beginning of solidification, are usually richer in the component having the higher melting point. The peripheral layers of the crystals and the spaces between the axes are filled in later and will be enriched by the component lowering the melting point of the alloy.

No true equilibrium is attained, for such cases, between the liquid and solid phases in solidification. Therefore, the diagram, which is drawn for equilibrium cases, cannot be used for determining compositions of the phases by intersections on the liquidus and solidus.

Upon faster cooling, the structure of solid-solution type alloys will be of a clearly defined dendritic character (Fig. 84) due to the

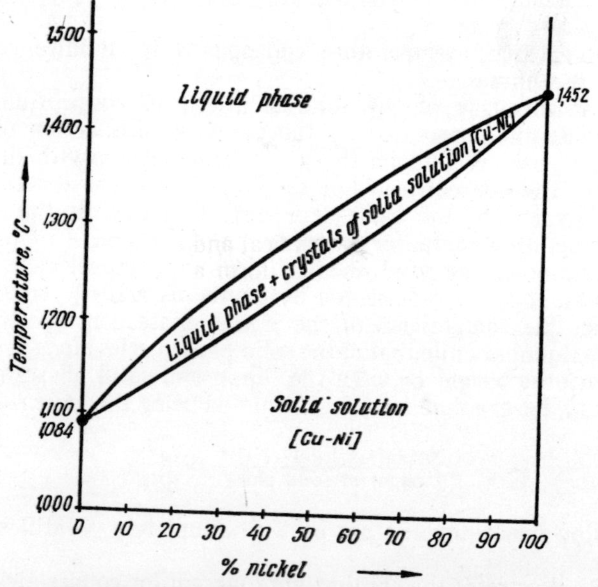

Fig. 83. The copper-nickel equilibrium diagram

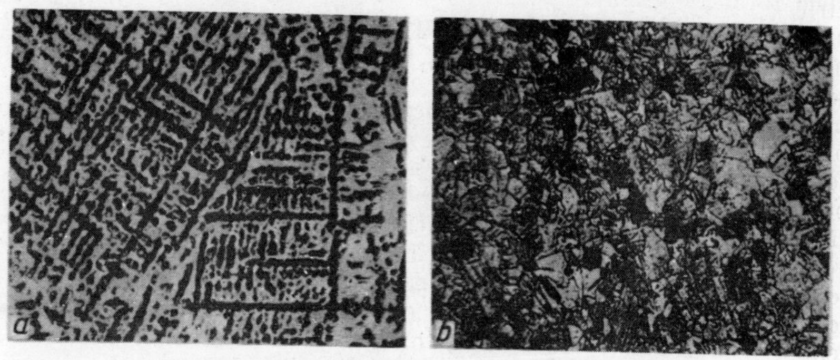

Fig. 84. Microstructure of a solid solution of 90 per cent Cu+10 per cent Zn:
a—dendrites of the solid solution (as-cast), × 3, and *b*—polyhedrons of the solid solution (after homogenising or diffusion annealing), × 100

lack of homogeneity in the composition of the crystals. Such inhomogeneity of the alloy within the crystals is called *dendritic segregation* or *coring*.

The larger the temperature interval between the liquidus and solidus, the greater the degree of dendritic segregation. It may be reduced somewhat by holding the alloy for a long period at a temperature high enough to ensure a suitable rate of diffusion. After such treatment (annealing), the dendritic structure of the cast alloy is no longer evident and polyhedral grain structure of the homogeneous solid solution may be observed (Fig. 84, *b*).

The variation in mechanical, physical and fabrication properties of alloys that form a continuous series of solid solutions is shown in Fig. 85. The diagrams indicate that solid solutions possess higher strength and hardness than their components. In many cases, however, this increased tensile strength is not accompanied by reduced ductility. On the contrary, ductility is often increased together with tensile strength and hardness (for example, in α brasses).

N. S. Kurnakov first noted that the formation of a solid solution always increases the electrical resistivity. This is associated with the considerable distortion of the electrical field of the solvent metal by the atoms of the solute metal. It must be noted, however, that the electrical resistivity of a solid solution remains practically unchanged upon an increase in temperature while that of pure metals substantially increases. Therefore, at high temperatures, the resistivity of pure metals may exceed that of solid solutions.*

Alloys which form homogeneous solid solutions are ordinarily easily worked (by rolling, forging, stamping, etc.) but have comparatively low casting properties (they have a tendency to form cracks and dispersed porosity). Such alloys are machined with difficulty in many cases due to their high

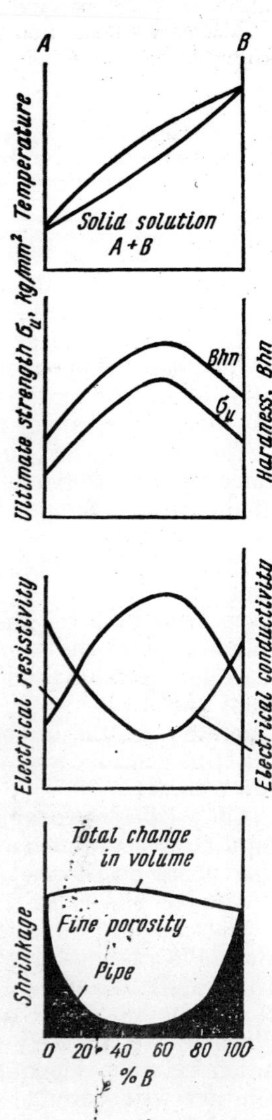

Fig. 85. Relationship between properties and composition of alloys that form a continuous series of solid solutions

* This indicates that the relationship deduced by N. S. Kurnakov is valid only at sufficiently low temperatures.

ductility. Substances which reduce the ductility are added to these alloys to improve their machinability.

Alloys forming homogeneous solid solutions are widely used as engineering materials.

b. Equilibrium Diagrams of Systems Whose Components Have Complete Mutual Solubility in the Liquid State and Limited Solubility in the Solid State and in Which the Solid State Solubility Decreases with the Temperature

Fig. 86, *a* shows how an equilibrium diagram of this type is drawn while Fig. 86, *b* is the complete diagram on which the obtained structures are indicated.

The lines *ae* and *be* are the liquidus. Crystals of a solid solution of component *B* in component *A* (α) begin to precipitate from the liquid alloy along line *ae*; the solid solution of component *A* in component *B* (β) precipitates along line *be*.

The lines *ad*, *dec*, and *bc* correspond to the solidus. The point *d* corresponds to maximum solubility of component *B* in component *A* at the eutectic temperature (t_e). Point *c* is the same for component *A* in component *B*.

In a similar manner, points *k* and *f* indicate the solubilities of component *A* in component *B*, and *B* in *A*, respectively, at normal room temperatures. The lines *dk* and *fc* thus show the variation in solubility with the temperature.

At point *e*, which conforms to the conditions of nonvariant equilibrium ($F=2+1-3=0$), the solid solutions α and β simultaneously separate from the liquid phase to form the eutectic ($\alpha+\beta$).

Let us consider now the solidification process of alloys 1, 2, and 3 (Fig. 86, *b*).

Alloy 1 begins to solidify at temperature t_1 and finishes at t_2. The solid alloy contains only crystals of the solid solution α and no further phase transformations occur at lower temperatures.

Alloy 2 begins to solidify at temperature t_3 and finishes at t_4. The composition of the liquid phase varies during solidification along the liquidus; that of the solid phase varies along the solidus. Thus, at temperature *t*, points *m* and *n* determine the compositions of the liquid phase and the solid solution crystals, respectively. When it is completely solid, alloy 2 will consist only of α solid solution crystals. Upon further cooling to temperature t_5, the solid solution will become supersaturated and, at lower temperatures, it will decompose and the surplus component *B* will be separated as β crystals whose amount increases as the temperature falls. Therefore, at temperatures below t_5, the alloy will consist of two phases $\alpha+\beta$ (Fig. 87). The composition of the α and β solutions varies along line *dk* for α crystals and along *cf* for β crystals upon a fall in temperature.

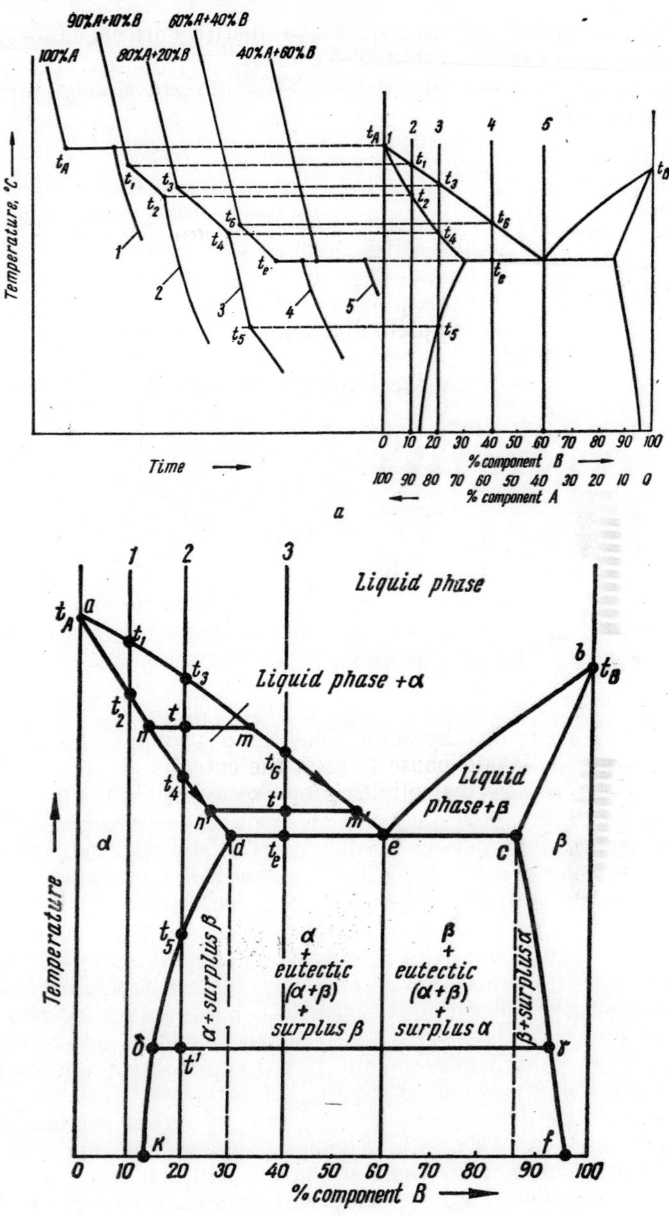

Fig. 86. Equilibrium diagram of a system of components that have complete mutual solubility in the liquid state and limited solubility in the solid state.
a—constructing the diagram, *b*—the finished diagram

For example, at temperature t', the composition of the α-phase is determined by point δ and that of the β-phase, by point γ. The quantitative relation between the weights of the α- and β-phases conforms to the lever rule:

$$\frac{\text{amount of } \alpha\text{-phase}}{\text{amount of } \beta\text{-phase}} = \frac{t'\gamma}{t'\delta}$$

On the basis of energetic and concentration fluctuations, nuclei of the β-phase are first formed on the boundaries of the original α-phase

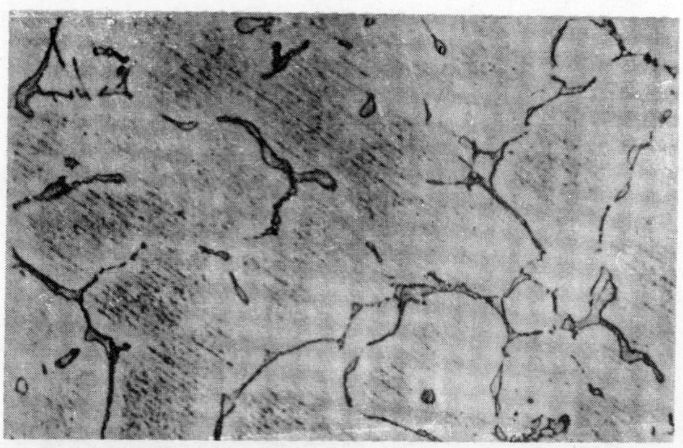

Fig. 87. Separation of the surplus phase upon decomposition of the solid solution (Al-Cu alloy), ×250

grains (Fig. 87). Rapid cooling (quenching) will suppress the formation of β-phase crystals and a homogeneous alloy may be supercooled to low temperatures. A quenched solid solution of this type, supersaturated with component B, is unstable and will decompose upon heating and, for some alloys, even at normal temperature. The rate of decomposition increases with the temperature to which the alloy is heated. This separation of the surplus phase due to decomposition of a supercooled solid solution is accompanied by considerable changes in the properties of the alloys. An increase in hardness is characteristic for this process which is called *age hardening* or *precipitation hardening*.

Alloy 3 (Fig. 86, b) begins to solidify at temperature t_6 and is completely solidified at temperature t_e.

In solidification, the composition of the liquid part of the alloy varies continuously along the liquidus, approaching the eutectic composition (point e); the composition of the solid phase varies along the solidus moving toward the maximum solubility (point d). The relation between the liquid and solid phases may be found at any temperature by the lever rule. At temperature t_e, the liquid phase reaches the eutectic composition and the alloy completely solidifies. In this, alpha and beta solid solutions, forming the grains of the eutectic, are simultaneously precipitated from the liquid phase.

The solidified alloy will consist of primary fully saturated (point *d*) crystals of alpha solid solution, precipitated in the temperature interval t_6-t_e, and the eutectic, formed at temperature t_e and comprising alpha and beta solid solutions of the extreme compositions (points *d* and *c*, respectively). The α crystals, both primary and those

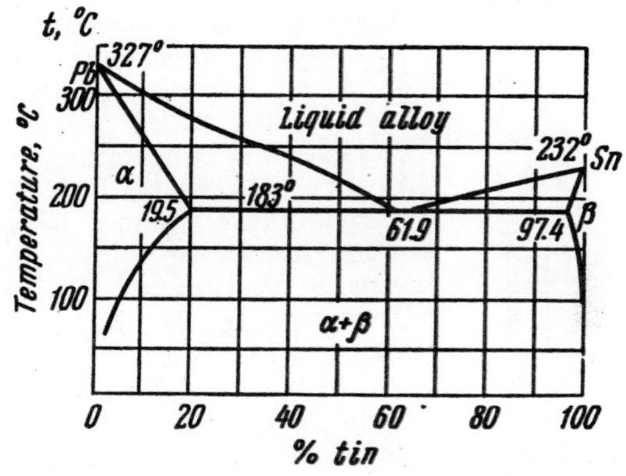

Fig. 88. The lead-tin equilibrium diagram

in the eutectic, decompose upon a further drop in temperature due to the decrease in solubility. As a result beta solid solution crystals precipitate from the α crystals. The compositions of the α-phase and the β crystals, both surplus and those in the eutectic, vary with the temperature along the lines *dk* and *cf*, respectively. After the complete solidification of an alloy, their compositions are indicated by points *k* and *f*.

An alloy of the composition corresponding to point *e* (60 per cent *B*) will contain only grains of the eutectic (α+β) after solidifying.

Hypereutectic alloys will solidify in exactly the same manner as the hypoeutectic ones considered previously except that primary β crystals will precipitate instead of primary α crystals. Equilibrium diagrams of this type are of great practical significance since the properties of the alloys (for example, alloy 2) can be substantially changed by heat treatment (hardening and ageing).

The Pb-Sn system has an equilibrium diagram of the above-described type; this diagram is shown in Fig. 88.

N. S. Kurnakov has shown that the properties of alloys of this type vary along a curve in the region of homogeneous solid solutions and along a straight line for regions of two-phase mixtures (Fig. 89).

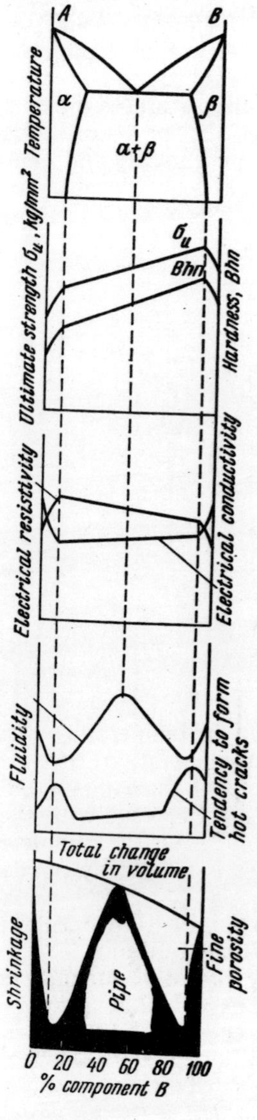

Fig. 89. Relationship between properties and composition in a system with limited solubility in the solid state

In the second case, the end points of the straight line correspond to the properties of fully saturated solid solutions.

It must be noted that the properties, in the regions in which two phases (α and β) exist, depend upon the degree of dispersity of the mixture (Fig. 89). The properties of highly dispersed mixtures will not conform to the straight line relationship.

Investigations conducted by A. A. Bochvar and co-workers have found a definite relationship between the composition of an alloy and its casting properties (Fig. 89). The larger the temperature interval of solidification, the less the fluidity of an alloy. The greater the distance between the liquidus and solidus (in a horisontal direction), the more tendency an alloy has for dendritic segregation. Fluidity is noticeably increased and the tendency to form hot cracks is reduced in alloys which form a eutectic in solidifying.

Shrinkage appears as a concentrated pipe in alloys which solidify in a narrow freezing range and, particularly, in eutectic alloys which solidify at a constant temperature. On the other hand, shrinkage in alloys, solidifying in a wide freezing range, is found chiefly in the form of dispersed porosity (Fig. 89) throughout the casting. This may necessitate rejection of the casting.

Therefore, to ensure good casting properties, the concentration of the components must exceed tneir maximum solubility in the solid state and should approach the eutectic concentration.

Alloys in the region of homogeneous solid solutions possess high ductility and, therefore, are easily rolled, forged, pressed, etc. Plastic deformation of alloys consisting of a mixture of several phases is much more difficult. Ductility is reduced to the greatest extent when the eutectic appears in the structure. Consequently, the maximum solubility at the eutectic temperature is the upper limit of concentration for wrought alloys.

Alloys of the type here described are extensively used in engineering.

c. Equilibrium Diagrams of Alloys Whose Components Have Complete Mutual Solubility in the Liquid State and Limited Solubility in the Solid State (Alloys with a Peritectic Transformation)

The equilibrium diagram for this type of alloy is shown in Fig. 90. This diagram differs from the preceding type (Fig. 86) in that the crystals of beta solid solution, precipitated at the beginning of solidification, react with the liquid alloy of a definite composition to

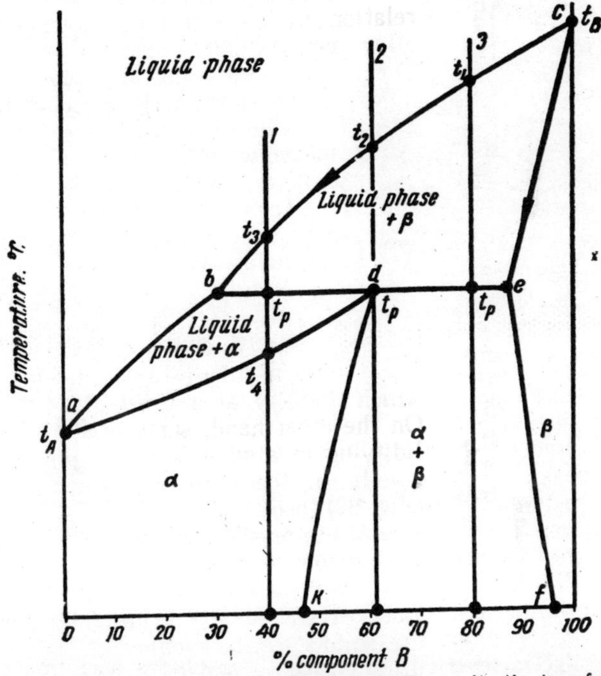

Fig. 90. Equilibrium diagram of an alloy subject to peritectic transformation

form new crystals of alpha solid solution. This transformation or reaction occurs at a constant temperature (as do eutectic transformations) and is called *peritectic transformation.*

In the equilibrium diagram (Fig. 90), the line *abc* is the liquidus and *adec* is the solidus. Point *d* represents the maximum solubility of component *B* in component *A* at temperature t_p; point *e* is the same for component *A* in component *B*. Points *k* and *f* represent maximum solubility at normal temperatures. Thus, lines *dk* and *ef* show the variation in solubility in the alpha and beta solid solutions upon cooling.

The diagram can be better explained by considering the solidification of alloys *1*, *2*, and *3* which contain 40 per cent *B* and 60 per cent *A*, 61 per cent *B* and 39 per cent *A*, and 80 per cent *B* and 20 per cent *A*, respectively.

The 80 per cent *B* alloy begins to solidify at temperature t_1 when crystals of beta solid solution precipitate from the liquid alloy. At temperature t_p (Fig. 90), the liquid phase has the composition conforming to point *b* and the crystals of solid solution are enriched by component *B* to the maximum concentration, shown by point *e*. Here the alloy completely solidifies according to a peritectic transformation that consists in an interaction between the liquid alloy of composition *b* with the fully saturated β crystals (point *e*). This forms a new alpha solid solution of the maximum concentration (point *d*):

Fig. 91. Two-phase structure formed as a result of peritectic transformation (Cu-Zn alloy), × 250

$$L_b+\beta_e \rightarrow \alpha_d+\beta_e$$

For any given composition of the alloy, the proportion of the reacting phases is characterised by the amount of excess β crystals and, therefore, a certain amount of β crystals remains unused in the peritectic transformation.

The structure of the solid alloy will be a peritectic mixture of two solid solutions (α+β) whose compositions will vary along the lines *dk* and *ef* at a further fall in temperature.

Fig. 91 illustrates the structure of an actual alloy after a peritectic transformation. The light crystals are α while the dark ones are β.

The peritectic reaction, in which three phases participate (liquid and crystals of alpha and beta solid solutions), corresponds to non-variant equilibrium, i. e., the process proceeds at constant temperature:

$$F=2+1-3=0$$

Alloy *2*, containing 61 per cent *B*, begins to solidify at temperature t_2 and finishes at the peritectic temperature t_p. The alloy completely solidifies at a constant temperature. This is accompanied by the peritectic transformation:

$$L_b+\beta_e \rightarrow \alpha_d$$

Thus, the result of the peritectic transformation (for the given alloy) will be the disappearance of both initial phases (liquid phase of composition *b* and crystals of the beta solid solution of composition *e*) and the formation of a single alpha solid solution. At temperatures below t_p, however, the alpha solid solution will be partially decomposed, resulting in the precipitation of surplus beta solid solution crystals.

The compositions of the alpha and beta solid solutions will vary along the lines *dk* and *ef* which indicate the maximum solubility at various temperatures.

After being cooled to room temperature, the alloy will consist of separated crystals of alpha and beta solid solutions, having the compositions *k* and *f*.

Alloy *1*, containing 40 per cent *B*, begins to solidify at temperature t_3 and crystals of beta solid solution precipitate from the liquid phase. The following peritectic transformation occurs when temperature t_p is reached:

$$L_b + \beta_e \rightarrow L_b + \alpha_d$$

The liquid phase is in excess after the peritectic transformation in this alloy. Thus, the solidification of this alloy is not completed at the temperature of peritectic transformation and crystals of alpha solid solution continue to precipitate from the liquid alloy upon further fall in temperature. The alloy completely solidifies at temperature t_4 and consists only of crystals of the alpha solid solution.

It is evident from Fig. 90 that alloys having a component *B* concentration less than 47 per cent, i. e., located to the left of point *k*, consist solely of alpha solid solution crystals. Alloys containing more than 96 per cent component *B* (to the right of point *f*) consist only of beta solid solution crystals. Peritectic transformations occur in a number of alloys which are important engineering materials. They include: Fe-C (Fig. 108), Cu-Zn (Fig. 257), Cu-Sn (Fig. 260), and others.

6-5. Equilibrium Diagram of a System in Which the Components Have Unlimited Solubility in the Liquid State and Form Chemical Compounds Upon Solidification

One or several chemical compounds may be formed in many metallic systems.

Chemical compounds are classified in the following three main types: 1) valency compounds, 2) electron phases, and 3) interstitial phases.

Valency compounds include Mg_2Si, Mg_2Sn, Mg_2Pb and other compounds of magnesium with elements of the fourth, fifth, and sixth groups in the periodic system. A characteristic feature of these compounds is the practically complete insolubility of the components.

Valency compounds are seldom formed in metal alloys. In the majority of cases, the components combine in intermetallic phases (chemical compounds) which do not follow simple valence rules.

Electron compounds (Hume-Rothery phases) are of prime importance in metallic systems. Compounds of this type have a definite ratio of the number of valence electrons to the number of atoms, i. e., a definite electron ratio. Thus, a whole series of compounds exist in which the ratio is $^3/_2$; in others it equals $^{21}/_{13}$ and $^7/_4$. Each of these ratios corresponds to a definite type of crystal lattice. All compounds with an electron ratio of $^3/_2$ have a body-centred crystal lattice. This type of compound includes CuZn *, Cu_3Al, Cu_5Sn, NiAl, FeAl, and others. They are known as the β-phases.

Compounds Cu_5Zn_8, Cu_9Al_4, and others, with an electron ratio of $^{21}/_{13}$, have a complex cubic lattice. They are called the γ-phase.

Finally, such compounds as $CuZn_3$, Cu_3Sn, Cu_3Si, and others with a ratio of $^7/_4$, have a close-packed hexagonal lattice and are called the ε-phase.

A large group of metals (Fe, Cr, Mo, W, Ti, V, and others) form chemical compounds (carbides, nitrides, etc.) with carbon, nitrogen, hydrogen and boron, all of which have a small atomic radius. These are called *interstitial compounds*. This group of compounds may be formed if the ratio of the atomic radius r_x of the nonmetal (C, N, H, or B) to the atomic radius r_m of the metal is less than or equal to 0.59 $\left(\frac{r_x}{r_m} \leqslant 0.59\right)$.

Interstitial compounds have the following types of chemical formulae: Me_4X [Mn_4N, Fe_4N, etc.], Me_2X [W_2N, Fe_2N, etc.], and MeX [TiN, TiC, WC, VC, etc.].

The atoms of interstitial compounds form one of the most typical cubic or hexagonal crystal lattices into which the atoms of the metalloid are inserted, occupying definite pores or vacant positions.

Interstitial compounds or phases differ from interstitial solid solutions in that their crystal lattices differ from those of the metals of which they are formed. They have certain typically metallic properties (electrical conductivity which is reduced with an increase in temperature). They also have an exceptionally high hardness and melting point.

Solid solutions, usually with excess metal atoms, are easily formed on the basis of interstitial compounds.

a. Equilibrium Diagram of a System Whose Components Have Complete Mutual Solubility in the Liquid State and Form Stable Chemical Compounds After Solidification

In this diagram, the components A and B form a chemical compound $A_n\ B_m$ which has a constant melting point, shown on the diagram by

* In compounds copper provides one valence electron, zinc provides two, aluminium three, tin four, while nickel and iron do not provide any.

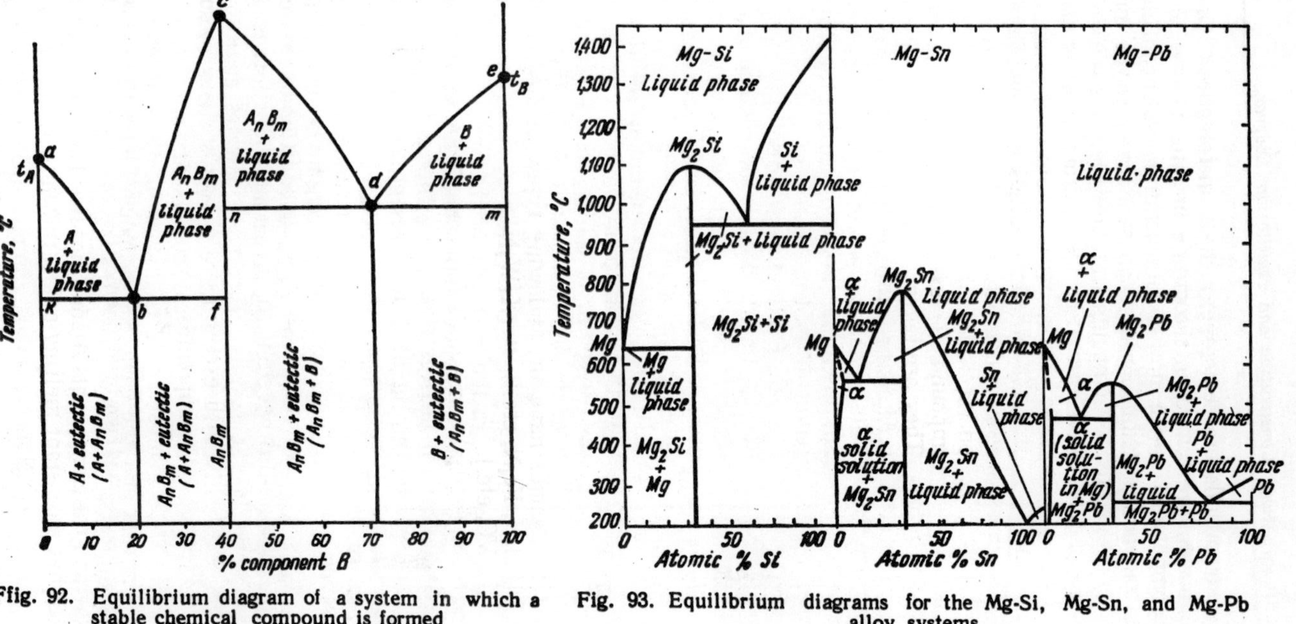

Ffig. 92. Equilibrium diagram of a system in which a stable chemical compound is formed

Fig. 93. Equilibrium diagrams for the Mg-Si, Mg-Sn, and Mg-Pb alloy systems

point c (Fig. 92). Under a microscope, a chemical compound is observed as separate polyhedral grains, the same as in a pure metal.

The effect of the existence of a stable chemical compound on the equilibrium diagram is that an ordinate, corresponding to the composition of this alloy, divides the diagram into two parts (Fig. 92). We obtain what may be considered as two diagrams of the first type, A-A_nB_m and A_nB_m-B, joined along the ordinate corresponding to the chemical compound composition.

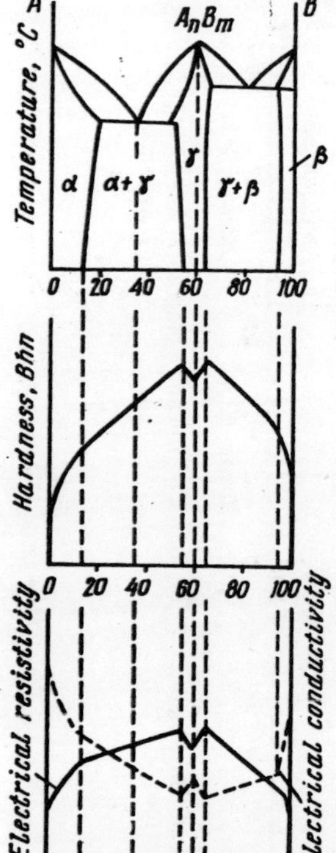

Fig. 94. Relationship between properties and composition in a system in which a stable chemical compound is formed

Alloys A-A_nB_m begin to solidify at the temperatures indicated by the line abc. A crystals begin to separate on line ab and crystals of the compound A_nB_m along line bc. Solidification is completed at the eutectic temperature (line kbf).

A_nB_m-B alloys begin to solidify along line cde. On line cd crystals of compound A_nB_m separate and B crystals along line de. In this case, solidification is completed along the eutectic line ndm.

It is evident from the diagram (Fig. 92) that two eutectics are formed. One, represented by point b, consists of a mixture of A crystals and compound A_nB_m; the second, represented by point d, consists of compound A_nB_m and B crystals.

In the first part of the system (A-A_nB_m), hypoeutectic alloys begin to solidify with the precipitation of primary A crystals from the liquid alloy; crystals of the compound A_nB_m precipitate in hypereutectic alloys. In the second part of the system (A_nB_m-B), hypoeutectic alloys contain primary crystals of compound A_nB_m in addition to the eutectic (A_nB_m+B); hypereutectic alloys contain B crystals. Fig. 93 illustrates equilibrium diagrams of three actual systems in which stable chemical compounds are formed. It must be noted, however, that in most cases, the components and the chemical compound form limited regions of solid solutions. In these cases, the diagram will be complex and will consist of two diagrams of the eutectic type having limited solubility in the solid state (Fig. 94). The γ region is a solid solution on the basis of the chemical compound or a compound of variable composition. Since the typical equilibrium diagram given in Fig. 94 is identical to a diagram previously explained (Fig. 86), there is no necessity for following the solidification of various alloys in this system.

Fig. 94 also shows the relation between properties and composition for alloys of this type. Characteristic features of chemical compounds are their exceptional hardness, high electrical resistivity, and unsuitability for plastic deformation; they are very brittle.

On the properties vs composition curves, there is a sharp bend in the straight line at compositions corresponding to the chemical compound. The point, at which the curve bends sharply, corresponds to the composition of the chemical compound and is called the *singular point*. The singular point helps to exactly locate the position of the chemical compound on the equilibrium diagram.

b. Equilibrium Diagram of a System of Components that Form an Unstable Chemical Compound upon Solidification and Which Decomposes upon Reheating, Before Melting, by a Peritectic Reaction, Forming a Solid Solution and a Liquid Phase

Fig. 95 illustrates the diagram obtained in this case. The chemical compound A_nB_m is located at the ordinate 80 per cent B. This compound

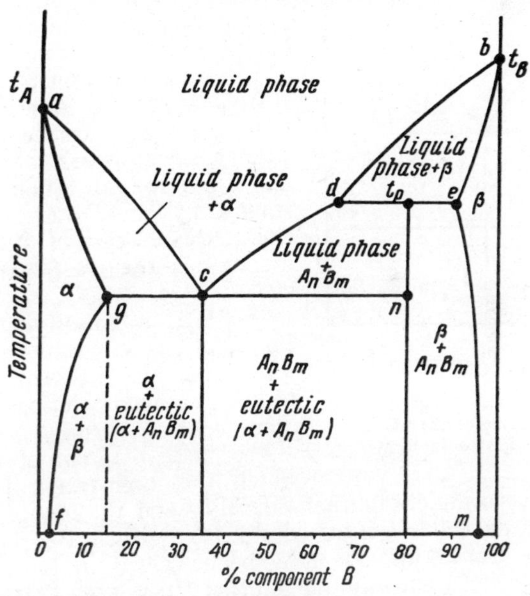

Fig. 95. Equilibrium diagram of a system in which an unstable chemical compound is formed

is formed at temperature t_p as a result of a peritectic transformation of the liquid phase, having a composition indicated by point d (L_d), and the beta solid solution of maximum concentration (point e):

$$L_d + \beta_e \rightarrow A_nB_m$$

A chemical compound is formed only in alloys of the indicated composition. Alloys of higher concentration (located to the right of 80% B) will contain excess crystals of β solution, in addition to the compound A_nB_m; those of lower concentration (to the left of 80% B) will contain the eutectic $(\alpha+A_nB_m)$.

6-6. Equilibrium Diagram of a System Whose Components are Subject to Allotropic Transformations

Phase transformations occur in the solid state of systems in which one or both components are subject to allotropic transformations, for example, in systems of which Fe, Co, Mn, Ti and Zr are components.

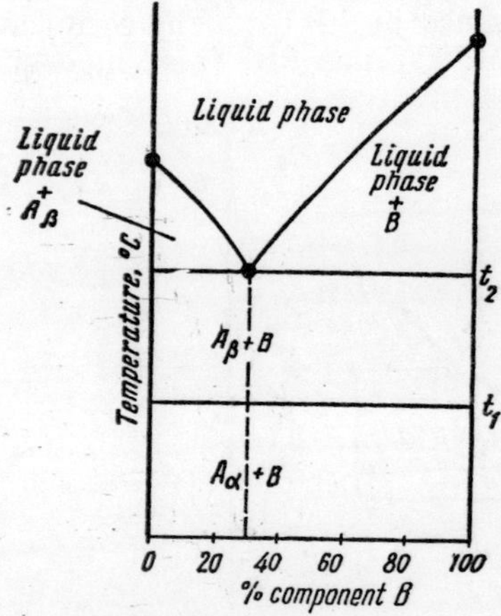

Fig. 96. Equilibrium diagram of a system in which one of the components has two allotropic forms

Fig. 96 shows an equilibrium diagram for the case when the components of the system solidify as pure metals and the component A has two allotropic forms, α and β.

The transformation of component A from allotropic form α to form β takes place at a constant temperature t_1, since three phases α-A, β-A and B are in equilibrium during the transformation $(F=2+1-3=0)$. A characteristic case of this type of diagram is the Fe-S system (Fig. 203).

If alloys, in solidifying, form solid solutions of components that may exist in different allotropic forms, various relationships may occur between them.

The phase diagram of a case, in which the high-temperature modifications have complete mutual solubility and the low temperature modifications have but limited solubility, is illustrated in Fig. 97.

Solid state transformations may be described in the same way as in solidification from a liquid solution * (see page 114). It is evident that the line ac corresponds to the beginning of decomposition of the beta solid solution with the precipitation of α' solid solution

* In real systems, solid state transformations take place in a more complex manner.

crystals. On line *cb*, α'' solution will precipitate. The beta solid solution will decompose into a mixture of the α'- and α''-phases along line *dce* at a constant temperature ($F=0$).

The process is similar to a eutectic transformation except that the initial phase is a solid solution. This is called a *eutectoid* transformation and the mixture of crystals ($\alpha'+\alpha''$) is called the *eutectoid.*

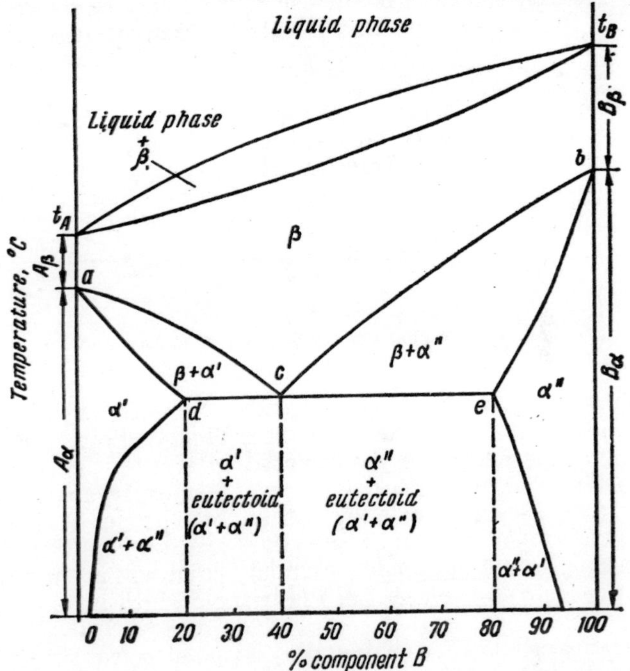

Fig. 97. Equilibrium diagram of a system in which the high-temperature allotropic forms of the components have complete solubility, and the low-temperature forms—limited solubility

It should be noted that solid solutions differ from liquid solutions in that they have a larger tendency to supercool and produce nonequilibrium states which, naturally, cannot be shown on the equilibrium diagram where all phases are in equilibrium.

6-7. Diagrams of Ternary Systems

Since ternary and more complex alloys are widely used in engineering, it is necessary to study the equilibrium diagrams for multiple-component systems. Such diagrams enable more expedient alloy compositions to be selected and help us to understand the various processes that occur in such systems. The Russian school of physical metallurgists, under the leadership of Academician N. S. Kurnakov, have done extensive research in this field.

Many ternary diagrams have been constructed and they are widely used in actual practice. The diagram of a ternary system is a three-dimensional model constructed on a base which is an equilateral triangle. The concentration of the components may be shown in the triangle in the same scale. This is called a concentration triangle.

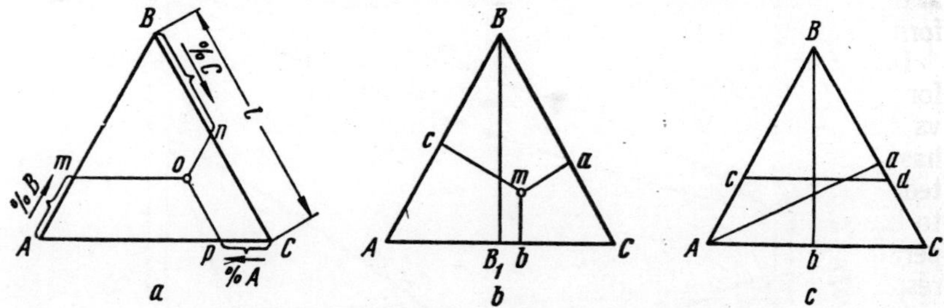

Fig. 98. Determining the composition of ternary alloys

The components of the alloy are located at the corners of the triangle while the sides represent the three binary alloys. Any point within the triangle specifies the composition of a ternary alloy.

A property of an equilateral triangle is frequently used to determine the composition of an alloy within the triangle. This geometrical rule states that the sum of three lines ($on+om+op$) (Fig. 98, *a*), drawn parallel to the three sides, from any point within the triangle, is equal in length to one side l of the triangle which is taken as 100 per cent.

To find the composition of the alloy at point o, we simply draw three lines from this point parallel to the sides of the triangle. The corresponding lengths of the lines show the percentages of components A, B, and C in the alloy.

The composition may also be determined by another method. In this case, the altitude of the triangle is taken as 100 per cent and the geometrical theorem which is used states that the sum of three lines drawn perpendicular to the three sides from a point in an equilateral triangle, is a constant equal to the altitude ($BB_1=ma+$ $+mc+mb$) (Fig. 98, *b*).

On the basis of the properties of equilateral triangles, it is not difficult to show that:

1) All alloys whose composition lies on a straight line connecting an apex of the triangle with the opposite side have a constant composition ratio of two components, for example, alloys on the line Aa have a constant ratio of components B and C (Fig. 98, *c*).

2) All alloys which lie on altitudes of the triangle (Fig. 98, *c*) have the same composition for two of the components, for example, alloys on the altitude *Bb* contain equal amounts of components *A* and *C*.

3) All alloys which lie on a line parallel with one side of the triangle (*cd*) have the same content of the component located at the apex opposite which this line is drawn (*B* in Fig. 98, *c*).

The composition of ternary alloys is represented on a plane while a three-dimensional model is required to show temperature transformations.

In constructing a ternary diagram, the first step is the same as for binary alloys where cooling curves are plotted in temperature vs time coordinates for alloys of various composition. When this has been done, the points representing the various alloys are plotted in the composition triangle. Perpendiculars are constructed to the plane of the triangle at the corresponding points and the temperatures of the critical points for each alloy are located on the corresponding perpendicular. Surfaces passing through the critical points, in the same way as lines were drawn through the critical points in binary diagrams, will provide a three-dimensional representation of the ternary diagram.

Ternary diagrams are classified in accordance with the same features as for binary alloys, i. e., on the basis of the solubility of the components in the liquid and solid states, presence of chemical compounds, etc.

Fig. 99 illustrates a ternary diagram for a system in which the components have complete mutual solubility in the liquid state and form a simple mechanical mixture when they solidify.

Binary systems *A-B*, *B-C*, and *A-C* are constructed on each side of the composition triangle (Fig. 99).

The surface $t_1E_1' E'E_3' t_1$ is a part of the liquidus and corresponds to temperatures at which component *A* begins to precipitate from the liquid alloy, surface $t_2E_1' E'E_2' t_2$ is the same for component *B* and surface $t_3E_2' E'E_3' t_3$ refers to precipitation of component *C*. The lines $E_1' E'$, $E_2'E'$, and $E_3' E'$ represent the binary eutectics in the ternary alloy.

All three surfaces composing the solidus intersect at the common point E' which is the ternary eutectic.

The projections of the lines of the binary eutectics on the plane of the composition triangle give three two-dimensional curves E_1E, E_2E, and E_3E (Fig. 99, *b*).

We may follow the solidification of an alloy having a composition represented by point *a* to better understand this diagram. It is evident from Fig. 99 that at point *R* the alloy is in a liquid state. When the temperature is lowered to point *1*, the liquid alloy becomes saturated with component *A* which begins to precipitate at this point. At this, the liquid alloy will become richer in components *B* and C and its composition will change along the surface of the liquidus in the direction $1b'$.

The projection of the curve $1b'$ on the composition triangle will be a straight line ab passing through the apex A *. When the composition of the liquid alloy reaches point b, it will be saturated with

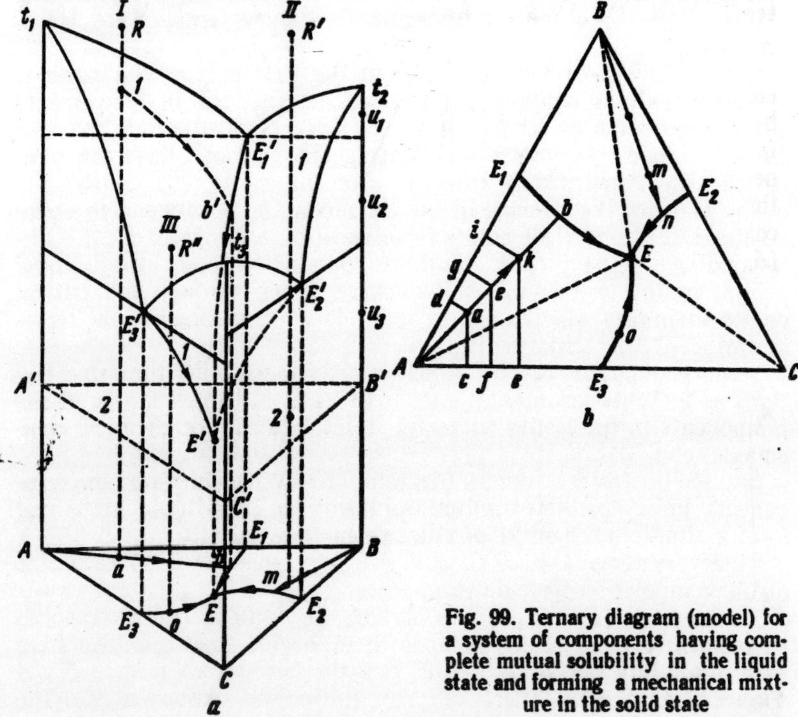

Fig. 99. Ternary diagram (model) for a system of components having complete mutual solubility in the liquid state and forming a mechanical mixture in the solid state

components A and B. Therefore, both of these components will solidify at a further drop in temperature and form a binary eutectic. This binary eutectic ($A+B$) is formed along a temperature interval ($F=3+1-3=1$). In this temperature interval (formation of the eutectic), the composition of the liquid alloy will vary along line $E_1' E'$, from point b' to the ternary eutectic E'. In the projection on the composition triangle, this change will follow along curve E_1E, from point b to point E (Fig. 99, b).

* Though the composition of the liquid alloy varies, the proportions of the components B and C remain constant since their absolute amount in the alloy does not change. This requires the composition of the liquid alloy to vary along a straight line passing through an apex of the triangle.

When the liquid alloy reaches the concentration represented by point E (E'), the ternary eutectic ($A+B+C$) is formed. This takes place at a constant temperature ($F=3+1-4=0$) corresponding to point *2* (Fig. 99) in the eutectic plane ($A'B'C'$) passing through point E'. The completely solidified alloy will consist of A crystals, binary eutectic $A+B$, and the ternary eutectic $A+B+$ $+C$.

Another alloy with the composition represented by point *m* will solidify in exactly the same manner. After solidification this alloy (alloy *II*) will consist of $B+(B+C)+$ $(A+B+C)$. The latter two phases are the binary and ternary eutectics, respectively.

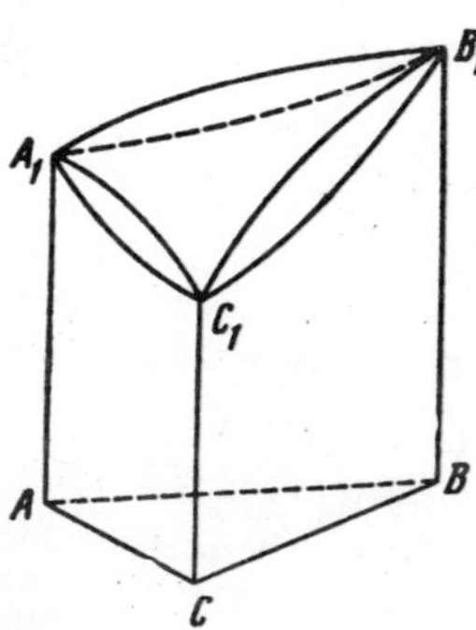

Fig. 100. Ternary diagram (model) for a system of components having complete mutual solubility in both the liquid and solid states

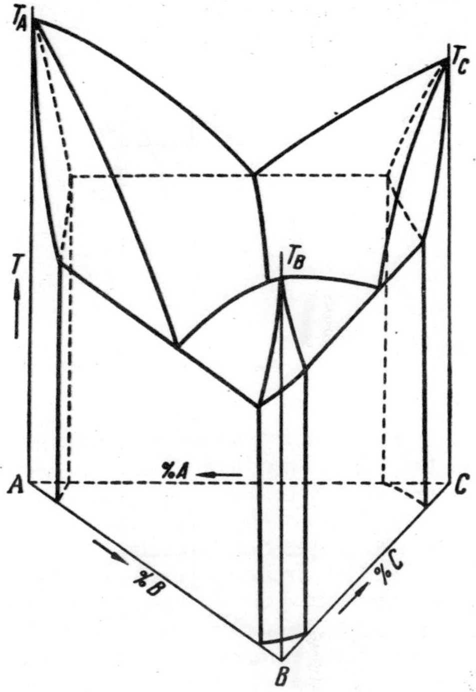

Fig. 101. Ternary diagram (model) for a system of components having limited solubility in the solid state

Alloys located on the lines E_1E and E_2E, for example, alloy *III* in Fig. 99, will consist, after solidifying, of the binary and ternary eutectics. Alloys on the lines AE, BE, and CE will contain no binary eutectics; they will consist only of an excess component (A, B or C) and the ternary eutectic ($A+B+C$). The alloy at point E will consist only of the ternary eutectic ($A+B+C$).

A ternary diagram for components having complete mutual solubility in both the liquid and solid states is shown in Fig. 100. This diagram has two surfaces, liquidus and solidus, between which the ternary solid solution is solidified. The solid alloy consists of this ternary solid solution.

Fig. 101 illustrates a ternary diagram of a system having limited solubility in the solid state.

Quite often, only certain known sections of a ternary diagram are used to show transformations. There are several methods for constructing such sections:

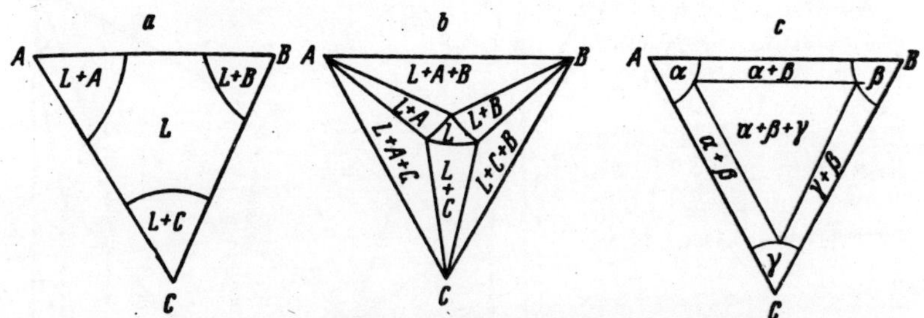

Fig. 102. Isothermic sections of a ternary diagram

1. Horizontal sections of the ternary diagram are made, either in the form of isothermal sections, which represent the phase and structural composition of all the alloys at a definite temperature, or as projections of certain surfaces and lines on a horizontal surface (the composition triangle).

2. Vertical sections are made either parallel with one side of the composition triangle or passing through one of its apexes. Sometimes, special sections are constructed. These sections show the phase and structural composition for all temperatures but only for definite compositions of ternary alloys.

The type used most frequently is a vertical section of a ternary diagram with a constant composition of one component or with constant proportions between two of the components.

Fig. 102 illustrates isothermic sections of a ternary diagram (Fig. 99) which are horizontal planes made at a temperature above the temperatures for the binary and ternary eutectics (*a*) and at a temperature slightly above the ternary eutectic point (*b*). Fig. 102, *c* shows a horisontal section below the ternary eutectic point for alloys forming limited solid solutions (Fig. 101).

Several vertical sections in planes parallel with side *AB* are shown in Fig. 103.

In the section *ab*, points *a*, *K*, *M*, *N*, and *b* are intersections of the plane in question with corresponding volumes of the ternary diagram (Fig. 99). In our section, the two surfaces of the liquidus

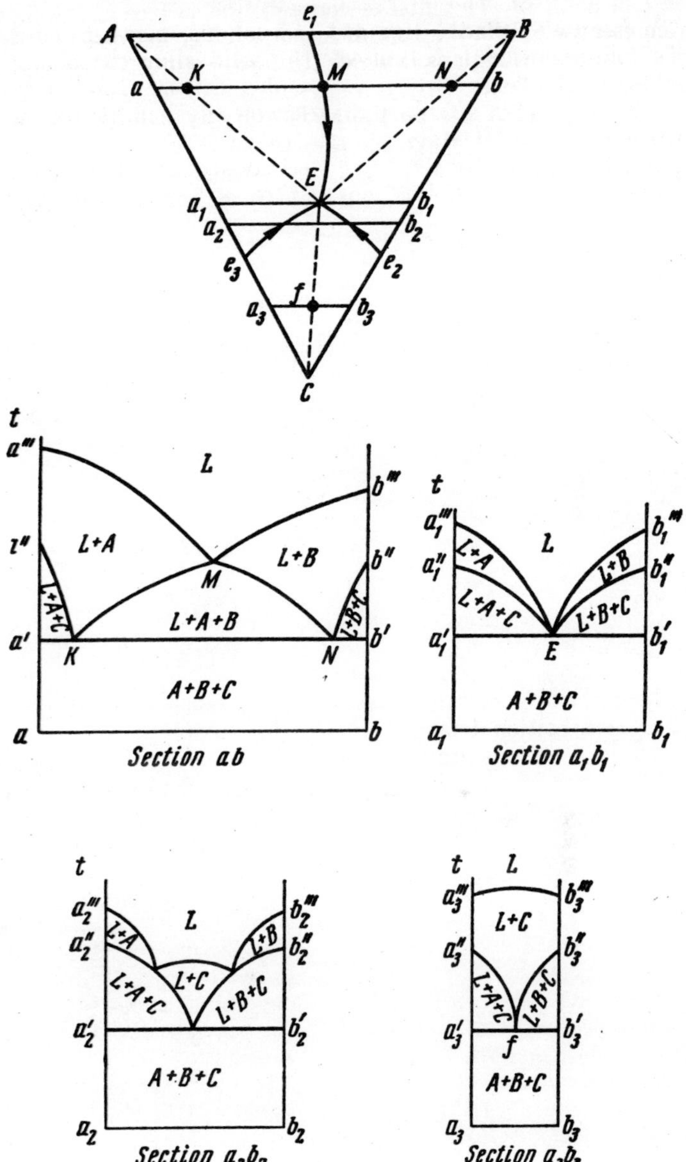

Fig. 103. Vertical sections of a ternary diagram

intersect in point M. The intersection with the surface of the liquidus is represented by the lines $a'''M$ and Mb'''. Further down, the vertical section cuts the regions in which the binary eutectics precipitate from the liquid alloys. This is represented by the lines $a''K$ and Nb''. Points K and N are on the surface of the solidus, i.e., at the temperature of solidification of the ternary eutectic. They can therefore be connected by a straight line parallel with the base.

Sections a_1b_1, a_2b_2, and a_3b_3 were obtained in a similar manner.

Chapter 7

THE IRON-CARBON EQUILIBRIUM DIAGRAM

Alloys of the iron-carbon system include steel and cast iron, which are of the most vital importance to modern industry due to their extensive, versatile applications.

Alloys with a carbon content up to 2.0 per cent are called *steels*; those with a carbon content exceeding 2.0 per cent are called *cast irons.**

Earliest research on the iron-carbon equilibrium diagram was carried out by the great Russian scientist D. K. Chernov.** He first showed, in 1866, that when steel is heated (or cooled), transformations, associated with structural changes, occur at definite temperatures (depending on the composition of the steel). Chernov called these temperatures "peculiar points" (critical points) and designated them with the letters *a* and *b*.

At the present time Chernov's points *a* and *b* are designated as A_1 and A_3, respectively. Chernov's discovery of critical points in steel has served as a basis for the development of the science of metals.

The works of Academicians A. A. Baikov, I. V. Gutovsky, and N. P. Chizhevsky, as well as A. A. Rzheshotarsky were a great contribution to the study of the iron-carbon system.

The investigations of such famous foreign scientists as H. C. Sorby, F. Osmond, W. Roberts-Austen, and P. Goerens helped to construct the iron-carbon equilibrium diagram as we know it today.

7-1. Iron

The melting point of iron is 1,539° C. There are two allotropic forms of iron: α- and γ-iron.

Curves showing the variation in free energy with the temperature for α- and γ-iron are given in Fig. 104. The α-iron free-energy curve

* This classification of steel and cast iron, though accepted, is not exact since there are alloy steels containing more than 2.0 per cent C and alloyed cast irons containing less than 2.0 per cent C.

** D. K. Chernov, "A Critical Review of the Articles by Messrs Lavrov and Kalakutsky on Steel and Steel Gun Barrels and Research Made by the Author on the Same Subject", Transactions of the Russian Engineering Society, 1868.

has a gradually changing inclination. That of the γ-iron curve increases sharply at 910° C and intersects the α-iron curve a second time at 1,400° C.

Therefore, α-iron exists at temperatures up to 910° C and in the interval from 1,400° to 1,539° C; γ-iron is stable at temperatures from 910° to 1,400° C. In the interval from 1,401° to 1,539° C, α-iron is often called δ-iron.

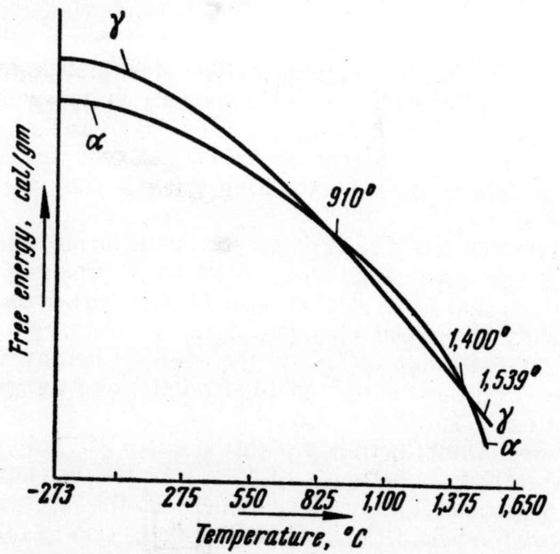

Fig. 104. Free-energy vs temperature curves for α- and γ-iron

The cooling curve for pure iron is illustrated in Fig. 105.

The crystal lattice of α-iron is of the body-centred cubic type with a lattice constant of 2.86 Å. Up to 768° C, α-iron is ferromagnetic. The maximum solubility of carbon in α-iron is 0.025 per cent (at 723°), at 20° C, the solubility is only 0.0025 per cent.*

The solid solution of carbon in α-iron is called *ferrite*. Under a microscope ferrite is seen as homogeneous polyhedral grains (Fig. 106).

The mechanical properties of ferrite (0.06 per cent C) are characterised by the following values:

$$\sigma_u = 25 \text{ kg/mm}^2; \quad \sigma_s = 12 \text{ kg/mm}^2;$$
$$\delta = 50 \text{ per cent}; \quad \psi = 80 \text{ per cent.}$$

Gamma (γ) iron has a face-centred cubic crystal lattice with a constant of 3.63 Å (at 910° C). It is very weakly magnetic (paramagnetic). The carbon solubility of γ-iron at 1,130° reaches 2.0 per cent.

* According to more exact data which have not been entered in the diagram, the carbon solubility at 20° C is about 10^{-5} per cent.

The solid solution (interstitial type) of carbon in γ-iron is called *austenite*. Fig. 107, *a* illustrates the crystal lattice of austenite.

The carbon atom occupies the centre of the face-centred cubic lattice of γ-iron. This carbon is in the form of ions which have been

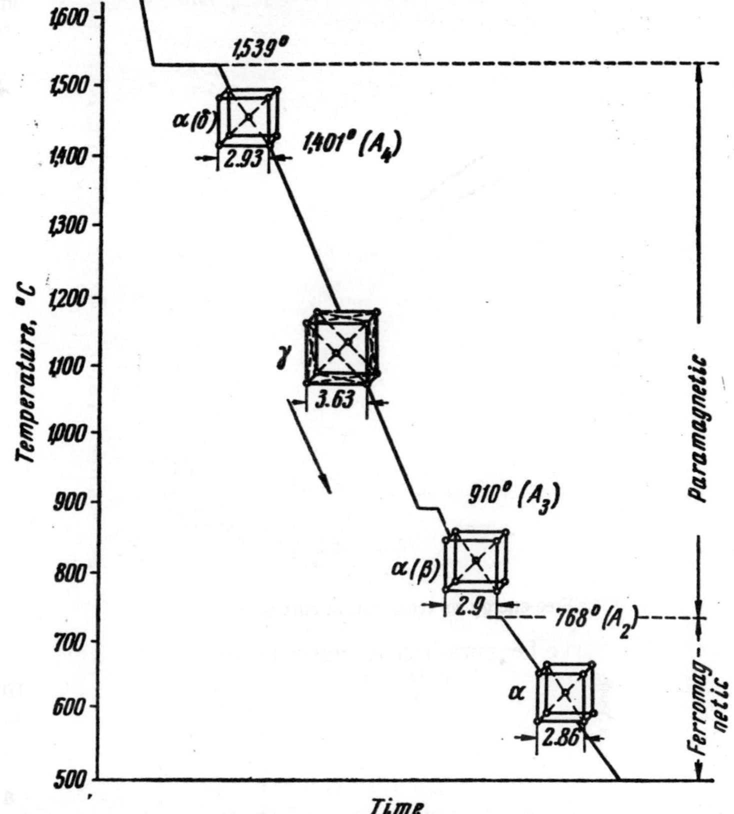

Fig. 105. The cooling curve for pure iron

ionised twice (having given up two electrons for common use) while the iron atom has been ionised but once (having given up one electron).

The microstructure of austenite is shown in Fig. 106, *b*. The structure of austenite was first observed by A. A. Baikov.

In addition to solid solutions (ferrite and austenite), iron and carbon form a chemical compound, iron carbide Fe_3C which has been named *cementite*. Cementite has a carbon content of 6.67 per cent.

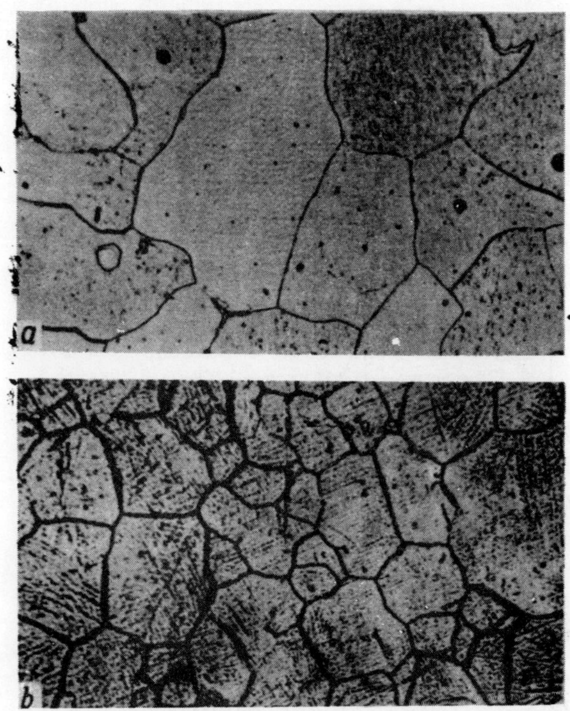

Fig. 106. Microstructure of iron:
a—ferrite, ×500, *b*—austenite, ×250

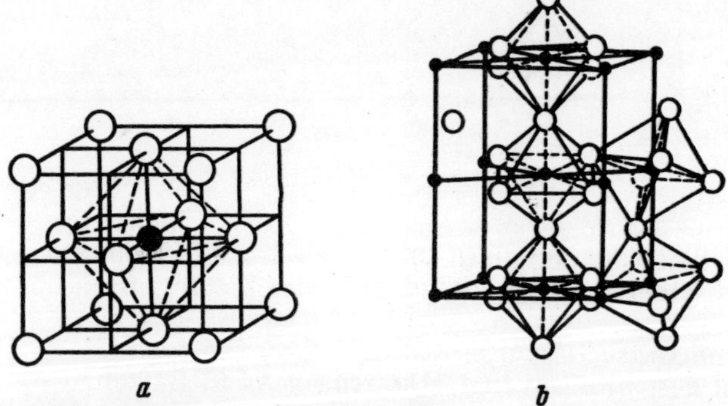

Fig. 107. Schematic representations of crystal lattices:
a—austenite, *b*—cementite

Cementite has an orthorhombic lattice with close-packed atoms (Fig. 107, *b*). Its lattice constants are: a=4.518 Å and c=5.069 Å. Six atoms of iron form an octahedron with the carbon atom at its centre. The bond of the lattice is of the metallic type (Fig. 107, *b*). The melting point of cementite has not been exactly determined but is approximately equal to 1,550° C. Up to a temperature of 217° C, cementite is ferromagnetic. A characteristic feature of cementite is its high hardness ($Bhn \geqslant 700$) and low ductility. Solid solutions may be formed on the basis of cementite but the limits of solubility have not yet been established.

Under definite conditions, cementite decomposes to form free carbon as graphite.

7-2. The Iron-Carbon Equilibrium Diagram

A modern version of the iron-carbon equilibrium diagram is illustrated in Fig. 108 *.

The iron-carbon equilibrium diagram concerns transformations that occur in alloys having compositions from pure iron to cementite (6.67 per cent C). Point *A* (1,539° C) on the diagram is the melting point of pure iron and point *D* (~1,550° C) is the melting point of iron carbide (cementite, Fe_3C). Points *N* (1,401° C) and *G* (910° C) correspond to the allotropic transformations of α-iron into γ-iron. Point *E* shows the solubility limit of carbon in γ-iron at 1,130° C (2.0 per cent).

The upper left-hand portion of the diagram (Fig. 108) represents the allotropic transformations $\gamma \rightleftarrows \alpha(\delta)$ at high temperatures. At temperatures on the line *AB* (Fig. 108), crystals of the solid solution of carbon in α-iron (δ-solution) begin to precipitate from the liquid alloy. The line *HJB* represents peritectic transformation taking place at a constant temperature. The result of the peritectic reaction is the formation of the solid solution of carbon in γ-iron, i. e., austenite.

To simplify further study of the iron-carbon diagram we shall ignore this part of the diagram in the future since it is of no practical importance.

The alloys begin to solidify when they reach a temperature corresponding to the line *ACD* (liquidus) ** (Fig. 109). Complete solidification corresponds to the solidus temperatures *AECF*. Austenite is precipitated from the liquid alloy along line *AC* and cementite along line *CD*. At point *C* with a temperature of 1,130° C and at

---

* Only the metastable equilibrium diagram of iron-carbon alloys (Fe-Fe_3C) will be considered in this section. The stable diagram (Fe-graphite) is shown in Fig. 247.

** See the simplified diagram.

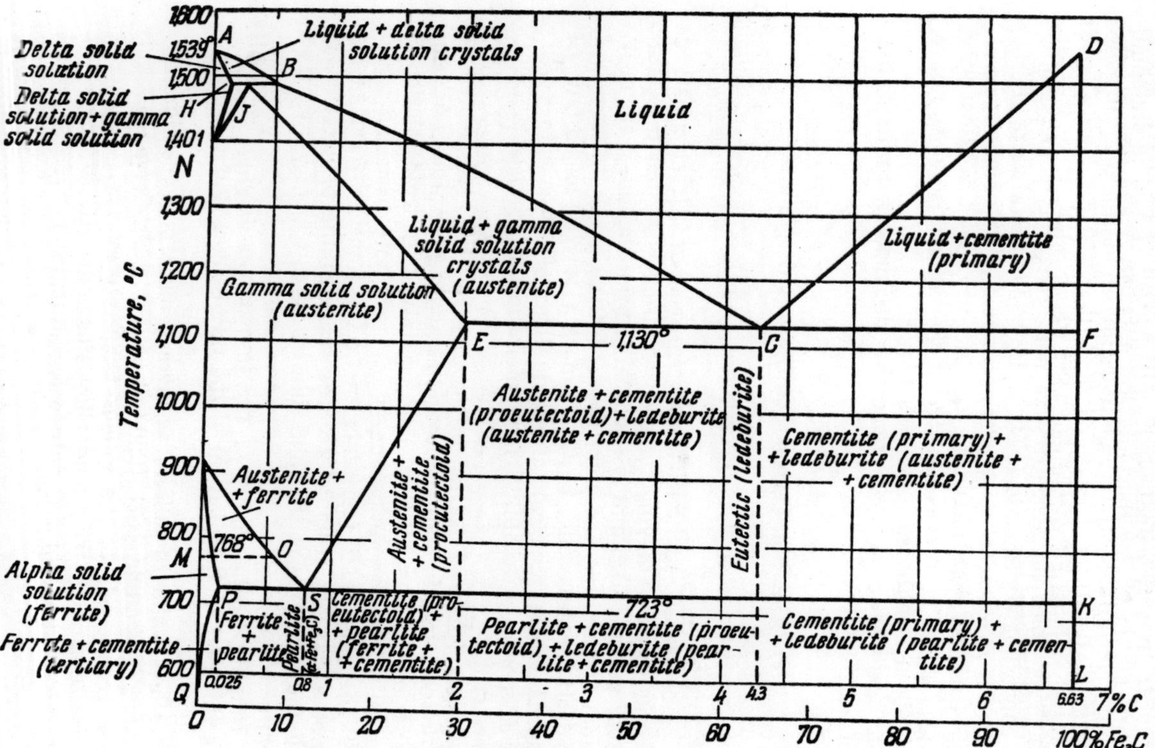

Fig. 108. The iron-carbon equilibrium diagram

a carbon content of 4.3 per cent austenite and cementite are simultaneously precipitated from the liquid alloy to form a eutectic called *ledeburite*. All alloys containing more than 2.0 per cent C (cast irons) and completely solidifying at 1,130° C (line *ECF*) form the eutectic (ledeburite).

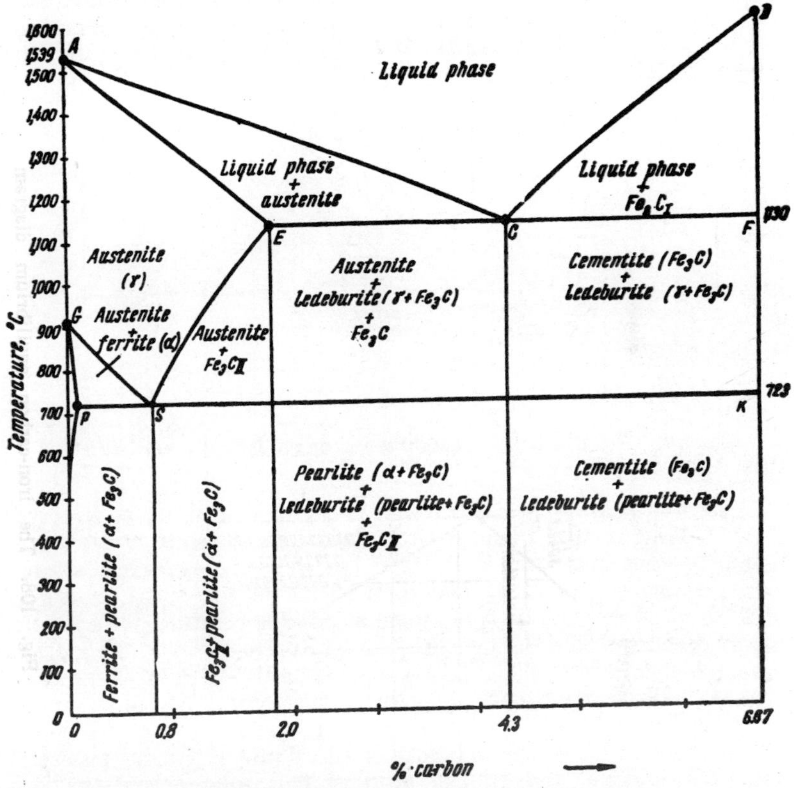

Fig. 109. Simplified iron-carbon equilibrium diagram

A discussion of the primary solidification (transformation from the liquid to the solid state) of steels containing 0.75 per cent C and cast irons containing 3, 4.3, and 5.75 per cent C will clarify the interpretation of the diagram.

Steel containing 0.75 per cent C begins to solidify at the temperature t_1 (Fig. 110). Here the solid solution of carbon in γ-iron, austenite, begins to precipitate from the liquid alloy. The amount of austenite continuously increases as the temperature falls and its composition varies along the solidus (AE) while that of the liquid phase varies along the liquidus (AC). For example, at a temperature t_2, the composition of austenite corresponds to point a; that of the liquid phase is indicated by point b. This steel completely solidifies at temperature t_3 and will consist only of austenite. Steel of any composition will solidify in exactly the same manner and will consist of only austenite, at the moment when the process has been completed.

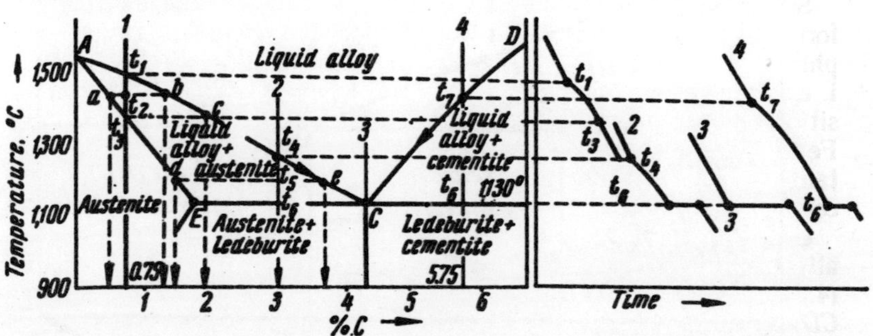

Fig. 110. The iron-carbon equilibrium diagram (primary crystallisation)

Cast iron containing 3.0 per cent C begins to solidify at temperature t_4 lying on the liquidus (AC). Here too austenite crystals precipitate from the liquid alloy in a continuously increasing amount as the temperature falls. The composition of the austenite and of the liquid phase may be determined at any temperature, for example, t_5 by drawing a line through this temperature parallel to the abscissa so that it intersects the liquidus and the solidus. Point d indicates the austenite composition and point e, the composition of the liquid phase which is in equilibrium at temperature t_5.

The proportion of the amounts of the liquid and solid phases at any temperature between the liquidus and solidus is determined by the lever rule. Thus, at the temperature t_5 the quantitative proportion of these phases is expressed by the ratio

$$\frac{t_5 d}{t_5 e}$$

Cast iron containing 3.0 per cent C completely solidifies at temperature t_6 (1,130° C). At this temperature, the liquid phase, enriched in carbon up to 4.3 per cent (the eutectic concentration), is finally solidified. From it, saturated austenite crystals (2.0 per cent C) precipitate simultaneously with cementite to form the eutectic mixture, ledeburite. Therefore, directly following solidification, the cast iron will consist of primary austenite crystals and ledeburite, formed at 1,130° C (t_6). Cast iron of any composition between 2.0 to 4.3 per cent C (*hypoeutectic cast irons*) will solidify in exactly the same way as has been described above.

An alloy containing 4.3 per cent C (*eutectic cast iron*) differs from the previous alloys in that it solidifies at a constant temperature (point *C* at 1,130° C). Above this temperature, the alloy will be in a liquid state and when the temperature falls to 1,130° C (point *C*), the alloy solidifies at this constant temperature forming ledeburite.

Solidification of the eutectic alloy at a constant temperature conforms to the phase rule. In a two-component system containing three phases (liquid phase, austenite, and cementite) $F=2+1-3=0$, i. e., the system is nonvariant. All three phases have definite compositions (liquid phase—4.3 per cent C, austenite—2.0 per cent C, and Fe_3C—6.67 per cent C) and the transformation temperature is constant—1,130° C. A horizontal step on the cooling curves (Fig. 110) corresponds to the formation of the eutectic (ledeburite).

Cast iron containing 4.3 per cent C will consist of only ledeburite after solidification. The solidification of *hypereutectic cast irons* (4.3 to 6.67 per cent C), as is evident from Fig. 110, begins along line *CD* by the precipitation of cementite (Fe_3C) from the liquid alloy. A cast iron containing 5.75 per cent C, for example, begins to solidify at the temperature t_7 where cementite crystals (Fe_3C) begin to precipitate from the liquid alloy. Since cementite is the phase richer in carbon, its precipitation will change the composition of the liquid alloy, upon a further fall in temperature, toward a reduction in carbon content along the liquidus *CD*. At 1,130° C, the liquid alloy will reach the eutectic composition (4.3 per cent C) and will solidify at this (constant) temperature to form ledeburite. The completely solid alloy will consist of primary cementite crystals and ledeburite. All hypereutectic alloys will solidify in a similar manner.

Thus, all steels consist of austenite after the primary crystallisation, while all cast irons consist of ledeburite and primary crystals of excess austenite (hypoeutectic cast irons containing from 2.0 to 4.3 per cent C) or cementite (hypereutectic cast irons containing from 4.3 to 6.67 per cent C).

We will next consider transformations known as secondary crystallisation which occur in the solid state. In the iron-carbon alloys, these are associated with the transformation of γ-iron to α-iron and the decomposition of austenite. The line *GS* in the equilibrium diagram (Figs. 109 and 111) indicates the beginning of austenite decomposition and the precipitation of ferrite from the austenite. The critical points along the *GS* line are designated as Ac_3 in heating and as Ar_3 in cooling. The line *SE* indicates the temperatures at which austenite begins to decompose with the precipitation of excess carbon as cementite. This latter is called *secondary* or *proeutec-*

toid cementite to distinguish it from the cementite that precipitates from the liquid phase. Temperatures along the *SE* line are designated as *Acm* points. Point *S*, corresponding to 0.8 per cent C, shows the minimum temperature (723° C) at which austenite may exist in a state of equilibrium. At point *S* austenite decomposes with simultaneous precipitation of ferrite and cementite which form the *eutectoid mixture* known as *pearlite*.

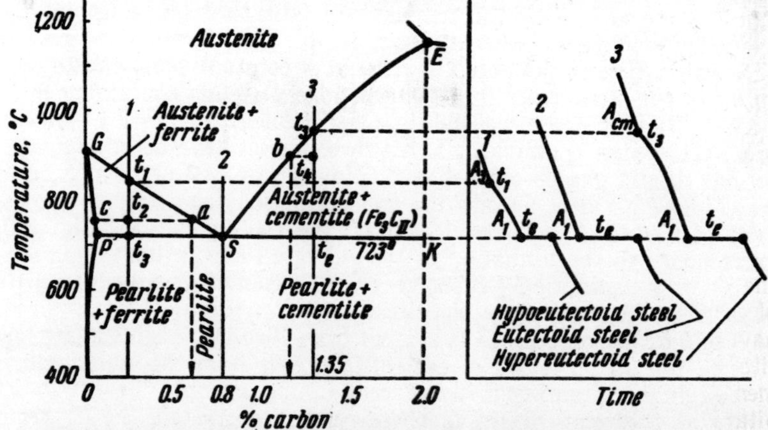

Fig. 111. The iron-carbon equilibrium diagram (transformations in steels)

The formation of pearlite proceeds at a constant temperature since, in the presence of three phases of constant composition (austenite, ferrite, and cementite), the number of degrees of freedom of the system will be zero ($F=2+1-3=0$). Consequently, there is a horizontal step on the cooling curve at the pearlite formation temperature (Fig. 111).

In structure, pearlite consists of thin alternating plates or lamellae of cementite and ferrite called *lamellar pearlite* (Fig. 112, *a*). As a result of a special heat treatment the so-called *granular or divorced pearlite* (cementite) may be formed (see page 188). It consists of rounded globules of cementite in a ferrite field (Fig. 112, *b*).

The decomposition of austenite with the formation of pearlite corresponds to the line *PSK* (723° C) for all iron-carbon alloys. The temperature (critical point) at which pearlite is formed in cooling is designated as Ar_1; the transformation of pearlite into austenite in heating occurs at a temperature designated as Ac_1.

Point *P* in Figs. 109 and 113 indicates the maximum solubility of carbon in α-iron at 723° C (0.025 per cent). In the same way as the line *SE* running from point *E* indicates the reduction in the solubility of carbon in γ-iron (austenite) with a drop in temperature,

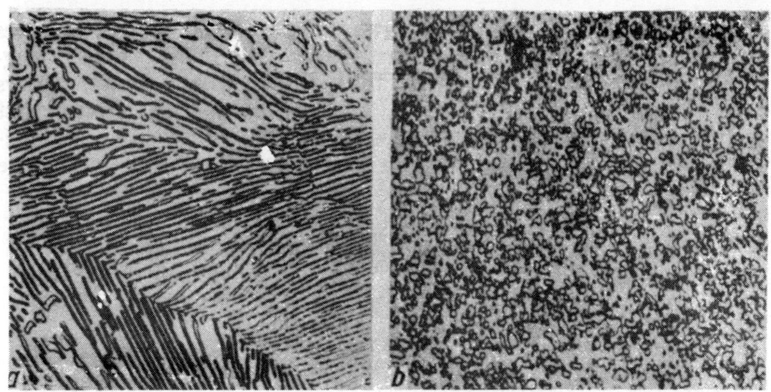

Fig. 112. Pearlite, × 500:
a—lamellar pearlite, b—granular pearlite

the line PQ runs from P and indicates the reduction of the solubility of carbon in α-iron (ferrite) upon cooling the alloy. Therefore, alloys having a carbon content between points P and Q will consist of ferrite and excess (tertiary) cementite which usually precipitates along the boundaries of the ferrite grains (Fig. 114). The composition of the ferrite is indicated by the line PQ. After being completely cooled, alloys of a composition to the left of point Q will consist of only the solid solution of carbon in α-iron, i.e., of ferrite grains.

We will next follow transformations that occur in steels containing from 0.025 to 0.8 per cent C (hypoeutectoid steels) when they are cooled from the solid solution (austenite) region.

The diagram (Fig. 111) shows that no changes occur in the alloy until the temperature drops to the line GS. In this range, the alloys have a sing-

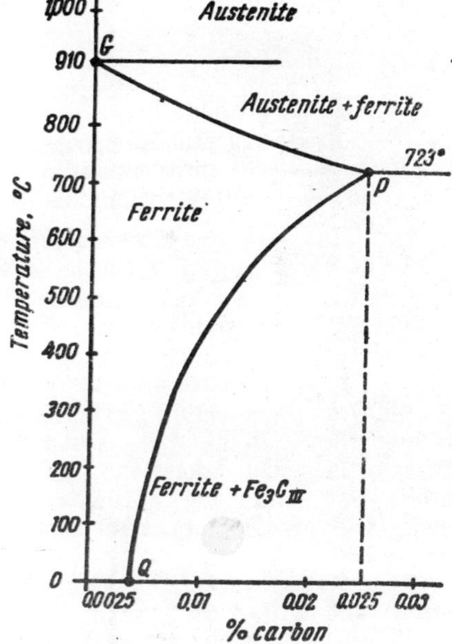

Fig. 113. Solubility of carbon in α-iron

le-phase, austenite structure. Ferrite precipitates from the austenite at temperatures below line *GS*. Thus, a two-phase state exists below line *GS*: austenite (γ) and ferrite (α).

Since the ferrite precipitating from the austenite upon further cooling contains very little carbon, the carbon concentration of the remaining austenite will continuously increase along the line *GS*. For example, in cooling a 0.2 per cent C steel, the austenite at temperature t_2 will have a carbon content corresponding to point *a* (Fig. 111) while that of the ferrite will correspond to point *c*. Upon further cooling the carbon content of the austenite will increase and at a temperature of 723° C (Ac_1), it will reach 0.8 per cent.

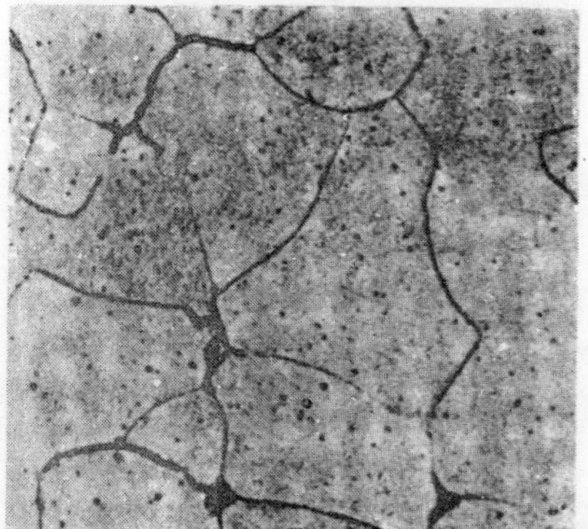

Fig. 114. Tertiary cementite precipitated along the boundaries of the ferrite grains, ×600

Austenite of the eutectoid composition will decompose at the constant temperature of 723° C and pearlite will be formed.

Thus, hypoeutectoid steels (containing less than 0.8 per cent C) will consist of ferrite and pearlite when they are completely cooled. The higher the carbon content of such steels, the more pearlite and the less ferrite they will contain. This is well illustrated in the photomicrographs of Fig. 115.

Other processes occur when steels containing more than 0.8 per cent C (hypereutectoid steels) are cooled. The line *ES* indicates the change in the solubility of carbon in austenite (γ-iron) upon cooling. Only austenite exists above line *ES*. When the temperature drops to the line *ES*, the austenite is saturated with carbon which precipitates as cementite upon further cooling. Therefore, at temperatures below the line *ES*, a two-phase state will exist, consisting of austenite (γ) and secondary or proeutectoid cementite Fe_3C_{II} (Fig. 111).

The carbon content of the austenite is reduced upon the precipitation of cementite along the line *ES*. At a temperature of t_4, the carbon content in the austenite will be 1.2 per cent (point *b*) when a hypereutectoid steel containing 1.35 per cent C is cooled.

At the temperature 723° C (Ac_1) the austenite will contain 0.8 per cent C and will decompose at this constant temperature into the ferrite-cementite eutectoid mixture which we call pearlite.

Thus, completely cooled hypereutectoid steels (0.8 to 2.0 per cent C)

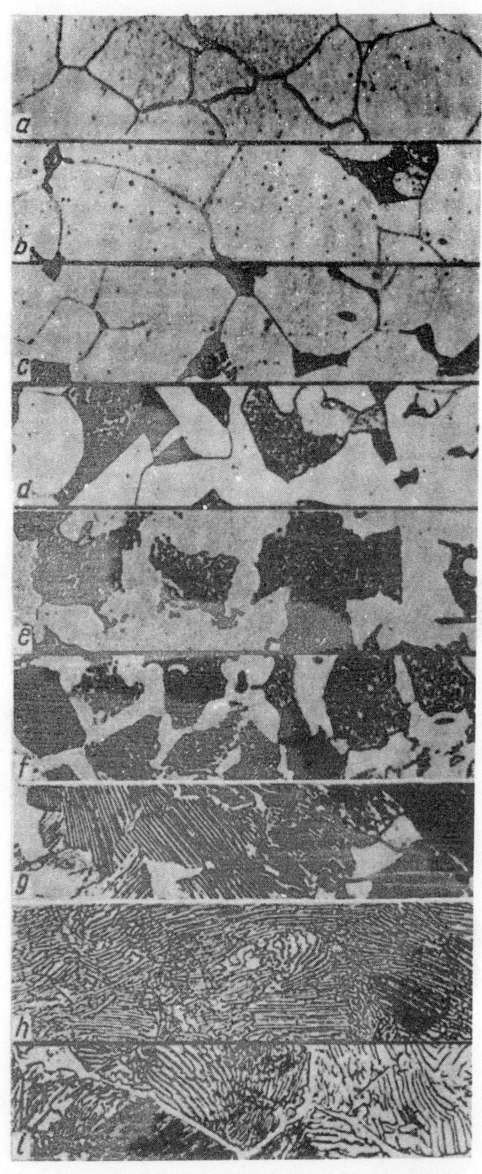

Fig. 115. Microstructures of steels:
a—0.06 per cent C, *b*—0.13 per cent C, *c*—0.15 per cent C, *d*—0.2 per cent C, *e*—0.26 per cent C, *f*—0.5 per cent C, *g*—0.62 per cent C, *h*—0.82 per cent C, *i*—1.3 per cent C, ×600 (after Hanemann)

will have a structure consisting of pearlite and excess (proeutectoid) cementite (Figs. 115, *i* and 116).

We may now consider the transformations that occur in cast irons. As has been mentioned above, cast irons contain ledeburite, after primary crystallisation, which is a eutectic consisting of saturated austenite (2.0 per cent C) and cementite. Hypoeutectic cast irons (2.0 to 4.3 per cent C) consist, in this state, of ledeburite and primary austenite crystals, containing 2.0 per cent C. Hypereutectic cast irons (4.3 to 6.67 per cent C) consist of primary cementite in addition to the ledeburite.

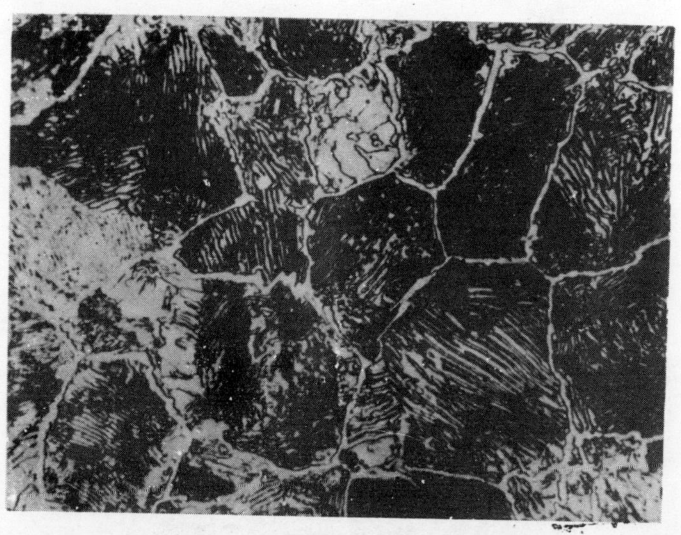

Fig. 116. Microstructure of hypereutectoid steel (excess cementite has precipitated as a network along the grain boundaries), ×500

Upon further cooling, the solubility of carbon in the austenite is reduced (along line *SE*) and the primary crystals of the solid solution (austenite) will partly decompose, together with the austenite in the eutectic, with the precipitation of secondary cementite crystals. The composition of the austenite will change along the line *SE*. When the line *PSK* (723° C) is reached, the austenite, depleted of carbon to the eutectoid concentration (0.8 per cent C), decomposes into ferrite and cementite which form the eutectoid (pearlite).

Therefore, upon being completely cooled, hypoeutectic cast irons will have a structure consisting of pearlite, ledeburite (pearlite+ +cementite) and secondary cementite. Fig. 117, *a* illustrates the structure of hypoeutectic cast iron. Grains of pearlite are seen here,

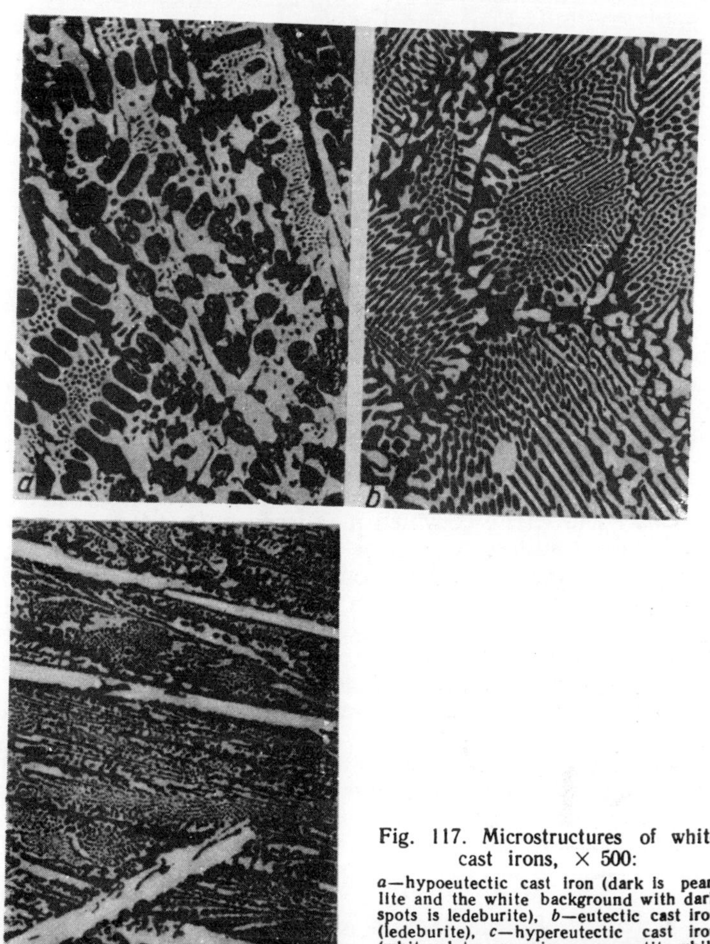

Fig. 117. Microstructures of white cast irons, × 500:

a—hypoeutectic cast iron (dark is pearlite and the white background with dark spots is ledeburite), *b*—eutectic cast iron (ledeburite), *c*—hypereutectic cast iron (white plates are primary cementite while the main background is the eutectic—ledeburite)

surrounded by the eutectic (ledeburite), consisting of pearlite and cementite. The precipitates of secondary cementite usually blend with the eutectoid and cannot be distinguished in the microphotograph. Eutectic cast iron consists of the eutectic—ledeburite—only. The ledeburite consists of pearlite and cementite (Fig. 117, *b*).

Hypereutectic cast irons consist of ledeburite (pearlite+cementite) and primary cementite (Fig. 117, *c*). In this case, the primary cementite precipitates in the form of long plates or needles.

Chapter 8

PHASE TRANSFORMATIONS IN THE IRON-CARBON SYSTEM

8-1. Formation of Austenite (Transformations That Occur in Heating Steel)

Austenite may be formed from pearlite only after it is heated slightly above the equilibrium temperature (Ac_1) so that the free energy of the austenite is less than that of the ferrite-cementite mixture (pearlite) (Fig. 118).

This austenite formation is a crystallising process and proceeds by nucleation and subsequent growth of the austenite crystals.

The composition of the austenite greatly differs from that of the cementite and ferrite of which it is formed. Therefore, the pearlite-to-austenite transformation is of the diffusion type and is associated with the movement of carbon atoms over considerable distances.

Extensive research has shown that the austenite nuclei always appear on the boundaries between crystals of ferrite and cementite (Fig. 119). These boundaries become rich in carbon when steel is heated above the equilibrium temperature (Ac_1).

The higher the temperature to which the pearlite is heated (above the critical point), the larger the difference will become between the free energies of pearlite and austenite, the smaller the critical size of austenite nuclei, the higher the nucleation rate, and the higher the rate of linear crystal growth will be.

If, for example, the temperature is raised from 740° to 800° C, the rate of nucleation will increase 280 times and that of crystal growth—82 times (after M. E. Blanter).

The number of nuclei that appear at a given temperature increases with a larger carbon content in the steel and with higher dispersion of the cementite (carbides) in the initial structure.

The process of transforming the ferrite-cementite mixture into austenite proceeds at the highest rate if the initial structure is finely lamellar pearlite, at a lower rate if the pearlite is coarsely lamellar, and slowest of all for granular pearlite. This is explained by the fact that higher carbide dispersion increases the interface with the ferrite. This, in turn, increases the rate of austenite nucleation and crystal growth.

The pearlite-to-austenite transformation in heating eutectoid steel may easily be followed on the schematic diagram proposed by S. S. Steinberg and shown in Fig. 120. In its initial state (before heating), the steel consists of cementite crystals surrounded by ferrite (Fig. 120, *I*). Upon heating slightly above the equilibrium point Ac_1, regions of austenite are formed (Fig. 120, *II*) which dissolve the cementite. The layer of austenite adjacent to the cementite will, therefore, become more saturated with carbon.

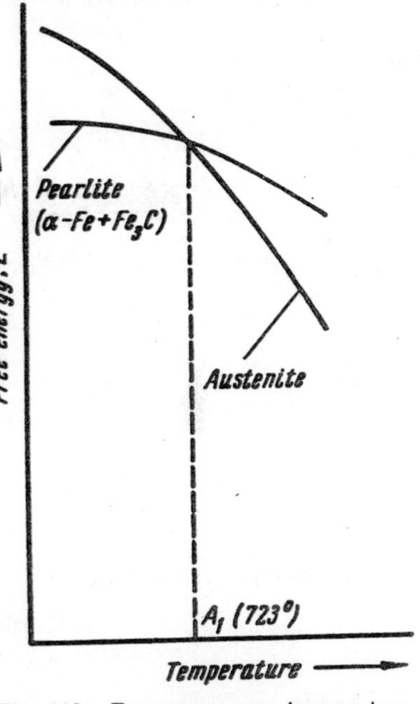

Fig. 118. Free-energy vs temperature curves for pearlite and austenite (schematic diagram)

The austenite nuclei develop due to the dissolution of cementite and transformation of the ferrite. The rate of austenite crystal growth due to ferrite is always higher than that due to the dissolution of the cementite. Therefore, a certain amount of cementite remains in the austenite after all of the ferrite has been transformed (Fig. 120, *III*). If the steel is held further at this temperature (or if the temperature is raised), this remaining cementite is also dissolved in the austenite (Fig. 120, *IV*).

After the ferrite has disappeared (Fig. 120, *III*) and even after the cementite is completely dissolved (Fig. 120, *IV*), the austenite will not be homogeneous in carbon content.

Regions of austenite adjacent to the cementite particles (Fig. 120, *III*) and the places the particles occupied before being dissolved (Fig. 120, *IV*) are always richer in carbon than regions occupied by the ferrite. Additional time or higher temperatures, which increase the rate of carbon diffusion, is required to homogenise the carbon concentration throughout the austenite crystals (Fig. 120, *V*). Consequently, the pearlite-austenite transformation consists of three stages: formation of the austenite nuclei and their growth by taking up the cementite and ferrite, dissolution of the cementite, and finally, equalising the austenite composition throughout the crystals (homogenisation of austenite).

The diagram in Fig. 121 shows the time at which the pearlite-to-austenite transformation begins and ends for various temperatures. It is evident that the rate of transformation is higher for higher temperatures. At temperatures above 800° C, the rate is so high that it cannot be determined experimentally.

Upon continuous heating at various rates (lines *1*, *2*, and *3*), pearlite is transformed into austenite (Fig. 121), not at a constant temperature,

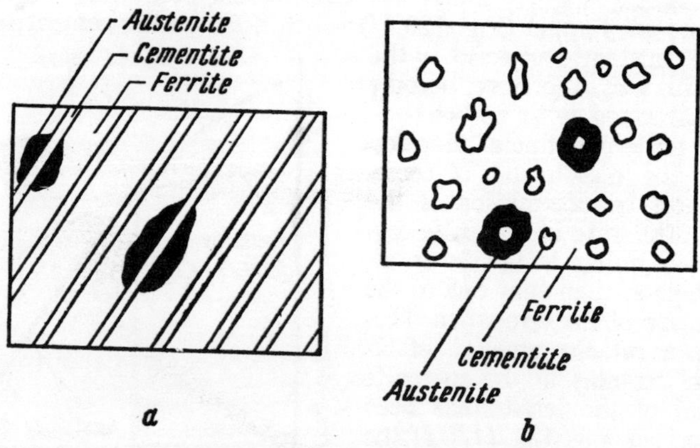

Fig. 119. Formation of austenite on the boundary between ferrite and cementite: *a*—lamellar pearlite, *b*—granular pearlite

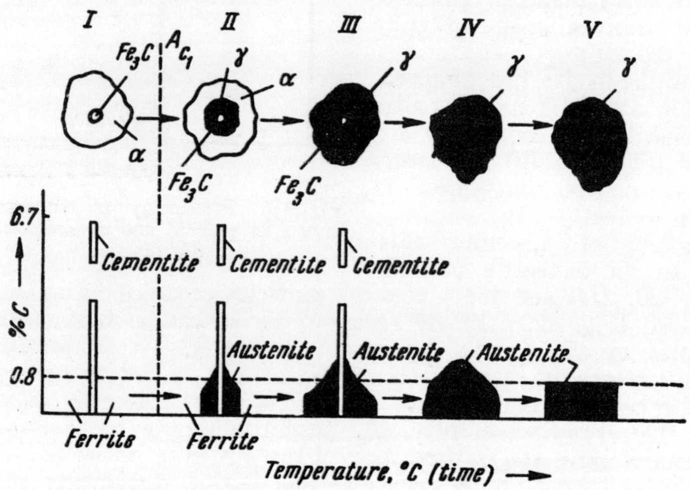

Fig. 120. Schematic diagram illustrating the formation of austenite from the ferrite-cementite mixture in eutectoid steel (after S. S. Steinberg)

but in a certain temperature interval (t-t_1, t_2-t_3, and t_4-t_5). The higher the heating rate, the higher will be the temperature of the transformation.

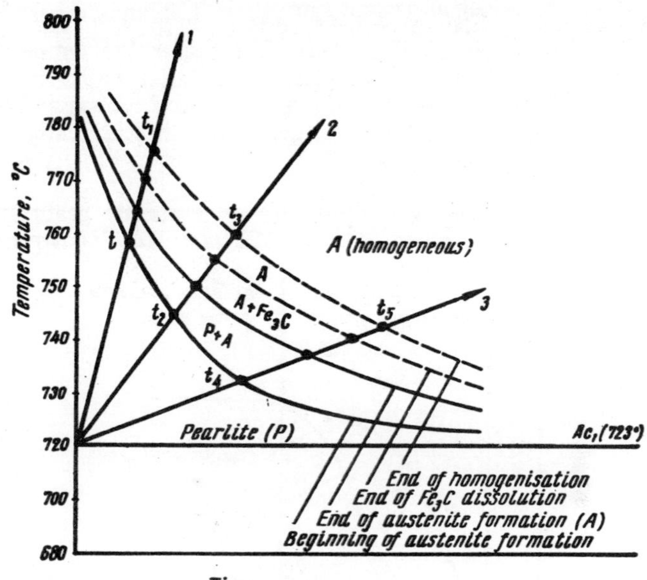

Fig. 121. Pearlite-to-austenite transformation at constant temperature and continuous heating

8-2. Austenite Grain Growth in Heating

The effect of austenite grain growth at an increase in temperature must be taken into consideration in selecting the rate at which steel is to be heated. The size of the austenite grains greatly affects the mechanical properties and, particularly, the impact strength of the steel after heat treatment.

A grain of austenite appears immediately upon the completion of the pearlite-to-austenite transformation. It is usually called the *original austenite grain*. Its size depends on the number of nuclei appearing in a unit time in a given volume and their rate of growth. The more the carbide phase is dispersed in the previous structure, the smaller the original grain of austenite will be.

The austenite grain is ordinarily very small (500 to 1,000 sq microns) at the moment it appears but it grows rapidly if held for appreciable time at the given temperature or at increasing temperatures.

Austenite grain growth is spontaneous and is due to the tendency to reduce the free energy of the system by decreasing the grain surface.

Grain growth develops by the enlargement of certain grains at the expense of other smaller grains which are, consequently, less stable thermodynamically. Holding the austenite at a constant temperature does not at first cause detectable grain growth. This is the so-called incubation period (Fig. 122). The higher the temperature, the less time required for the incubation period after which intensive growth begins. In this second period, considerable nonuniformity of grain size is observed. After this, grain growth practically ceases and holding the austenite further at this temperature will only equalise grain size.

The effect of the heating temperature on austenite grain size is illustrated by the data given in Figs. 123 and 124. Higher carbon

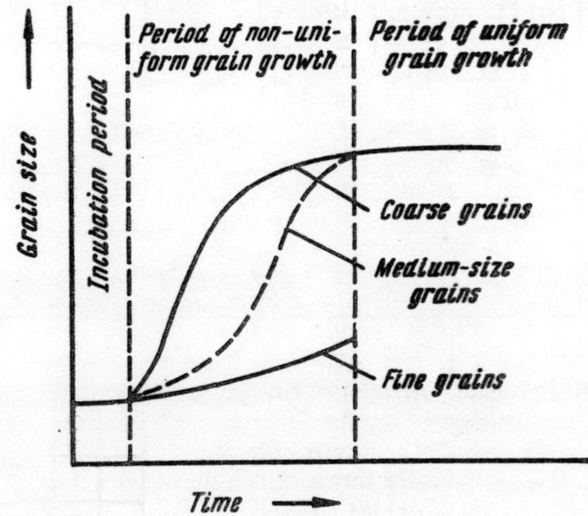

Fig. 122. Schematic diagram illustrating austenite grain growth vs time relationships at constant temperature

content in a steel increases its tendency to grain growth (Fig. 123) if grain size is determined at the same temperatures in a single-phase region.

All other conditions being equal, the tendency of cast steel to austenite grain growth is less than for hot-worked steel. This may be explained by the presence of impurities at the grain boundaries which almost entirely envelop the grains in cast steel and occur only as separate inclusions in hot-worked steel.

This tendency to grain growth may differ even for steels of the same grade but of different melts.

In this respect, two types of steel are distinguished: 1) inherently fine-grained and 2) inherently coarse-grained.

Inherently fine-grained steel is characterised by its small tendency toward austenite grain growth in heating; *inherently coarse-grained steel*, on the opposite, is disposed to considerable grain growth under the same conditions (Fig. 125).

The tendency toward grain growth (of inherent grains) is determined by the method of deoxidation of the steel and the deoxidisers employed. Steels, deoxidised by ferromanganese only (rimming steel) or by ferromanganese and ferrosilicon, belong to the group of inherently coarse-grained steels. Steels, additionally deoxidised by aluminium or those containing vanadium, molybdenum, and titanium, are of the inherently fine-grained type.

It has been proved by experiments * that aluminium, dissolved in steel, contributes to an increase in the number of austenite nuclei, appearing in heating steel, and inhibits their rate of growth. This explains why finer grain structure is obtained after introducing aluminium. Theoretical research has shown that aluminium reduces the boundary and surface energy and increases the activation energy, i.e., it impedes atomic exchange in phase transformations (in austenite nucleation and grain growth). Many investigators, however, explain the favourable effect of aluminium by the fact that it easily forms oxides (Al_2O_3) and nitrides (AlN) which are disposed in a dispersed state on the boundaries of the austenite grains and impede their growth in a purely mechanical manner.

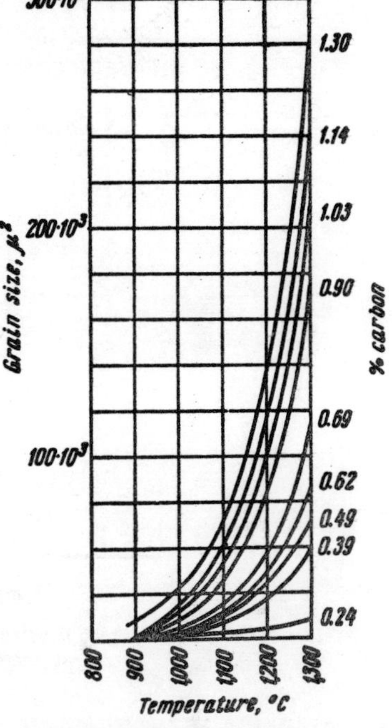

Fig. 123. Effect of temperature and carbon content on austenite grain growth (after Gayev and Gendler)

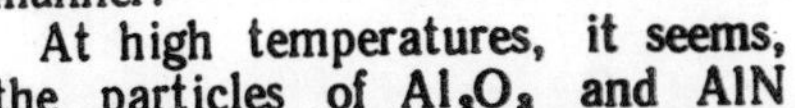

At high temperatures, it seems, the particles of Al_2O_3 and AlN are being dissolved at the grain boundary layers of the austenite, and the grains begin to grow rapidly (Fig. 125). This, however, has not been proved by experiment. It must be emphasised again that the terms "inherently coarse-grained" and "inherently fine-grained" do not mean that the given steel always has a coarse or a fine grain size. They only indicate that in heating, a coarse-grained steel will acquire a coarse grain structure at a lower temperature than will an inherently fine-grained steel.

At a sufficiently high temperature, an inherently fine-grained steel may have austenite grains of even larger size than an inher-

---

* *Journal of Engineering Physics*, Issue I, Volume XXIV, Moscow, 1954.

ently coarse-grained steel. This is quite evident from the curves in Fig. 125. For this reason, the term *actual grain* is employed to

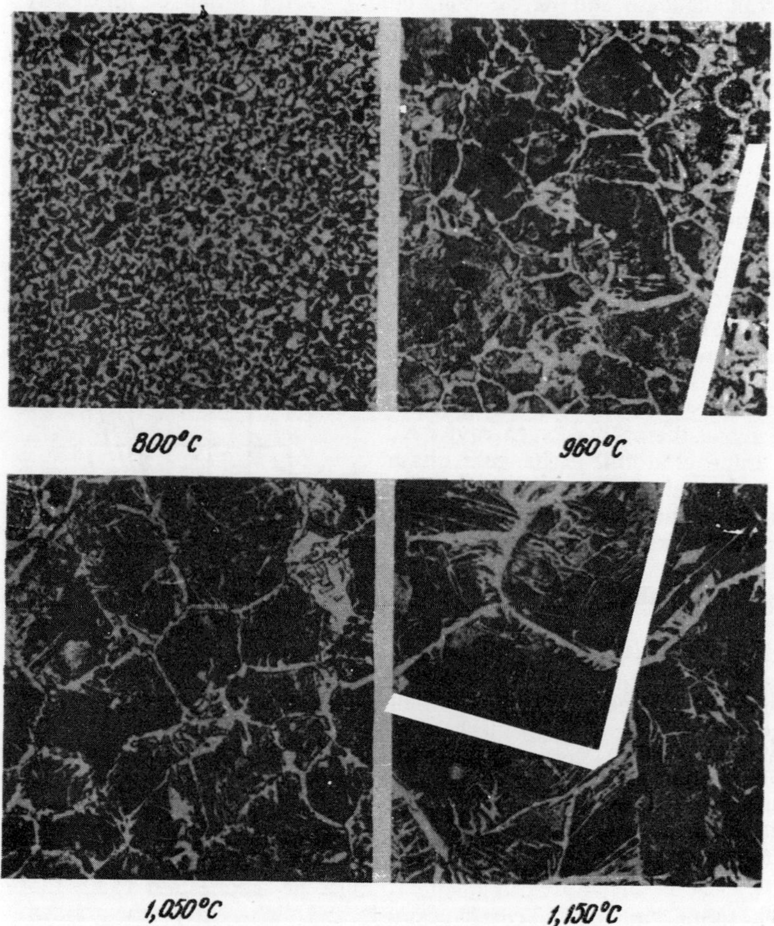

Fig. 124. Effect of temperature on grain growth of hypoeutectoid steel containing 0.45 per cent C (Ac_1=725° C, Ac_3=780° C), ×100 (after Druzhinin)

indicate the grain obtained in steel at a given temperature or as a result of a certain type of heat treatment.

The grain size of the structural components of steel is closely related to that of the prior austenite grains. Consequently, actual grain size, observed at normal temperatures, depends only on the temperature and time in heating; upon subsequent cooling, the grain size does not change. Therefore, actual grain size is determined exclusively by the temperature to which steel is heated in heat treatment and the tendency toward grain growth of the given steel in heating.

Actual grain size of steel determines its final properties. Coarsening of the grain has a comparatively small effect on tensile strength, hardness, or relative elongation, but sharply reduces impact strength.

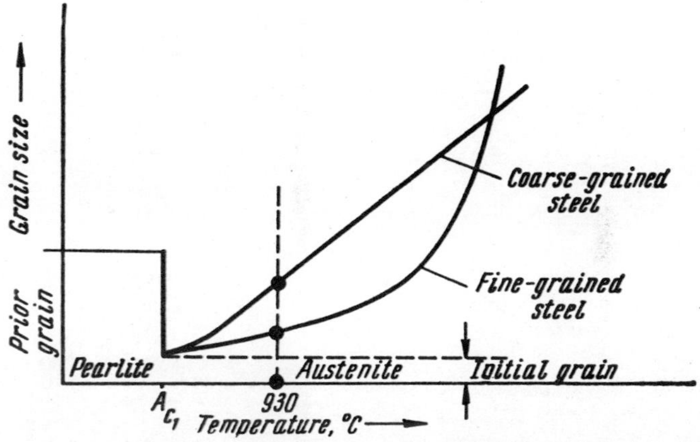

Fig. 125. Schematic diagram of austenite grain growth in inherently fine- and coarse-grained steels

It should be noted that when hypoeutectoid steels are heated to temperatures considerably exceeding the point Ac_3, in addition to austenite grain growth, excess ferrite may precipitate, on subsequent cooling, as long plates (or needles) cutting across the pearlite grains (Fig. 167, *a*). This is called a *Widmanstätten structure*. The formation of a Widmanstätten structure in overheating steel further reduces impact strength. Overheated steel is characterised by a coarse-grained fracture since the fracture passes along the boundaries of the large grains. If the austenite grains are small, the fracture will have a fine-grained structure.

The size of the inherent grain mainly affects the fabrication properties of steel. Thus, for example, inherently fine-grained steel may be heated to higher temperatures without the fear of overheating, i.e., appreciable coarsening of the grains.

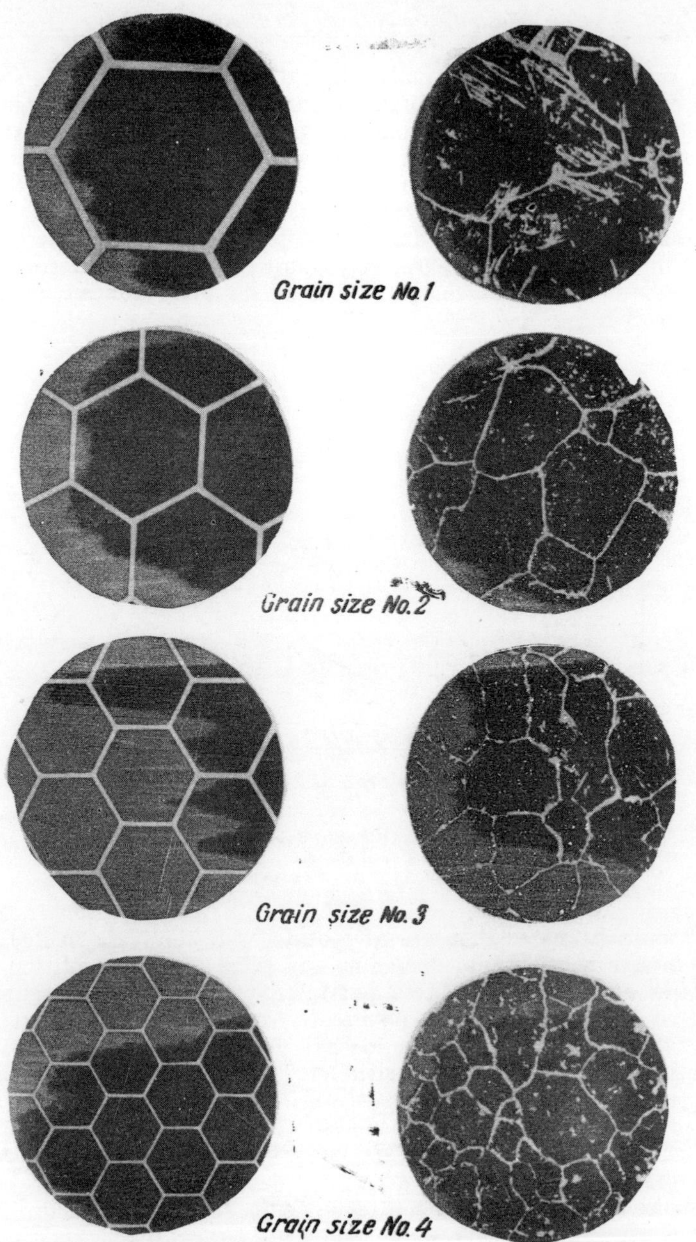

Fig. 126. Standard chart of steel

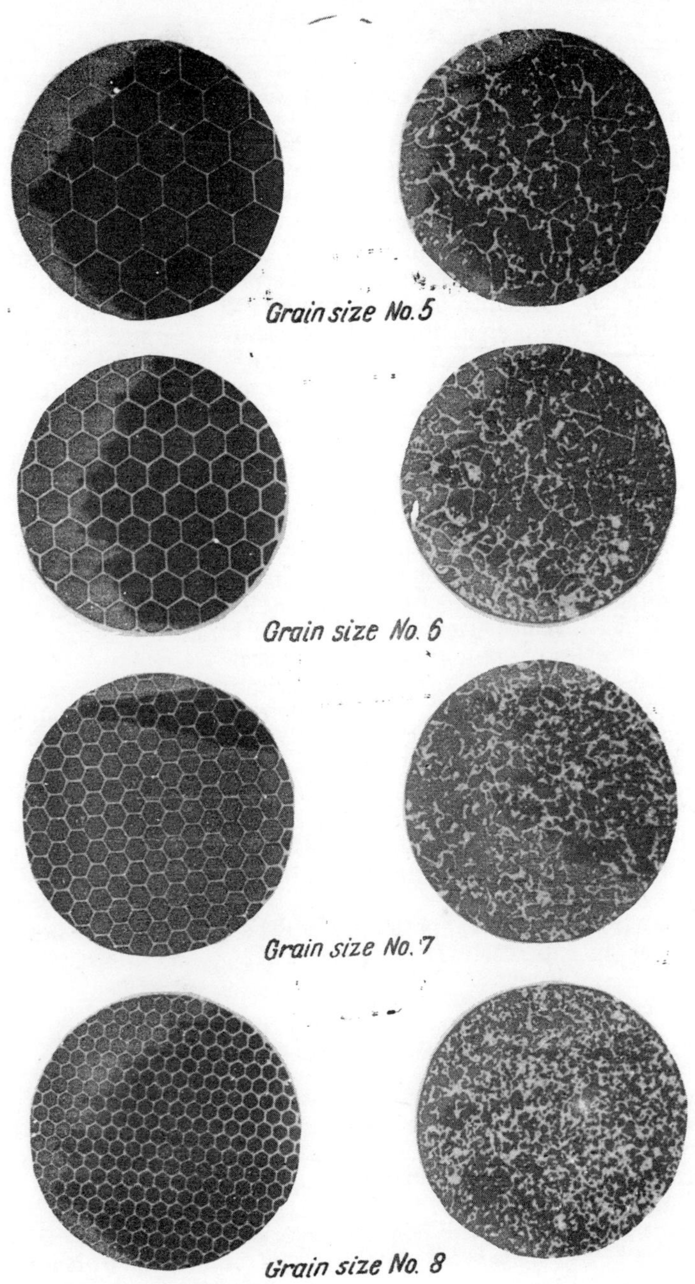

grain size (GOST 5639-51), ×100

DETERMINATION OF GRAIN SIZE

A method specified by GOST 5639-51 is employed to determine inherent (or actual) grain size. By this method, the specimen is carburised (saturated with carbon) at 930° C for eight hours. The carbon content in the austenite of the surface layer will reach a value exceeding the eutectoid concentration. As a result of subsequent slow cooling, proeutectoid cementite will be precipitated along the boundaries of the austenite grains, forming a continuous network which is retained even after the austenite is transformed into pearlite. Thus, the size of the original austenite grains is revealed by the cementite network in the hypereutectoid zone of the specimen (Fig. 126).

The grain size of hypoeutectoid steel may be determined after a treatment which consists in heating steel to 930° C and subsequently cooling it in still air (normalisation). The boundaries of the prior austenite grains are indicated, in this case, by a network of excess ferrite (Fig. 124). Austenite grain size may be determined, as well, by the oxidation method in which a microsection is heated, then cooled rapidly in oil (quenched) and finally slightly polished. After polishing off the continuous layer of oxides, light grains of martensite will be revealed, surrounded by a dark network of oxides along the boundaries of the former austenite grains.

The grain size is measured by comparison, under a microscope with a magnification of ×100, with standard charts shown in Fig. 126. Steels with a grain-size number from 1 to 5, inclusively, are coarse-grained; Nos. 6, 7, and 8 are fine-grained steels.

For most steels, the heating temperature in heat treatment does not exceed the temperature of the standard test (930° C). Therefore, if the standard test shows that the steel is of the inherently coarse-grained type, the actual grain size obtained as a result of heat treatment will also be coarse.

Likewise, a fine actual grain size will be obtained after heat treatment of an inherently fine-grained steel.

8-3. Transformation of Austenite into Pearlite (Isothermal Decomposition of Austenite)

Austenite is unstable below point Ar_1 (723° C) since its free energy is higher than that of the ferrite-cementite mixture (see Fig. 118). Therefore, the transformation of austenite into pearlite (ferrite-cementite mixture) is possible only upon a certain degree of supercooling in reference to the equilibrium point Ar_1.

The pearlite transformation of supercooled austenite is of a crystallisation type and proceeds by a diffusion mechanism. This is evident from the fact that practically homogeneous austenite, as far as carbon concentration is concerned, decomposes into a mixture of phases which have sharply different carbon concentration—ferrite (almost pure iron) and cementite (6.67 per cent C).

The kinetics of the austenite transformation into the ferrite-cementite mixture, in accordance with the degree of supercooling, is described by isothermal austenite decomposition diagrams (usually called TTT diagrams as they relate the *transformation* of the austenite to the *time* and *temperature* conditions to which it is subjected). Other names for these diagrams are S-curves and C-curves. A study of the isothermal decomposition of supercooled austenite provides the most comprehensive understanding of the kinetics involved.

For this purpose, small samples of steel are heated to a temperature at which austenite is stable (above the critical point) and then rapidly cooled to temperatures of 700°, 600°, 500°, 400°, 300° C, etc. The supercooled austenite is held at this temperature (isothermally) until the austenite is completely decomposed. The degree of decomposition is determined by microscopic, magnetic, dilatometric, or other methods.*

The results of the research are plotted as a curve showing the amount of decomposed austenite against the time elapsed from the beginning of decomposition.

As may be seen in Fig. 127, *a*, no decomposition is observed experimentally during an initial period of time (S_1, S_2, S_3, and S_4). This is called the *incubation* period and, following it, austenite begins to decompose into the ferrite-cementite mixture. At first, the rate of decomposition increases rapidly and then gradually slows down. After a certain period of time (F_1, F_2, F_3, and F_4), the process is completed or stopped.

If such curves are plotted for a large number of subcritical temperatures (Fig. 127, *a*), they may be used as a basis for constructing the TTT curve (Fig. 127, *b*). For this purpose, the lengths of time from the start (S_1, S_2, S_3, and S_4) to the end (F_1, F_2, F_3, and F_4) of decomposition are transferred to a diagram with times ** plotted along the abscissas and temperatures along the ordinates. Finally the corresponding plotted points are joined by smooth curves.

In the TTT diagram obtained (Fig. 127, *b*) curve *1* characterises the beginning of austenite decomposition and curve *2* indicates the time required for its complete decomposition.

The region to the left of curve *1* determines the length of the incubation period. This region corresponds to supercooled austenite in which no appreciable decomposition has yet occurred. The length of the incubation period and that of the whole decomposition characterises the stability of the austenite. Fig. 127 shows that austenite

* See *Handbook of Physical Metallurgy and Heat Treatment of Steels*, Metallurgizdat, Moscow, 1956.

** For convenience the time scale is logarithmic since the decomposition interval may vary in a wide range—from fractions of a second to minutes and even hours.

stability first rapidly decreases with an increase in supercooling. It reaches a minimum at about 500° to 550°C and then begins to increase again.

Decreased austenite stability (increased rate of transformation) upon an increase in the degree of supercooling is explained by the

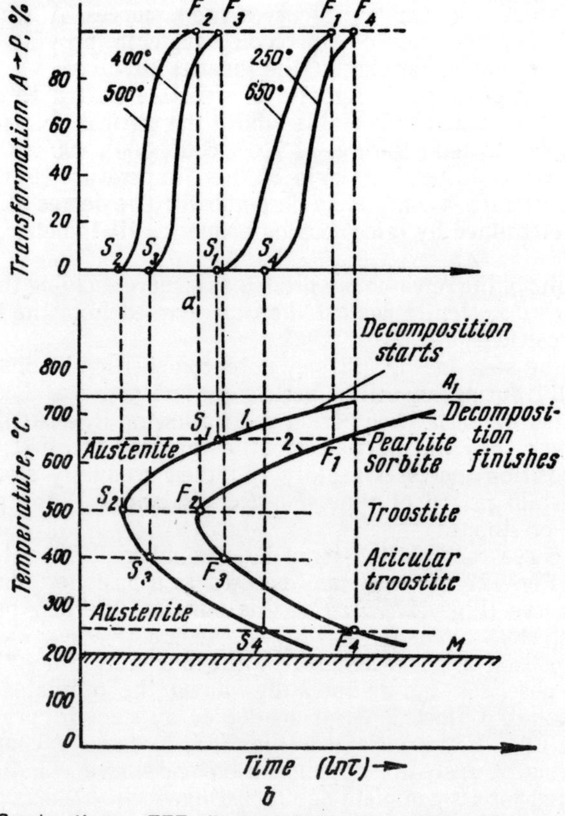

Fig. 127. Constructing a TTT diagram (isothermal decomposition of austenite)

increase in the difference between the free energy of austenite and pearlite. Increased supercooling reduces the size of the critical nucleus capable of growth and increases the number of regions in the austenite where the new phase may nucleate. This, naturally, decreases the stability of the austenite. The increase in austenite stability observed at high degrees of supercooling is due to a reduction in the rate of formation and growth of the new phase because diffusion is inhibited. At subcritical temperatures below point M, diffusion is completely suppressed and the formation of a ferrite-cementite mixture is excluded. In this case, we have diffusionless austenite transformation into a hardened steel structure called *martensite* which is a *supersaturated solid solution of carbon in α-iron.*

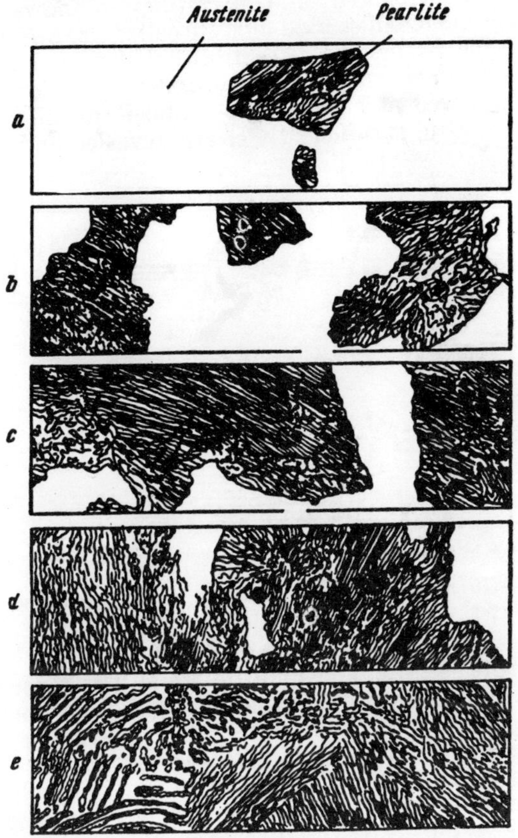

Fig. 128. Kinetics of the pearlite transformation at 700° C, ×750. Transformation time:

a—400 sec, *b*—1,150 sec, *c*—1,320 sec, *d*—1,450 sec, *e*—4,000 sec (after E. Bain)

Decomposition products of austenite in the temperature range from Ar_1 to the martensite point M are mixtures of ferrite and cementite. In the range from Ar_1 to 550°C austenite will decompose into a ferrite-cementite mixture with a lamellar structure.

Fig. 128 illustrates the kinetics of the pearlitic decomposition of austenite in this temperature range. The lower the decomposition temperature (higher degree of supercooling), the more dispersed the ferrite-cementite mixture will be. At low degrees of supercooling, a clearly differentiated ferrite-cementite mixture, pearlite, is obtained. A finer mixture, called *sorbite*, is obtained at higher degrees of supercooling. At subcritical temperatures in the region of 500° to 550°C, an even more dispersed mixture called *troostite* is obtained. In troostite, the ferrite and cementite particles cannot be resolved at the usual accepted magnifications. Under an optical microscope, troostite is observed as a dark mass, difficult to differentiate. The structure of troostite is sufficiently clearly revealed under an electron microscope.

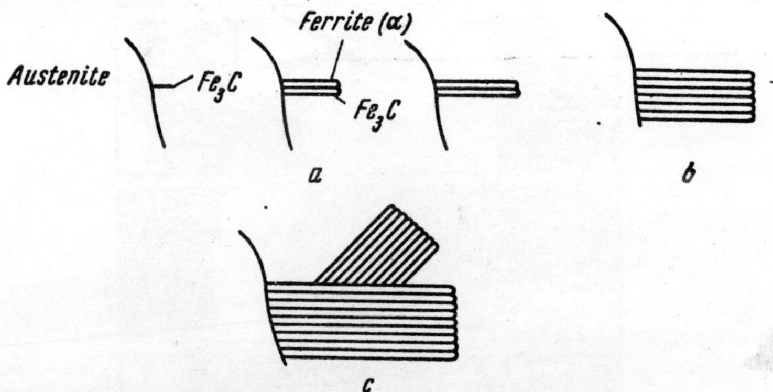

Fig. 129. Schematic representation of nucleation and growth of the pearlite grain or nodule

Thus, pearlite, sorbite, and troostite are all ferrite-cementite mixtures having a lamellar structure and distinguishable from each other in eutectoid steel only by their degrees of dispersion.

In eutectoid crystallisation, where two phases are precipitated simultaneously, one of the phases leads. The leading phase in pearlite decomposition of austenite, upon low degrees of supercooling (in the range Ar_1 to 550°C), will be cementite which nucleates on the boundaries of the austenite grains (Fig. 129). Here, concentration fluctuations, required for forming cementite, more easily occur and less energy is required for nucleation. The cementite nuclei grow due to diffusion of carbon from the adjacent austenite. The austenite, surrounding the pearlite plate, loses its carbon and is more easily transformed into ferrite (Fig. 129,*b*). The ferrite formed does not dissolve carbon to any appreciable extent, so it (the carbon) is rejected to the ferrite-austenite boundary where it facilitates the formation of the next cementite plate. This forms what is called a pearlite colony. Ferrite and cementite plates grow simultaneously. Repetition of this process forms the pearlite nodule or grain. A cementite nucleus with a different orientation appears on the boundary of the first pearlite grain and starts a new grain (Fig. 129). The process continues until the grains impinge upon each other and all of the austenite is transformed into pearlite.

Fig. 130. Kinetics of the bainite decomposition of austenite, ×750. Transformation time:
a—400 sec, *b*—500 sec, *c*—850 sec, *d*—900 sec, *e*—2,500 sec (after E. Bain)

Below the temperature interval of minimum austenite stability (500° to 550°C) a ferrite-carbide mixture of acicular (needle-like) structure is formed.

Fig. 130 illustrates the kinetics of austenite decomposition in this temperature range. The decomposition products in the upper temperature interval frequently have a feathery structure ("cut straw");

in the lower interval, the structure is acicular (Fig. 131). These structures are called upper and lower acicular troostite*. In investigating the structure of acicular troostite (bainite) under an electron micros-

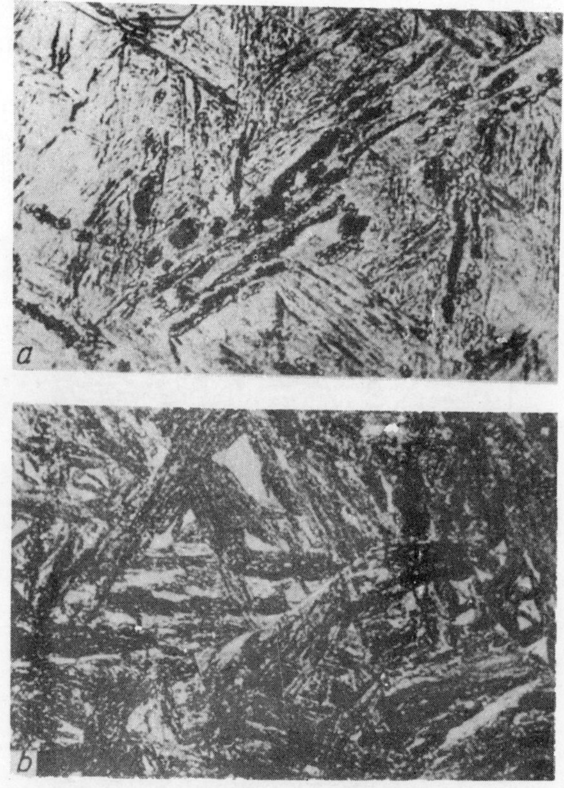

Fig. 131. Acicular troostite (bainite). Steel containing 0.35 per cent C, 1 per cent Cr, and 4 per cent Ni, ×500. Isothermal decomposition temperature: *a*—400°C, *b*—300°C

cope, A. I. Gardin found that troostite has a lamellar structure as do the other products of pearlite transformations. Acicular troostite (bainite) differs from pearlite (sorbite, troostite) in that the plates in the former are considerably shorter.

* They are called upper and lower bainite in English and American metallurgical literature.

The intermediate transformation into acicular troostite has certain features which are inherent to both pearlitic (diffusion type) and martensitic (diffusionless) transformations.

The transformation of austenite into acicular troostite is characterised by the incubation period, measurable rate of the process, and the possibility of supercooling. These are all features of pearlitic transformation. On the other hand, austenite does not completely decompose in the intermediate range, the amount of transformed austenite depends on the temperature, and the decomposition products (in the lower temperature interval) have an acicular structure—all typical features of martensitic transformation (see Sec. 8-5).

Intermediate or bainite transformation is a special type of austenite decomposition by diffusion which occurs under conditions when the diffusion process is so inhibited that cementite nucleation and growth is impossible though cementite is the leading phase in the pearlitic transformation. On the other hand, martensitic transformation is also impossible under these conditions.

The most probable mechanism that can be conceived for the intermediate (bainite) transformation, on the basis of extensive research in this field, is the following. In the transformation process, carbon is first redistributed within the solid solution by diffusion. This results in the formation of austenite regions both rich and poor in carbon in comparison with the initial concentration. The austenite regions with a low carbon content subsequently undergo a martensitic $\gamma \rightarrow \alpha$ transformation (see Sec. 8-5). Either carbides are separated in the regions rich in carbon or these regions are not transformed while held at constant temperature.

The $\gamma \rightarrow \alpha$ transformation is associated with the appearance of a specific microrelief on the surface of unetched microsections, typical of martensitic transformation.

The alpha solid solution formed as a result of the diffusionless martensitic transformation is supersaturated with carbon. Therefore, iron carbide is separated from it following the $\gamma \rightarrow \alpha$ transformation as in the tempering of austenite (see Sec. 8-6).

The precipitation of carbides from the enriched regions of austenite is accompanied by a reduction in their carbon content and their transformation into a supersaturated α-solution by the martensitic mechanism.

In the isothermal decomposition of hypoeutectoid and hypereutectoid steels, certain deviations are observed from the above-described transformation of eutectoid steel. In the upper temperature interval, either excess (proeutectoid) ferrite or excess (proeutectoid) cementite is precipitated first. The amount of proeutectoid ferrite (or cementite, as the case may be) decreases as the temperature is lowered and at a certain degree of supercooling, decomposition begins directly with the nucleation of troostite. The precipitation of proeutectoid ferrite (or cementite) is shown on the TTT diagram by an extra branch of the curve (Fig. 132). Since the amount of proeutectoid ferrite (or cementite) is reduced when the temperature is lowered, the pearlite will contain

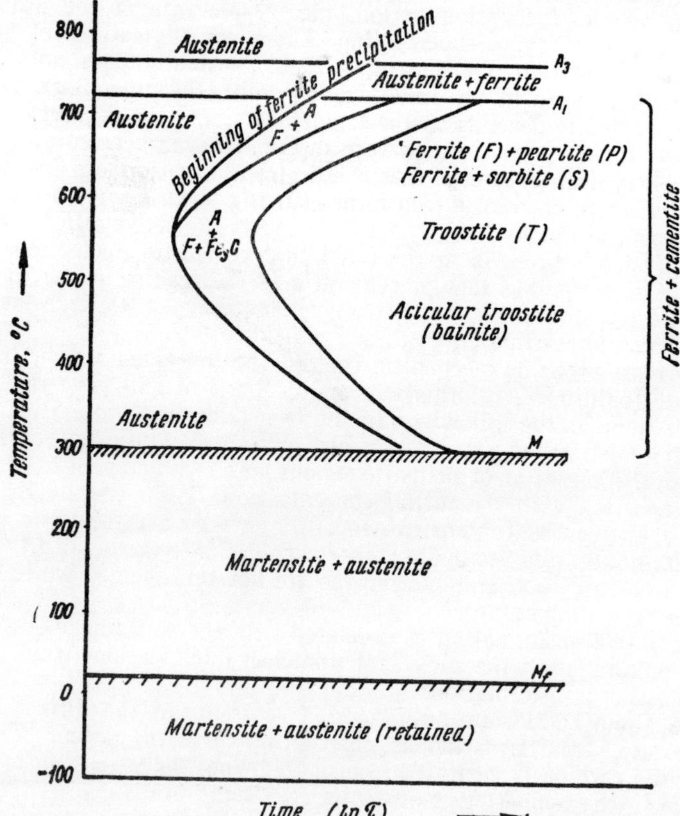

Fig. 132. TTT curve for austenite decomposition in a hypoeutectoid steel (schematic)

not 0.8 per cent C but less (for hypoeutectoid steels) or more (for hypereutectoid steels). Such a eutectoid (pearlite) having a carbon content differing from the equilibrium concentration (0.8 per cent C, point *S* on the iron-carbon diagram) has been called a *quasi-eutectoid* by A. A. Bochvar.

The lower the temperature of austenite decomposition (i.e., the higher the degree of supercooling), the harder the ferrite-cementite mixture will be.

The stability of austenite depends on the chemical composition of the steel, its grain size, the homogeneity of the austenite, and other factors. An increase in the carbon content of hypoeutectoid steels increases austenite stability. Upon a further increase in carbon content, austenite stability is reduced in the range of pearlite transformation and increased in the range of intermediate (bainite) transformation.

All lack of homogeneity in austenite, in particular, nonuniform concentration of carbon, the presence of undissolved carbides and other inclusions, reduces its stability. Austenite of larger grain size will be more stable.

Additions of alloying elements, especially nickel, manganese, chromium, and others, sharply increase austenite stability.

8-4. Transformation of Austenite Upon Continuous Cooling

Upon continuously lowering the temperature, austenite decomposition will occur at a constant temperature, near to point Ar_1, only under equilibrium conditions (see Fig. 108), i.e., at infinitely low cooling rates. As a rule, austenite is supercooled below the equilibrium point Ar_1 when it is cooled at actual rates and its decomposition proceeds, not at a constant temperature, but in a certain temperature interval. This is quite clear in Fig. 133 where cooling curves have been superimposed on the TTT curve.

The higher the rate of cooling, the higher the degree of supercooling of austenite will be in reference to the equilibrium point Ar_1, i.e., the lower will be the temperature interval in which austenite decomposes (Fig. 133).

At high cooling rates (Fig. 133, curve V_4), not all of the austenite has enough time to decompose at a high temperature with the formation of the ferrite-cementite mixture. A part of it is supercooled to point M and is transformed into martensite.

At an even higher cooling rate, austenite decomposition into the ferrite-cementite mixture by diffusion will be impossible. All of the austenite will be supercooled to point M and will be transformed into martensite (Fig. 133, curve V_5).

The above proposition is illustrated clearly in Fig. 134 where the diagrams indicate the effect of the cooling rate on the temperature of austenite decomposition in eutectoid carbon steel and also the amounts of the structural components present in the cooled steel.

The temperature at which supercooled austenite begins to decompose is designated as Ar' to distinguish it from the equilibrium point Ar_1.

As in isothermal decomposition of austenite, the lower the decomposition temperature, i.e., the higher the rate of cooling, the more dispersed (finer) the ferrite-cementite mixture will be. Thus, at a low cooling rate, pearlite will be formed; at a higher rate—sorbite; and at a still higher rate—troostite. At a certain, still higher rate, martensite appears in addition to troostite. With further increase in the cooling rate, austenite will be transformed, to a greater or lesser degree, into martensite. As can be seen in Fig. 134, austenite is never completely transformed into martensite. Therefore a certain amount of untransformed austenite, called *retained austenite*, is always found in hardened steel in addition to the martensite.

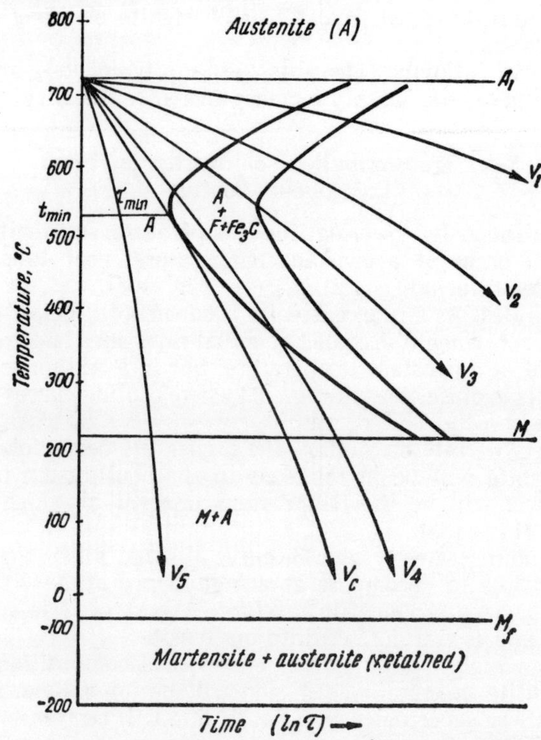

Fig. 133. Superimposition of cooling curves on a TTT curve (austenite decomposition). Cooling rates $V_1<V_2<V_3<V_4<V_5$

The more dispersed the ferrite-cementite mixture is, the higher its hardness will be. Hardness increase is especially notable when martensite appears in the steel.

The minimum cooling rate (see Fig. 133, curve V_c) at which all of the austenite is supercooled to point M and is transformed into martensite is called the *critical cooling rate.*

In Fig. 133 the critical cooling rate is tangent to the curve indicating the beginning of austenite decomposition. The magnitude

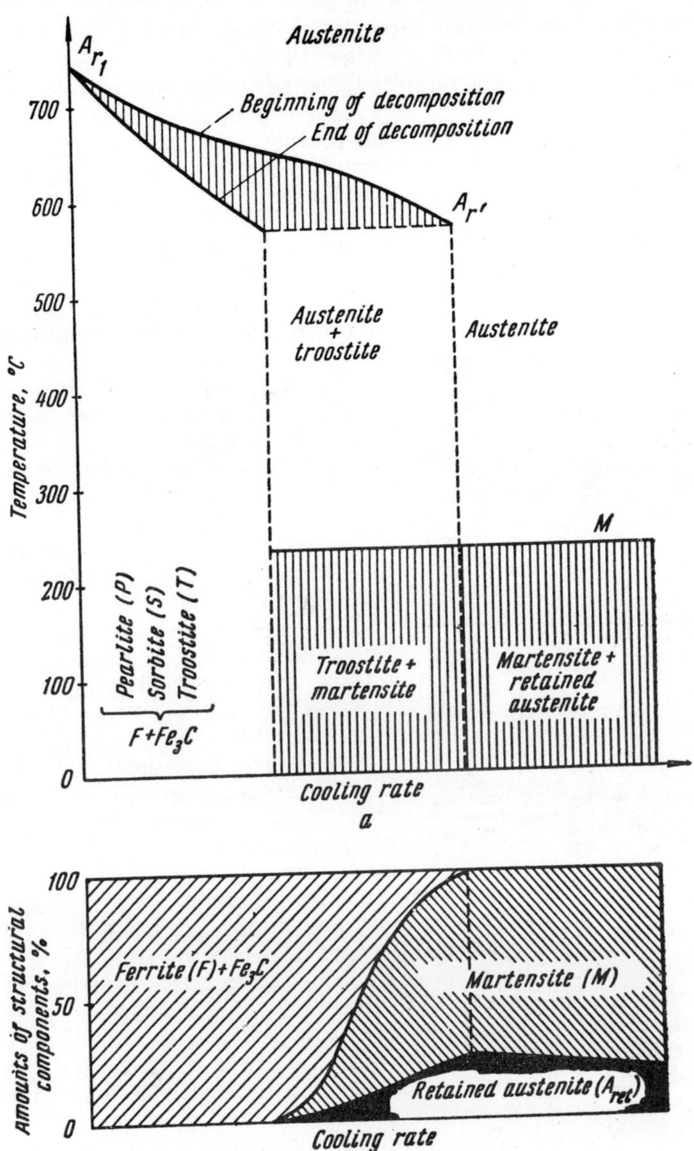

Fig. 134. Schematic diagram showing the effect of the cooling rate on the decomposition temperature of austenite in a eutectoid carbon steel (*a*) and on the amounts of structural components (*b*) (after A. A. Popov)

of the critical cooling rate depends on the stability of the austenite. The higher its stability (the farther to the right the transformation curve is on the TTT diagram), the less the critical cooling rate will be.

As a first approximation, the critical cooling rate may be determined from the formula proposed by Grange and Kieffer:

$$V_c = \frac{A_1 - t_{min}}{1.5\tau_{min}} \text{ degrees per sec}$$

in which A_1 is the temperature of the equilibrium point;

t_{min} is the temperature of minimum austenite stability;

τ_{min} is the length of the incubation period at t_{min} (see Fig. 133).

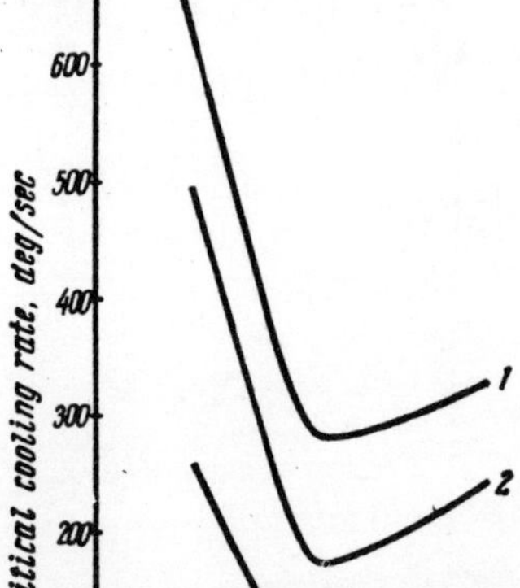

Fig. 135. Effect of carbon content and grain size on the critical cooling rate for steel (after French): 1—fine-grained steel with low manganese content, 2—medium-grained steel (0.4 to 0.5 per cent Mn), 3—coarse-grained steel (0.6 to 0.7 per cent Mn)

The critical cooling rate may vary in a wide range (Fig. 135) in accordance with the carbon content of a steel and the size of the austenite grains. Eutectoid steel has the lowest critical cooling rate. Most alloying elements increase the stability of austenite and thus lower the critical cooling rate.

The diagrams of isothermal austenite decomposition (TTT diagrams) provide only tentative data on continuous cooling transformation which are most frequently employed in heat-treating practice. For this reason, so-called thermokinetic diagrams are being more and more extensively used, in the last years, in addition to TTT diagrams, to establish proper heat-treating techniques.

Thermokinetic diagrams (Fig. 136) are plotted in temperature-time coordinates (as are TTT diagrams). They comprise a series of cooling curves on which the regions of pearlitic, intermediate (bainite), and martensitic transformation have been plotted.

Diagrams of the isothermal and thermokinetic decomposition of supercooled austenite for a hypoeutectoid steel (0.45 per cent C) are shown in Fig. 136. These diagrams show that at low cooling rates, austenite in carbon steels decomposes only by diffusion forming the ferrite-cementite mixture (pearlite) or various degrees of dispersion (pearlite, sorbite, and troostite). At high cooling rates, diffusion is suppressed and austenite is subject to the martensitic transformation.

There is no zone of intermediate transformation on thermokinetic diagrams for carbon steels.

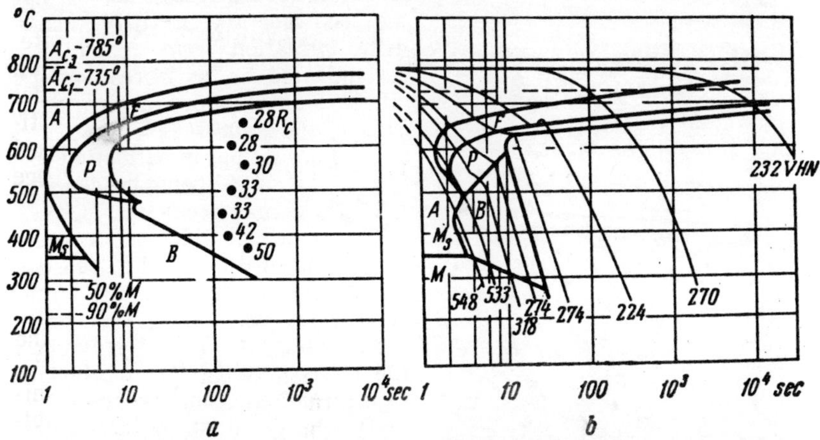

Fig. 136. Diagrams of the isothermal (*a*) and thermokinetic (*b*) decomposition of austenite for a steel containing 0.45 per cent C

For a complete quantitative description of the austenite transformation, two diagrams are required: 1) isothermal transformation, which enables one to control heat-treating processes associated with austenite transformation at a constant temperature (for example, isothermal quenching) and 2) thermokinetic, which indicates processes occurring in continuous cooling.

8-5. Martensitic Transformation in Steel

Martensite is *the supersaturated solution of carbon in α-iron.*

At room temperature, in an equilibrium state, the solubility of carbon in α-iron does not exceed 0.025 per cent. The carbon content of martensite may be the same as of the original austenite.

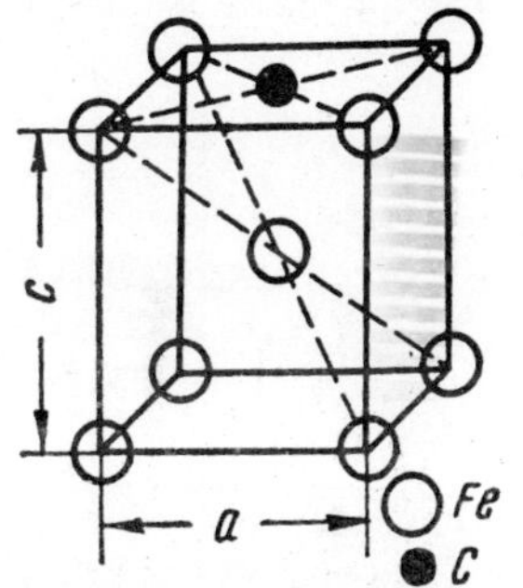

Fig. 137. Crystal lattice of martensite (schematic)

The crystal lattice of martensite is tetragonal (Fig. 137). This was first established by the research of N. Y. Selyakov, N. T. Gudtsov and G. V. Kurdyumov. The more the carbon content of the martensite, the larger the ratio of the lattice constants c/a (Fig. 139). Martensite has a characteristic acicular or needle-like structure (Fig. 138) formed as a result of intersection with the surface of the microsection of martensite crystals (martensite plates) which are flat discs, thicker in the centre than at the edges. The chief features of martensite are its high hardness and low ductility. Martensite formation from austenite is accompanied by an increase in specific volume (about 3 per cent). This is the main reason why large stresses are set up in hardening that distort the article being hardened so that cracks may occur.

Fig. 138. Martensite and retained austenite, × 800

The principal difference between the martensitic and pearlitic transformations is that the former is diffusionless.

According to G. V. Kurdyumov, the mechanism of the martensitic transformation consists in an orderly change from a face-centred cubic lattice to a body-centred one. Neighbouring atoms are displaced, relative to each other, without interchanging their positions, over distances not exceeding interatomic distances.

Kurdyumov is of the opinion that martensitic transformation may be considered as one similar to allotropic transformation, i.e., a phase

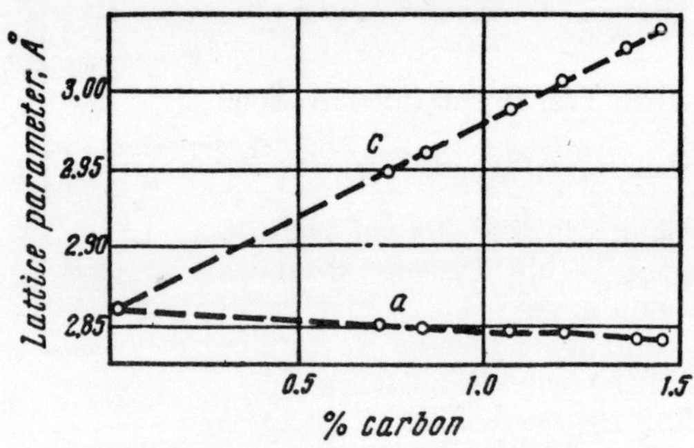

Fig. 139. Variation of lattice parameters of martensite with carbon content (after G. V. Kurdyumov and E. Z. Kaminsky)

transformation in a single-component system. Therefore, the general laws of phase transformation are valid for this case and martensitic transformation develops by nucleation and subsequent crystal growth as any other phase transformation.

The main features of the martensitic transformation are: 1) high rates of nucleation and crystal growth at low temperatures, 2) limited crystal growth; crystals grow rapidly to a certain limited size after which growth ceases, and 3) rapid attenuation of the transformation upon stopping the cooling process.

The extremely high rate of martensite crystal growth * at comparatively low temperatures may be explained by the fact that it proceeds with coherency between the austenite and martensite lattices and by the small movements of the atoms and low interfacial energy barrier (see Fig. 69).

Nucleation and growth of the martensite crystal results in the appearance of energy of elastic distortion of martensite formations. Therefore, coherent growth proceeds at a high rate until the increasing elastic energy causes plastic deformation which violates the coherency of the austenite and martensite lattices.

Coherent growth is violated not only by slip on the boundaries of martensite formations but also on the boundaries of the original austenite grains since these boundaries are distinguished for their disorderly atomic arrangement and adjacent grains have different orientation.

Noncoherent growth of the martensite crystal, which requires atomic displacement over considerable distances, is impossible due to the low temperatures.

Let us now discuss the conditions required for the formation of martensite and factors which affect the martensitic transformation. Martensitic transformation may occur only at cooling rates not less than the critical value which provides for supercooling austenite to the martensite point M where no diffusion processes occur.

Martensitic transformation occurs in a wide temperature range. It begins at a temperature corresponding to point M. When the cooling process passes through point M, austenite begins to transform into martensite. The lower the temperature, the more martensite will be formed.

The amount of martensite formed, according to the temperature, may be represented by the so-called martensite curve (Fig. 140). At a definite temperature for each steel, further transformation of austenite into martensite ceases. This temperature is usually denoted as M_f (Fig. 140).

The positions of points M and M_f do not depend upon the cooling rate (see Fig. 134) They are determined by the chemical composition of the austenite. More carbon in a steel will lower points M and M_f (Fig. 141).

---

* One martensite crystal is formed in from 0.5×10^{-7} to 5×10^{-7} second. The mean rate of growth of a martensite crystal is 10^6 mm per sec.

The transformation of austenite into martensite begins with the almost instantaneous formation of separate needles (plates) of martensite. The transformation develops not by the growth of the already formed crystals but by the appearance of newer and newer needles. The size of the martensite crystals is determined by the size of the original austenite grains. The larger the original grain, the larger the crystals will be.

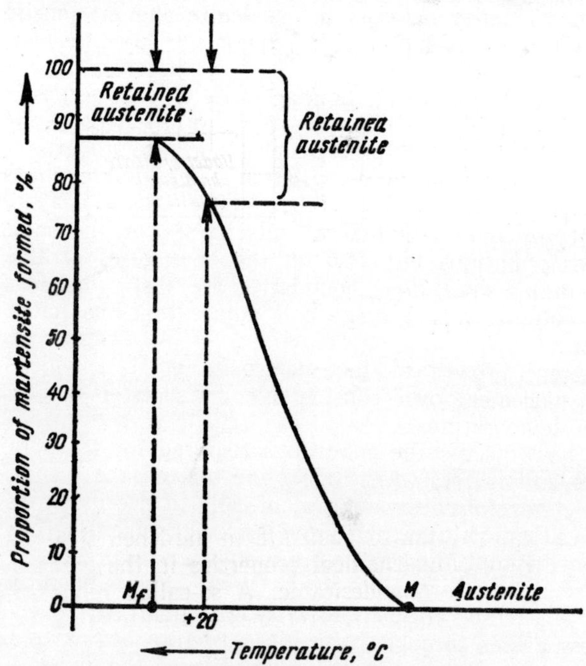

Fig. 140. Martensite curve (schematic)

The first martensite crystal formed has a length equal to the crosswise dimension of the austenite grain. Crystals, formed subsequently, are constrained in their development and are considerably smaller in size. The first crystals are usually oriented at angles of 60° or 120° to each other. Subsequent crystals are arranged either parallel to the first or in zigzags.

A characteristic feature of martensitic transformation is that it is practically never completed (see Fig. 140). Therefore, a hardened steel contains retained austenite, as a rule, in addition to martensite. The higher the carbon content, the more austenite will be retained (Fig. 142). The cooling rate below the martensite point also affects the amount of retained austenite. The lower this cooling rate, the more austenite will be retained in the hardened steel. If the steel is held isothermally after cooling below point M, the remaining untransformed austenite will be stabilised. If, after this, the temperature is lowered again, the transformation of austenite into martensite will be resumed, but only after a certain delay, and more austenite will be retained.

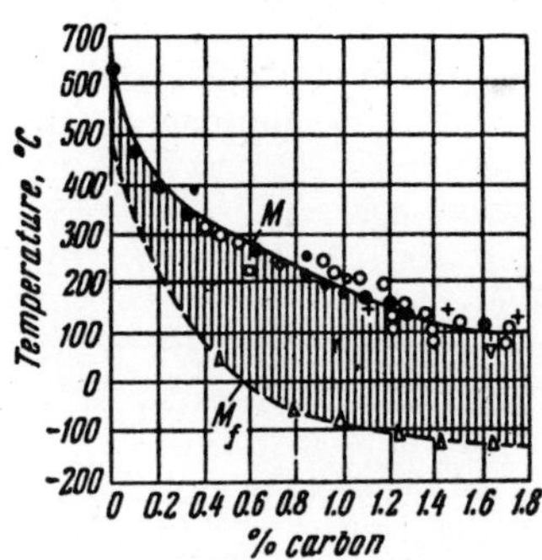

Fig. 141. Effect of carbon content on the martensite points M and M_f

Fig. 142. Effect of carbon content on the proportion of retained austenite

The presence of retained austenite in hardened steel has a detrimental effect on its mechanical properties in the great majority of cases and is therefore undesirable. A so-called *refrigeration treatment** is applied to reduce the retained austenite. It is based on the fact that a considerable proportion of retained austenite may be transformed into martensite if a hardened high-carbon steel is cooled to sub-zero temperatures.

To obtain more complete transformation of austenite into martensite, cooling must be continued to the M_f temperature (Fig. 141). Refrigeration treatment was first proposed by A. P. Gulyaev in 1937.

8-6. Tempering of Steel

Tempering is the process of heating steel (usually hardened steel) to a temperature below point Ac_1, and holding it for the proper time.

As mentioned above, hardened steel consists of two structural components—tetragonal martensite and retained austenite. Both of these components are unstable and tend to pass into a more stable state upon being heated.

* Also called sub-zero treatment and cold treatment.

This transformation of hardened steel to a more stable state takes place by diffusion and, therefore, its rate is determined mainly by the heating temperature.

To clarify the processes, upon which the tempering operation is based, it will be most expedient to consider phenomena occurring in martensite upon continuous heating.

No detectable changes take place when hardened steel is heated in the range from 80° to 100°C. In the range from 100° to 350°C, martensite decomposes with the precipitation of carbon as dispersed iron carbide particles.*

This reduces the number of carbon atoms in the martensite lattice which, consequently, becomes less tetragonal and approaches the cubic form characteristic of α-iron.

Martensite decomposes in tempering in two stages. The first, more rapid stage, proceeds at temperatures from 100° to 180°C. In this period, carbides are formed only of carbon atoms from zones surrounding the carbide nuclei while the more remote regions retain their original concentration. Carbide formation in the first stage does not require carbon diffusion over considerable distances. Thus, as a result of low temperature tempering, at least two alpha solid solutions (martensite) are present in addition to the carbides: one has a low and the other a high carbon concentration.

Further development of the tempering process proceeds, not by the growth of the solid solution regions, poor in carbon, but as a result of the formation of new regions of this concentration by further precipitation of carbides from the high-concentration regions. During tempering, the amount of high-concentration α solution is continuously reduced (due to carbide precipitation) while the amount of low-concentration solution increases.

The carbide particles precipitating during martensite decomposition have the form of thin plates, several-atom layers thick and several hundreds of angstroms in length. The carbide plates are coherently bonded to the α solution lattice. This decomposition mechanism is called the *two-phase* type.

---

* As to the question of the type of carbides formed in tempering steel, textbooks frequently give the diagram proposed by M. P. Arbuzov (*Journal of Engineering Physics*, Vol. XIX, Issue 10, 1949) who is of the opinion that cementite is formed at all stages in the decomposition of martensite. Foreign investigators (see, for example, M. Manes, A. D. Demick, M. M. Menster, E. M. Cohn, and L. I. Hafer, *Journ. Amer. Chem. Soc.*, Vol. 74, No. 24, 1952) and, more recently, Soviet investigators (see, for example, B. A. Apayev, *Physical Metallurgy and Metal Treatment*, 1957, No. 1) have shown that at low tempering temperatures, an intermediate hexagonal carbide ε Fe_xC (probably Fe_2C) is formed, and, at temperatures above 200° C, χ Fe_xC, which is transformed into cementite (Fe_3C) at temperatures from 350° to 400° C. The carbide phase χ Fe_xC occurs, apparently, only in steels containing more than 0.6 to 0.8 per cent C. Two carbide phases, ε Fe_xC and Fe_3C, occur in low carbon steels.

The second stage of decomposition occurs in the range from 180° to 350°C and comprises further slow precipitation of carbon, as carbides, from the solid solution (martensite).

The low rate of this transformation is due to the fact that carbon is precipitated from the solid solution in this period by enlargement of the carbide particles. This requires diffusion of the carbon within the solid solution. Such diffusion proceeds slowly at low temperatures. The coherency of the carbide and α solution lattices is partially retained in this period.

The result of martensite decomposition at temperatures below 350°C is the formation of *tempered martensite* which is a *supersaturated solid solution of carbon in α-iron* (of nonuniform concentration) with inclusions of highly dispersed iron carbide crystals, bonded coherently to the alpha solid solution lattice.

Tempered martensite retains its needle-like structure but is etched more intensively than untempered martensite (Fig. 143).

The carbon content of martensite (of the α solution) in a tempered carbon steel, depends on the temperature and time employed in tempering, as well as on the composition of the initial martensite.The diagram in Fig. 144 shows the effect of tempering temperature on the average carbon content of martensite (α solution).

It is evident from Fig. 144 that the carbon content in martensite is continually reduced with an increase in temperature. In the range from 350° to 400°C, decomposition of the solid solution is practically completed. With an increase in the holding time at a given temperature, carbon is intensively precipitated from the martensite, then the decomposition slows down and practically ceases at long holding times. Each tempering temperature corresponds to a definite carbon content of the martensite.

At temperatures from 250° to 350°C, in addition to martensite decomposition, the retained austenite also decomposes. The latter is transformed into tempered martensite, similar to that obtained in martensite decomposition at the same temperature and holding time. This process proceeds purely by diffusion.

After the decomposition of the martensite and retained austenite, steel, tempered at 350°C, will consist of elastically deformed crystals of alpha solid solution (ferrite) in which finely dispersed particles of cementite are distributed. This structure is called *secondary troostite.*

Elastic stresses are relieved in the range from 350° to 400°C in addition to the completion of carbon precipitation. These stresses appear as a result of volume changes at low temperatures, associated with decomposition of martensite and austenite in tempering. Separation of the carbide and ferrite lattices occurs in this same period. At higher temperatures, the coagulation and spheroidisation of the carbides proceed very intensively and they become globular in form.

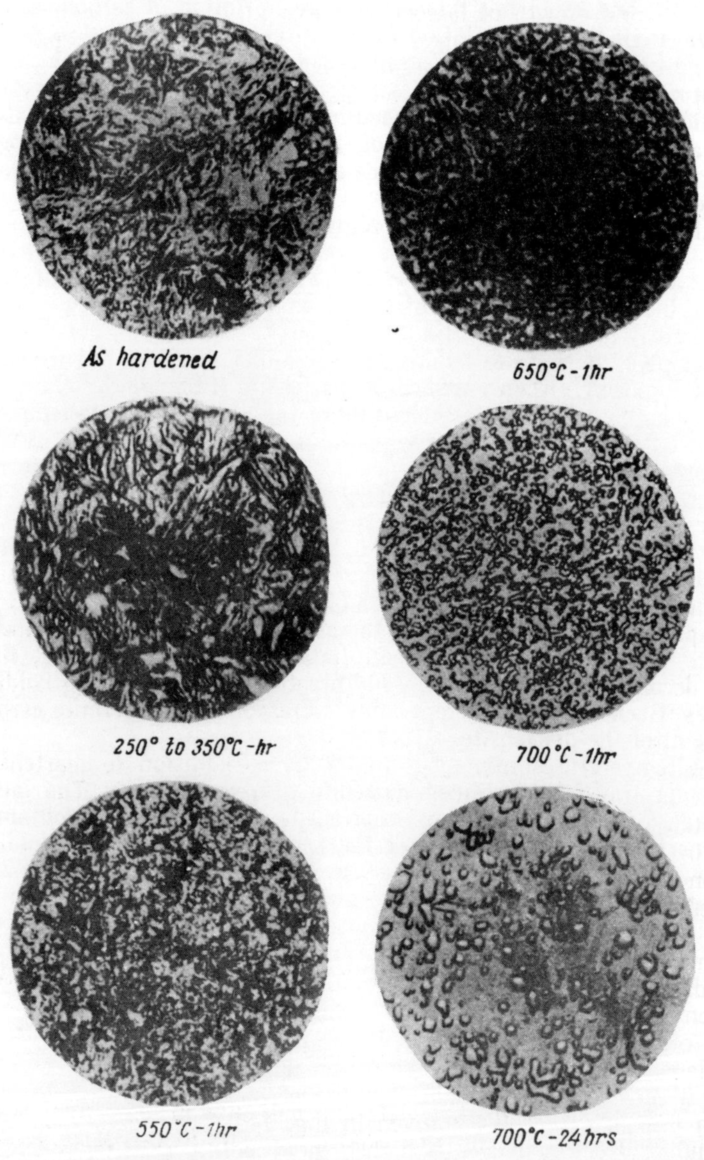

Fig. 143. Changes in steel structure after tempering. Steel containing 1 per cent C, × 500 (after B. S. Natapov)

The process of coagulation is due to the dissolution of smaller carbide particles and growth of larger particles.

When the small, thermodynamically unstable carbide particles are dissolved, a difference in carbon content is produced in the solid solution. The concentration is higher in the vicinity of dissolved particles and lower near the large undissolved particles. This leads to carbon diffusion from the higher- to lower-concentration regions. The solid solution becomes saturated in the neighbourhood of the large carbide particles and new carbides are precipitated and grow on the large particles which serve as a ready-made substrate.

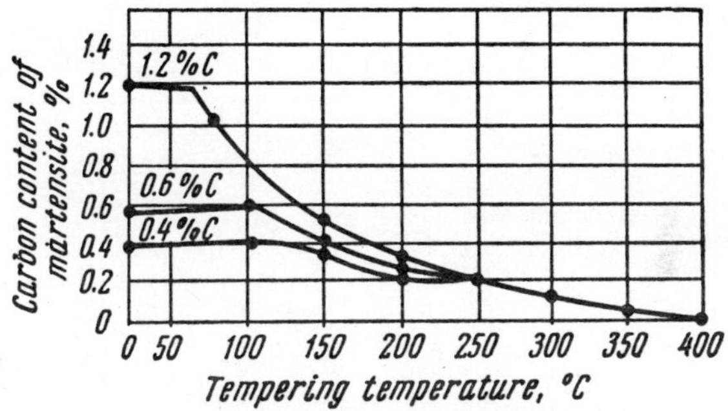

Fig. 144. Variation of the carbon content in martensite with the tempering temperature and carbon content of the steel (after G. V. Kurdyumov)

As a result of the coagulation of the carbides, secondary troostite is transformed into secondary sorbite at temperatures from 500° to 600°C and into pearlite at higher temperatures. The ferrite-cementite mixture, formed as the result of martensite decomposition, often retains its orientation when heated to 500-600°C. In this case, the secondary troostite and sorbite will have the acicular orientation inherent to martensite. A long tempering operation, between 650° and 700°C, will produce granular pearlite (Fig. 143).

In conclusion, it must be noted that secondary troostite and sorbite differ from the troostite and sorbite obtained in austenite decomposition in that the former have a granular and not lamellar structure.

A comparison of the mechanical properties of a steel having a lamellar structure (result of supercooled austenite decomposition) and one with a granular structure (result of martensite decomposition) with the same hardness is given in Fig. 145. These data show that granular structures have a higher yield point, relative reduction of area, and relative elongation than lamellar structures. The tensile strength of granular structures is somewhat lower.

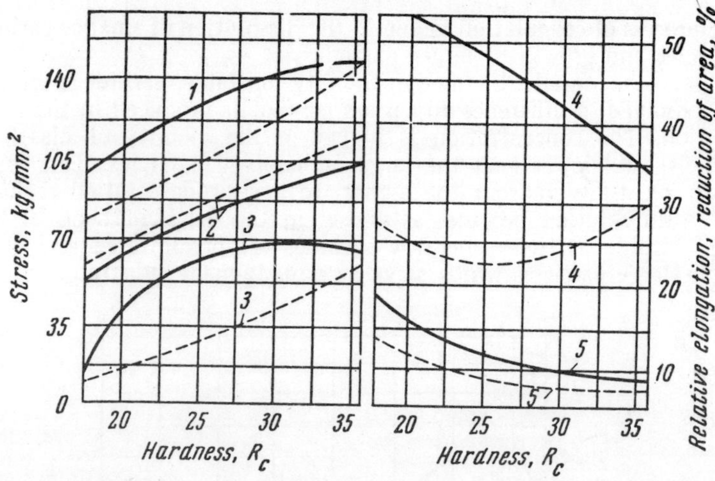

Fig. 145. Comparison of the mechanical properties of lamellar and granular structures (after E. Bain): —granular structure, ······· lamellar structure: 1—true tensile stress, 2—tensile strength, 3—yield point, 4—reduction of area, 5—elongation

8-7. Ageing

All the phase transformations discussed previously were associated with the allotropic transformation α-Fe$\rightleftarrows\gamma$-Fe.

In the iron-carbon system, however, as well as in other iron-base alloys (Fe-Mo, Fe-W, etc.), processes may take place that do not involve a change in the type of crystal lattice. Here, we have in mind transformations caused by partial decomposition of a supersaturated solid solution (see page 114). This type of transformation is usually observed in systems distinguished for their limited solubility in the solid state (see Fig. 86,*b*). If alloy *2* (Fig. 86,*b*) is heated to a region of the homogeneous alpha solid solution, rapidly cooled to obtain a supersaturated solid solution, and then reheated to comparatively low temperatures (ageing or tempering), the solid solution will be decomposed and the excess dispersed phase will precipitate from it.

The following stages may be noted in the ageing process. First, atoms of the excess component cluster in definite regions of the crystal lattice of the supersaturated solid solution. Then a new crystal lattice is formed, inherent to the precipitating phase. The crystallographic lattice of the new phase is coherently bonded to the lattice of the initial solid solution. Upon an increase in temperature (or holding time), the lattices are separated, i.e., coherent bonding is violated and independent, highly dispersed particles of the new phase are formed. Further on, the particles of the excess phase are enlarged (coagulated).

Both the coherent bond and the formation of finely dispersed particles of the new phase, increase the resistance of the alloy to plastic deformation, i. e., they increase the strength and hardness. This is often called *dispersion* or *precipitation hardening**. Coagulation of the excess phase is accompanied by a reduction in mechanical properties and is called overageing.

The following brief discussion will concern only the ageing of low-carbon steels.

There are two types of ageing: 1) ageing of a low-carbon steel cooled rapidly from a temperature above 650-700°C and 2) ageing after cold working (deformation ageing). Both types increase the hardness of steel and considerably reduce the ductility and toughness.

The probable reason for the first type of ageing is the varying solubility of carbon (see Fig. 111), nitrogen (see Fig. 188), and oxygen in α-iron. Heating to a temperature of 650-700°C results in dissolution of these elements in the ferrite and upon subsequent rapid cooling, a supersaturated alpha solid solution is obtained. Upon ageing at normal room temperature or in a range from 50° to 150°C, the solid solution decomposes. This increases the hardness and reduces the ductility and toughness of the steel.

Sheet steel is usually subject to deformation ageing after a small reduction (3 to 5 per cent), in the last passes in rolling (finishing passes), and in subsequent storage. In all probability, the main reason for deformation ageing is the decomposition of the solid solution with the precipitation of phases containing carbon, nitrogen, or oxygen. As a result of deformation ageing, the properties of low-carbon sheet steel and, in particular, its formability, are substantially lowered. Deformation ageing proceeds from 15 to 60 days at room temperature and several minutes at temperatures from 200° to 350°C. The tendency of steel to age is reduced if it is deoxidised (killed) with aluminium, titanium, or vanadium because these elements combine the nitrogen (or carbon) into nitrides (or carbides) which are insoluble in α-iron.

---

* Hardening with subsequent ageing or age hardening is widely used for aluminium alloys (Sec. 17-4), heat-resistant alloys (Sec. 13-11), and others.

Chapter 9

HEAT TREATMENT OF STEEL

Heat treatment is an important operation in the manufacturing process of many machine parts and tools. Only by heat treatment is it possible to impart the high mechanical properties to steel required for the normal operation of modern machinery and tools.

At the suggestion of Academician A. A. Bochvar, heat treatment is usually classified into four typical groups.*

The first group comprises first-order or recrystallisation annealing** which is employed to relieve internal stresses, reduce the hardness, and to increase the ductility of strain-hardened metal. It is also known as stress-relieving. At first, upon an increase in the heating temperature, the elastic distortions of the crystal lattice are eliminated (recovery). At higher temperatures, new grains form and begin to grow (recrystallisation). Due to recovery and recrystallisation, the metal is softened and regains its high ductility.

It is important to point out that the heating temperature for first-order annealing is not associated with phase transformation temperatures and only depends upon the recovery or recrystallisation temperatures peculiar to each metal.

The second group, second-order or full annealing, which involves phase recrystallisation, consists in heating alloys above the temperatures required for phase transformation. This is followed by slow cooling. Full annealing substantially changes the physical and mechanical properties and may refine a coarse-grained structure.

In the third group, hardening, alloys are heated above the phase transformation temperature and are then rapidly cooled (quenched) to retain at room temperature, either a phase stable at high temperatures or various stages of decomposition or transformation of this phase. Thus, hardening causes the formation of unstable (nonequilibrium) structures.

---

* The fifth heat-treatment group, termed by A. A. Bochvar as chemical heat treatment, is more properly to be considered separately from a methodological point of view.

** This type of annealing has been described in detail in Sec. 3-2.

The fourth group, tempering, involves the reheating of a hardened alloy (steel) to a temperature below that required for phase transformation so as to bring it nearer to an equilibrium state.

9-1. Annealing of Steel

Annealing is one of the most important widely-used operations in the heat treatment of steel.

The purpose of annealing is to obtain softness, improve machinability, increase or restore ductility and toughness, relieve internal stresses, reduce or eliminate structural inhomogeneity, refine grain size, and to prepare steel for subsequent heat treatment.

There are several types of annealing.

Full annealing consists in heating a hypoeutectoid steel 30-50°C above the critical point Ac_3, holding it at this temperature, and then slowly cooling (at a rate of 30-200°C per hour, depending on the composition).

The rate of heating for annealing rolled stock or forgings in heavy charges is as high as the given furnace can provide. The holding time at the annealing temperature for these conditions is taken as one-half to one hour per ton of the charge. Measures should be taken to prevent heavy scaling and decarburisation of the steel in annealing (see page 200).

Slow cooling is required in annealing to enable the austenite to decompose at low degrees of supercooling (see Fig. 133) so as to form a pearlite + ferrite structure in hypoeutectoid steel, a pearlite structure, in eutectoid steel, and a pearlite + cementite structure in hypereutectoid steel.

The higher the austenite stability, the slower the cooling must be to ensure austenite decomposition in the high temperature range. Thus, alloy steels, in which austenite is very stable, should be cooled much slower (30° to 100°C per hr) after annealing than carbon steel (150° to 200°C per hr).

The form in which the excess ferrite or cementite precipitates in hypo- and hypereutectoid steels depends on the cooling rate. At high rates, the excess components precipitate in the form of a network on the grain boundaries of the prior austenite (Fig. 146,*a*). Retarded cooling facilitates ferrite precipitation as separate clusters (Fig. 146,*b*).

The formation of separate ferrite regions in the structure of a steel is undesirable since, in subsequent heating for hardening, it is difficult to equalise the carbon content throughout the austenite and soft spots may remain after hardening.

Due to phase recrystallisation, grains of a steel are also refined after full annealing (Fig. 147).

Complete phase recrystallisation of hypereutectoid steel (full annealing) is required only in cases when hot-working (rolling or forging) was finished at high temperatures and the steel is, consequently, coarse-grained.

If hot-working is finished at a normal temperature (slightly above Ac_3), incomplete annealing will be sufficient to ensure the required properties.

Hypoeutectoid hot-worked steel (rolled stock, sheet, forgings, etc.), as well as castings of carbon and alloy steels, may undergo full annealing.

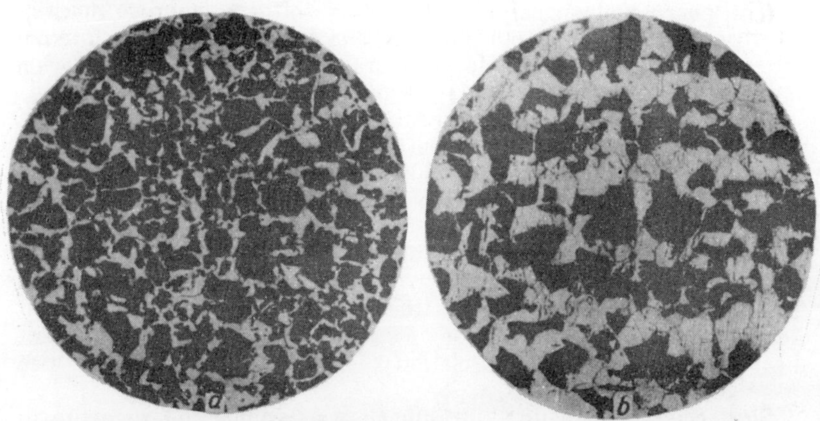

Fig. 146. The structure of hypoeutectoid steel (0.4 per cent C) after annealing, ×200:
a—rapid cooling, *b*—slow cooling

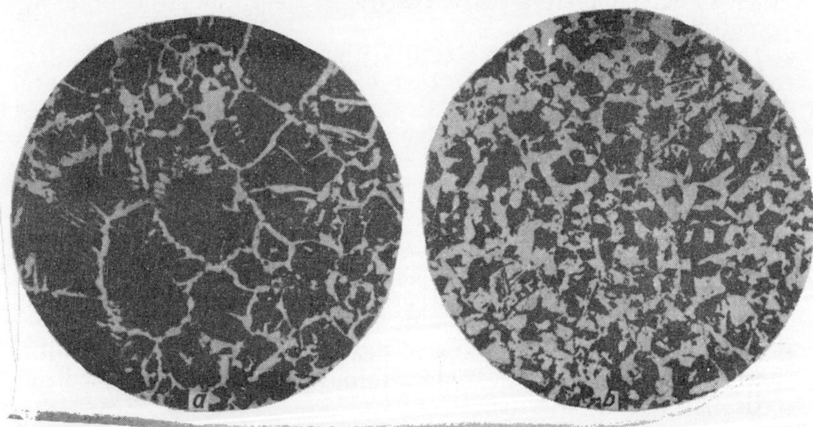

Fig. 147. Effect of full annealing on the structure of hypoeutectoid steel (0.4 per cent C), × 200:
a—after rolling, *b*—after annealing

Fig. 148 shows the effect of full annealing on the mechanical properties of rolled steels of various carbon content. Annealing reduces the tensile strength, yield point, and hardness of rolled (or forged) steel while elongation and reduction of area are increased.

When cast steel is annealed, its strength is also increased because a fine-grained structure is obtained and the Widmanstätten structure is eliminated.

Full annealing considerably improves the machinability and formability of steel.

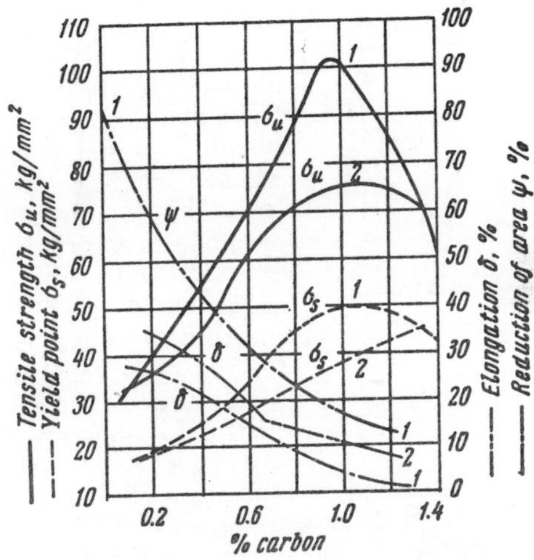

Fig. 148. Variation in the mechanical properties of rolled steel with the carbon content: (*1*) directly after rolling and (*2*) after full annealing followed by slow cooling

Incomplete annealing consists in heating steel to a temperature somewhat above the critical point Ac_1, holding it at this temperature, and then slowly cooling it. Incomplete annealing is used to relieve internal stresses and to improve the machinability of steel. It is associated with only partial recrystallisation; excess ferrite of hypoeutectoid steel or excess cementite of hypereutectoid steel does not pass over into the solid solution and is not recrystallised. Incomplete annealing is applied chiefly to eutectoid and hypereutectoid steels in which heating above point Ac_1 causes practically complete recrystallisation.

Sometimes, under conditions of excessively high heating temperatures and slow cooling, a cementite network appears in hypereutectoid steels. In this case, incomplete annealing will not ensure a high toughness and proper grain refinement. Prior to incomplete annealing, it will be necessary to eliminate the cementite network by heating above point *Acm* followed by rapid cooling (hardening or normalisation).

Isothermal annealing (Fig. 149). In this operation, steel is heated as for ordinary annealing and then cooled comparatively rapidly (in air or by a blast in a furnace) to a temperature below point Ar_1 (usually 50° to 100°C below in accordance with the character of the TTT curve for austenite decomposition). The steel is held isothermally at this temperature during a certain period of time to provide for complete austenite decomposition. This is followed by comparatively rapid cooling (Fig. 149).

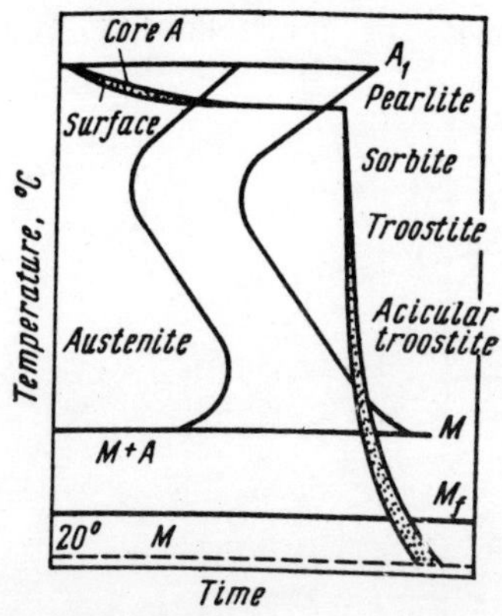

Fig. 149. Isothermal annealing curve

The main advantage of isothermal annealing is that it reduces the time required for heat treatment of the steel. This is especially true for alloyed steels which must be cooled very slowly to obtain the required reduction in hardness. Isothermal annealing produces good results in treating relatively small charges of rolled stock or small forgings.

Large forgings or charges of rolled stock have different cooling conditions on the surface and in the core so that the annealed structure will not be uniform when passing over to isothermal holding. Therefore, this type of annealing is not applied in such cases.

Annealing to obtain granular cementite (*spheroidising*) is extensively practised for high-carbon (tool) steels to transform lamellar pearlite into the granular type (Fig. 112). This process is performed by heating the steel slightly above the critical point Ac_1 (730-770°C) with subsequent holding at this temperature followed by slow cooling, at a rate of 25° to 30° per hour, to 600°C.

The annealing temperature should be only slightly above Ac_1. Heating to higher temperatures will make granular cementite difficult to obtain and will facilitate the formation of lamellar pearlite. Slow cooling is required to obtain austenite decomposition at low degrees of supercooling and so that the carbide particles coagulate to a sufficient extent. This will ensure low hardness after annealing the steel. Annealing procedures for steels containing from 0.7 to 0.9 per cent C (Grades У7 and У8) and from 1.0 to 1.3 per cent C (Grades У10, У12, and У13) are given in Fig. 150. The range of heating temperatures, at which granular cementite is formed, is very narrow in steels with near-eutectoid compositions. The optimum annealing temperature is 750°C for eutectoid steels and 770°C for hypereutectoid steels. Frequently, to heat the charge more uniformly throughout its height, the temperature is raised to 800°C at the beginning and end of the heating operation (Fig. 150).

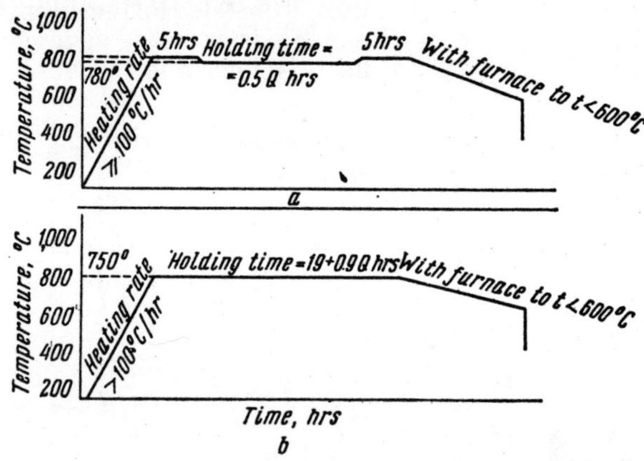

Fig. 150. Annealing steel to obtain granular cementite (spheroidising): *a*—steels У10 to У13, *b*—steels У7 and У8

The heating time is determined from the weight (*Q*) of the charge.

Spheroidised steel has a lower hardness and tensile strength and a correspondingly higher relative elongation and reduction of area than steel subject to normal annealing (Table 3).

Table 3

Mechanical Properties of Eutectoid Steel (various pearlite structures)

Structure \ Mechanical property	*Bhn*	σ_u, kg/mm²	δ, per cent	ψ, per cent
Sorbite-type pearlite	360	133	11	20
Lamellar pearlite	228	82	15	30
Granular pearlite	163	63	20	40

Steel having a granular cementite structure is easily machined.

Diffusion annealing (*homogenising*) is applied to alloy steel ingots and heavy complex castings for eliminating the chemical inhomogeneity within the separate crystals by diffusion.

Homogenising is carried out at temperatures from 1,100° to 1,200°C (the optimum temperature is 1,150°C) at which diffusion proceeds quite easily and, to some extent, equalises the composition of steels having developed dendritic segregations. The steel is heated to the maximum temperature of 1,150°C at the highest rate that the given furnace can provide for. Scaling is very intensive at high temperatures and this leads to excessive losses of metal. Holding time, therefore, should be at a minimum. It may be determined, in accordance with the weight of the charge, from the data given in Fig. 151.

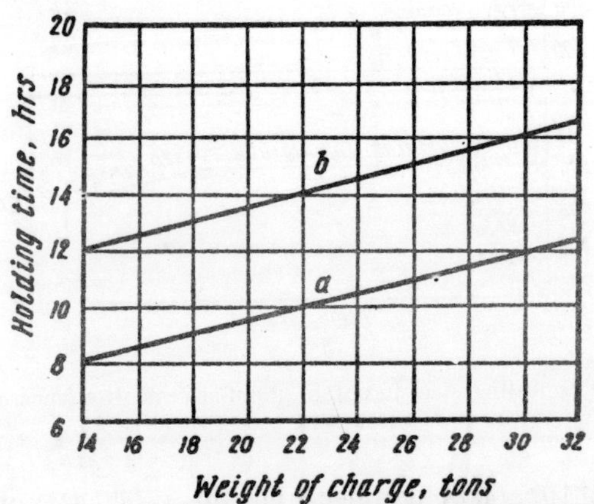

Fig. 151. Variation in holding time for homogenisation (1,150° C) with the weight of the charge (after V. M. Doronin):
a—for steels Grade 5XHM, 5XГM, and 30XГCA, *b*—for steels Grade 18XHBA, 25XHBA, 40XHMA (see Sec. 13-7)

Holding is followed by cooling with the furnace for 6 to 8 hours to 800-850°C and then further cooling in air. After homogenising, castings undergo full annealing to refine their structure.

9-2. Normalising of Steel

This process consists in heating steel to a temperature from 40° to 50°C above point Ac_3 (Acm), holding at this temperature for a short time, and subsequent cooling in air. Normalising is used to eliminate coarse-grained structures obtained in previous working (rolling, forging, or stamping, Fig. 152), to increase the strength of medium-carbon steels to a certain extent (in comparison with annealed steel), to improve the machinability of low-carbon steels, to improve the structure in welds, to reduce internal stresses, to eliminate the cementite network in hypereutectoid steels, etc.

Normalised carbon steel consists of pearlite and ferrite in hypoeutectoid steels and of pearlite and cementite in hypereutectoid steels.

More rapid cooling (in air), used in normalising, causes the austenite to decompose at lower temperatures. This increases the dispersity of the ferrite-cementite mixture (pearlite) and increases the

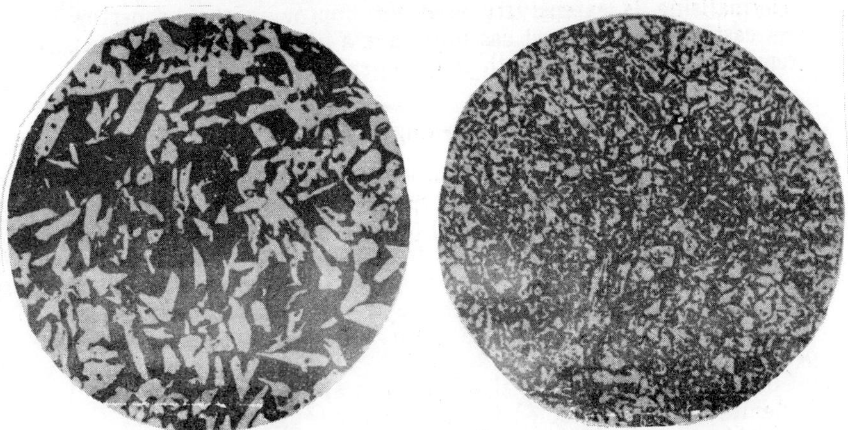

Fig. 152. Structures of hypoeutectoid steel (0.4 per cent C), ×200: *a*—after forging, *b*—after normalising

amount of eutectoid. Therefore, normalised steel has a higher strength and is harder than annealed steel (Table 4).

Table 4

Hardness of Annealed and Normalised Carbon Steels (after N. A. Minkevich)

Condition	Hardness, *Bhn*				
	Commercial iron	Structural steels			Tool steel
		low-carbon	medium-carbon	high-carbon	
Annealed	80 to 100	125	160	185	220
Normalised	90 to 100	140	190	230	270

Cooling in air employed in normalising alloy steels, in which the austenite is very stable, results in austenite decomposition at high degrees of supercooling. It is even possible to obtain a hardened steel structure, i.e., martensite, in this case. After being normalised, such steels will be very hard and must undergo high-temperature tempering at 550-650°C to enable them to be machined.

It is essential to note that two heat-treating operations, normalising and high tempering, require less time than annealing. Therefore, these two operations are often substituted for annealing in the treatment of alloy steels.

Normalising is frequently applied as a final heat treatment for items which are to operate at relatively high stresses.

Normalising is extensively used for improving the properties of steel castings. Normalised castings have a higher yield point, tensile strength, and impact strength than annealed castings.

9-3. Hardening of Steel

Hardening is a heat treating process in which steel is heated to a temperature above the critical point, held at this temperature, and then quenched (rapidly cooled) in water, oil, or molten salt baths.

Hypoeutectoid steels are heated from 30° to 50°C above point Ac_3 while hypereutectoid steels are heated above point Ac_1. In the first case, ferrite + pearlite, and in the second, pearlite + cementite, are transformed into austenite upon heating. A considerable part of the cementite is retained.

Cooling at a rate higher than the critical value should enable the austenite to be supercooled to the martensite point.

Hardened steel is in a stressed condition and is very brittle so that it cannot be employed for practical purposes. After hardening steel must be tempered to reduce the brittleness, relieve the internal stresses caused by hardening, and to obtain predetermined mechanical properties.

Hardening followed by tempering is intended for improving the mechanical properties of steel. The chief aim in hardening and tempering tool steel is to increase the hardness and wear-resistance, retaining sufficient toughness at the same time. For structural steels, the purpose is to obtain a combination of high strength, ductility, and toughness.

Selecting the hardening temperature. The hardening temperature of steel depends upon its chemical composition and, principally, upon its carbon content (Fig. 153).

Hypoeutectoid steels, containing pearlite and excess ferrite (Fig. 154,*a*), are hardened by heating to a temperature slightly (30-50°C) above point Ac_3. This will enable homogeneous austenite to be obtained which will be transformed into martensite (Fig. 154,*c*). Upon heating within the interval between the critical points (Ac_1-Ac_3), ferrite is partially retained with the austenite. The austenite is transformed into martensite after subsequent quenching but the ferrite remains unchanged. The result will be *incomplete hardening* since the segregates of free ferrite (Fig. 154,*b*) in the structure of the hardened steel reduce its hardness. The presence of ferrite, in addition to the brittle "framework" of martensite crystals, does not allow noticeable toughness to be obtained.

Ferrite zones lower the mechanical properties of steel not only after hardening, but after high tempering as well.

Overheating steel above its critical point will cause considerable austenite grain growth and coarse acicular martensite (Fig. 154,*d*) will be obtained after quenching. Such martensite has a low impact

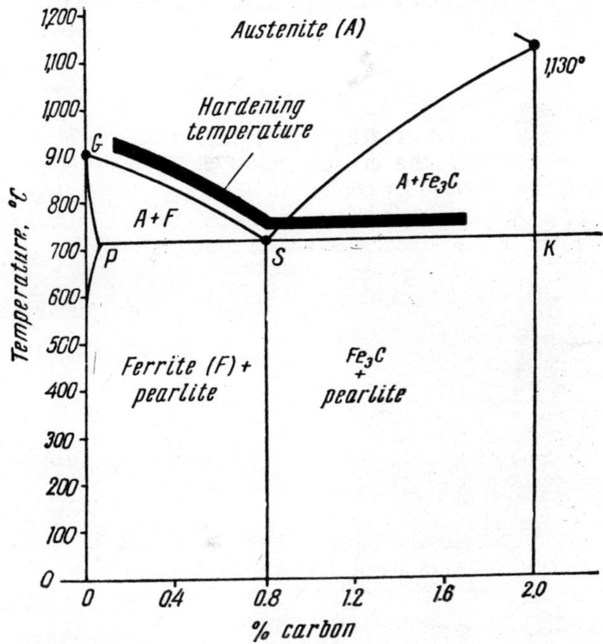

Fig. 153. Relationship between the hardening temperature of steel and its carbon content

strength. Overheating also increases the tendency of a steel to warp and crack during the quenching operation.

Hypereutectoid steels are heated in hardening to a temperature of Ac_1+ (30° to 50°C). A certain amount of cementite remains in the structure of the steel heated to this temperature (see Fig. 153), in addition to the austenite. Therefore, this cementite, which was not dissolved in heating, is retained in the structure of the hardened steel in addition to the martensite. Such a structure has a higher hardness and wear-resistance than that obtained upon quenching from a temperature above *Acm*, i.e., in the region of homogeneous austenite. This is explained by the reduction in the amount of retained austenite and because cementite is harder than martensite. Besides, heating to temperatures above *Acm* will inevitably lead to coarsening of the grain and warping of the part during quenching.

A necessary condition in hardening hypereutectoid steels is the presence of excess cementite as separate small grains. Excess cementite, having the form of a network (see Fig. 116), will increase the brittleness of hardened steel and promote the formation of hardening cracks.

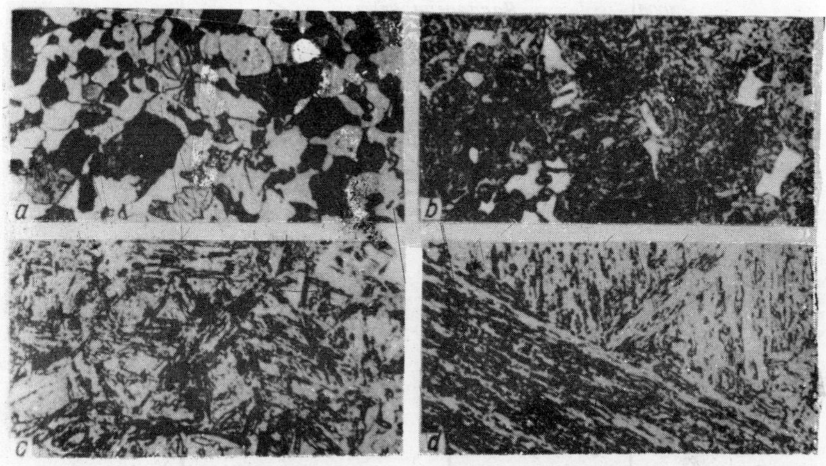

Fig. 154. Microstructure of hypoeutectoid steel (0.45 per cent C), × 400:
a—initial structure: pearlite+ferrite, *b*—after quenching from the range Ac_1-Ac_3 (770° C): martensite+ferrite (light regions), *c*—after hardening at a normal temperature (840° C): martensite; *d*—overheated (1,000° C): coarse acicular martensite

Fig. 155 shows the microstructure of hypereutectoid steel before hardening (Fig. 155,*a*), after normal hardening (Fig. 155,*b*), and after hardening with overheating (Fig. 155,*c*).

The heating rate and the heating time depend on the composition of the steel, its structure, residual stresses, the form and size of the part to be hardened, etc.

The more carbon and alloying admixtures in a steel, and the more intricate and larger the part being hardened, the slower it should be heated to avoid stresses due to temperature differences between the internal and external layers of the metal, warping, and even cracking.

The practically attainable heating rate depends upon the thermal capacity of the furnace, the bulk of the charged parts, their arrangement in the furnace, and other factors.

The diagrams in Fig. 156 illustrate possible heating methods. The first method (Fig. 156,*a*) provides a low rate of heating and may be

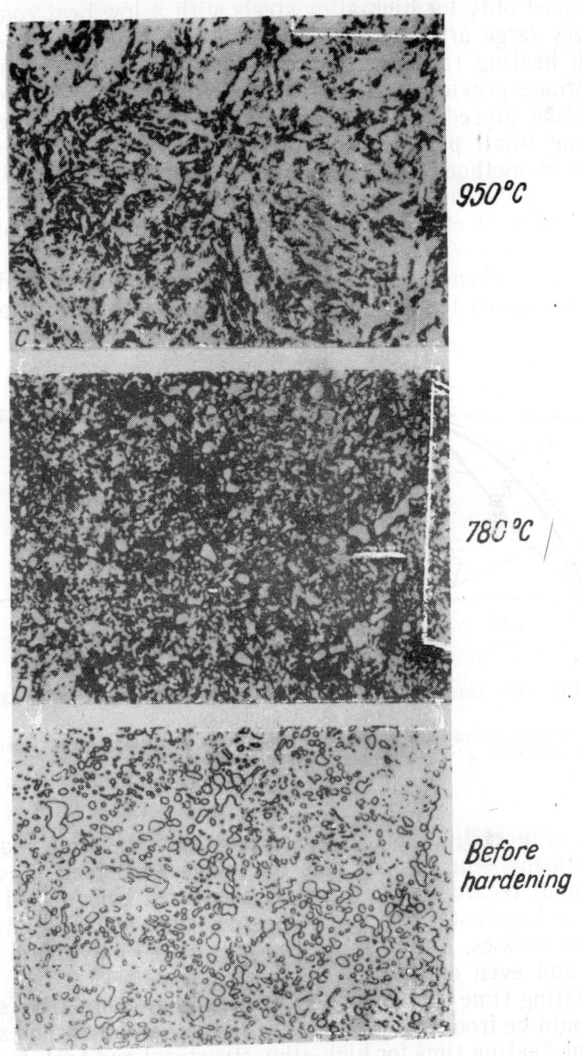

Fig. 155. Microstructure of hypereutectoid steel (1.2 per cent C), × 500:
a—initial structure: granular cementite, b—structure after properly conducted hardening: martensite and cementite, c—hardening with overheating: coarse acicular martensite

recommended only for high-alloy steels with a low heat conductivity or for very large articles.

A high heating rate may be achieved if the articles are charged into a furnace previously heated to the temperature specified by the heat treating procedure (Fig. 156,*b*). This method is applied chiefly for heating small parts in box furnaces or in continuous furnaces.

The third method (Fig. 156,*c*) is an example of forced heating. Here the articles are charged into a furnace with a temperature, at the moment of charging, even higher than the final heating temperature.

In all cases, whenever it is feasible, it is preferable to heat steel rapidly to the given temperature since this increases the output of the

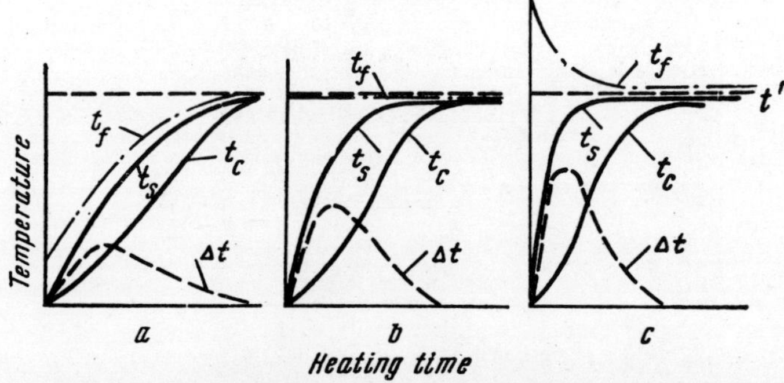

Fig. 156. Three possible procedures for heating parts in a furnace (after A. A. Shmykov):
t'—specified heating temperature, t_s—surface temperature of heated parts, t_c—temperature in the core of heated part, Δt—temperature difference between surface and core, t_f—furnace temperature

furnaces, reduces fuel consumption, and cuts the time required for heat treatment. The heating rate for low-carbon steels may be very high without regard for their initial condition. Unannealed high-carbon cast or forged steels should be heated somewhat slower to prevent additional stresses, due to rapid heating, which may cause excessive warping and even cracking.

The heating time for carbon tool steels and medium-alloy structural steels should be from 25 to 50 per cent more than for carbon structural steels. The heating time for high-alloy structural and tool steels should be from 50 to 100 per cent higher.

The heating rate is usually reduced, not by reducing the furnace temperature, but by preheating the articles.

The more uniform the heating is, the higher the permissible rate may be. For this reason, heating in salt baths (Table 5) may be conducted more rapidly than heating in box furnaces (Table 6).

Table 5

Composition of Salt Baths Used to Heat Steel Parts for Hardening

Composition	Melting point, °C	Range of practical application, °C	Features
100% NaCl	808	850 to 1,100	Decarburises steel
100% $BaCl_2$	960	1,000 to 1,400	Ditto
28% NaCl+72% $CaCl_2$	500	540 to 870	Decarburises steel above 820° C
50% Na_2CO_3+50% KCl	560	580 to 815	Decomposes above 875° C
50% NaCl+50% K_2CO_3	560	580 to 815	Ditto
33.3% $CaCl_2$+33.3% $BaCl_2$+33.4% NaCl	570	600 to 870	Decarburises steel above 820° C
50% $CaCl_2$+50% $BaCl_2$	600	650 to 900	Ditto
35% NaCl+65% Na_2CO_3	620	650 to 815	Decomposes above 815° C
22% NaCl+78% $BaCl_2$	635	700 to 870	Decarburises steel above 820° C
44% NaCl+56% KCl	665	700 to 870	Ditto

Table 6

Heating Times and Holding Times (at Working Temperatures) for Hardening Carbon and Structural Steels

Thickness or diameter of part, mm	Box furnaces (gas- or oil-fired)		Salt baths	
	heating, min	holding, min	heating, min	holding, min
25	20	5	7	3
50	40	10	17	8
75	60	15	24	12
100	80	20	33	17
125	100	25	40	20
150	120	30	50	25
200	160	40	65	35

Note: Heating time for alloy steels should be increased from 25 to 50 per cent.

Small items of any structural steel may be heated at the highest rate the furnace will provide.

When the specified heating temperature is reached, the parts to be hardened are held at this temperature until they are heated throughout, until all phase transformations are completed, and until the austenite composition becomes equalised throughout the full volume. The higher the heating temperature, the shorter the holding time may be.

Recommended heating and holding times for hardening structural steel parts of various cross-section are given in Table 6.

Table 7

Characteristics of Controlled Atmospheres (after A. A. Shmykov)

Type of atmosphere	Designation	Preparation	Composition (by volume) of dry atmosphere, per cent					
			CO	CO_2	H_2	CH_4	N_2	O_2, etc.
H_2-H_2O-N_2	ДА	Dissociation of NH_3 and drying with alumina gel or silica gel	—	—	75	—	25	—
		Ditto without drying (anhydrous NH_3)	—	—	75	—	25	—
	ПСА-08	Partial combustion of dissociated NH_3 at $\alpha=0.8$ and deep drying	—	—	7 to 20	—	93 to 80	—
CO-CO_2-N_2	ГГ-С	Dry producer gas from charcoal	30	2	6	1	61	—
	ГГ-ВО	Producer gas in an externally-heated retort at $t < 1{,}000°$ C	33	0.2 to 0.7	6	1	59.5	—
CO-CO_2-H_2-H_2O-N_2	ПС-06	Partial combustion of hydrocarbon gas at $\alpha=0.6$ and cooling to $+20°$ C	10	6	15	0.5	68.5	—
	ПС-09	Ditto, but at $\alpha=0.9$	1	10	1	—	88	—
	ПСС-06	Ditto, but at $\alpha=0.6$, and cooling to $+4°$ C	10	6	15	0.5	68.5	—

Continued

Type of atmosphere	Designation	Preparation	Composition (by volume) of dry atmosphere, per cent					
			CO	CO_2	H_2	CH_4	N_2	O_2 etc.
CO-CO_2-H_2-H_2O-N_2	ПСО-06	Ditto, but at $\alpha=0.6$, and cooling, removal of CO_2, drying to dew point $-40°$ C	10	0.1	16	1.5	72.5	—
	ПСО-09	Ditto, but at $\alpha=0.9$	2	0.1	2	—	96	—
	ГГ-С	Producer gas, removal of H_2S with bog ore, dried to $+4°$ C	25	5	12	1	57	0.3
	ГГ-О	Producer gas from anthracite, wooden blocks, etc., removal of CO_2, dried to $-40°$ C	25	0.2	12	1-2	62	0.3
	ПС-Э	Passing the combustion products of gas (without H_2S)+ +charcoal through red-hot charcoal	30	0.5	18	—	51.5	—
CO-CO_2-H_2-CH_4-N_2	КГ-ВО	Cracking hydrocarbon gases mixed with air at $\alpha=0.25$ and 0.3 and a gas temperature of 1,000° C	21	0.5	40	2	36.5	—
	КГ-H_2O	Pyrolysis and cracking of kerosene mixed with water vapour, removal of pitch and carbon black	16	0.5	60	10	10	0.5 O_2 (3 per cent C_nH_m)

PROTECTION AGAINST OXIDATION AND DECARBURISATION IN HEATING STEEL

Reaction of furnace gases (combustion products and air) with the surface of articles heated in flame or electric furnaces will lead to oxidation and decarburisation of the steel.

Steel is oxidised chiefly by its interaction with oxygen ($2Fe+O_2 \rightarrow 2FeO$), water vapour ($Fe+H_2O \rightarrow FeO+H_2$), and carbon dioxide ($Fe+CO_2 \rightarrow FeO+CO$). Oxidation in the heating process results in irretrievable losses of metal, deterioration in the condition of the ordinarily most highly stressed layers of metal and the necessity for subsequent descaling.

In the initial stage, steel is oxidised by the chemical reaction on its surface. After a film of oxide has formed on the surface, oxidation proceeds by the diffusion of oxygen atoms through the scale to the metal or by the diffusion of atoms (ions) of the metal through the scale to the surface. The latter is more often the case.

The oxidation rate increases with an increase in heating temperature. The higher the porosity of the scale formed, the more intensive oxidation will be.

Decarburisation at high temperatures is due mainly to the reactions between steel and hydrogen [$\gamma\text{-}Fe(C)+2H_2 \rightarrow Fe+CH_4$] or between steel and oxygen [$\gamma\text{-}Fe(C)+0.5O_2 \rightarrow Fe+CO$]. Decarburisation of the surface layers of steel reduces the hardness in the as-quenched condition as well as the wear resistance and fatigue strength.

Oxidation and decarburisation may be prevented if a protective gaseous medium is introduced into the furnace. This is called a *controlled or protective atmosphere.*

Tables 7 and 8 indicate the compositions of the most extensively employed controlled atmospheres and conditions for their application to protect steel against oxidation and decarburisation. The following atmospheres are most frequently used in heating medium-and high-carbon steels for hardening: ПСО-06, ГГ-ВО, ПСО-09, КГ-ВО, etc. (Table 8). If a controlled atmosphere is not available, the articles to be heated are packed in boxes with used carburising agents, in cast-iron chips, or various types of coatings are applied to the surfaces. High-speed steel tools, for example, are protected against decarburisation by heating them slightly (approximately [illegible]00°C) and then immersing them in a hot saturated solution of borax. This forms a crust of borax which becomes a continuous protective layer at high temperatures. Another method used is to heat the tools to 800-850°C and then to coat them with dehydrated borax powder before the final heating.

To prevent decarburisation when salt baths are used for heating (see Table 5), the baths should be carefully deoxidised by periodically adding ground ferrosilicon (1 to 1.5 per cent of the weight of the salt) or borax. Sometimes potassium ferrocyanide [$K_4Fe(CN)_6$] is used for this purpose.

Table 8

Applications of Controlled Atmospheres

Heat treatment	Metal to be treated	Treatment temperature, °C	Type of surface required	Recommended type of atmosphere
Annealing	Low-carbon steel	659 to 750	Bright	ДА, ПСА-08, ПСС-06, ПСО-09, ПС-06 with additional drying
	Medium- and high-carbon steel	6Ꮾ0 to 800	Bright, without decarburisation	ГГ-ВО, ПСО-06, ПСО-09
	Medium- and high-carbon alloy steel	700 to 870	Ditto	ГГ-ВО, ПСО-05, ПСО-09, ПС-Э
	High-speed steel	760 to 870	Ditto	ПСО-06, ПСО-09
	Chromium and chromium-nickel steel	980 to 1,150	Bright	ДА
Normalising	Low-carbon steel	870 to 1,000	Bright or clean	ПСА-08, ГГ-С, ПСС-06, ПС-06 with additional drying
	Medium- and high-carbon steel and alloy steel	800 to 1,100	Bright, without decarburisation	ПСО-06, ПСО-09, ГГ-ВО, ГГ-О, КГ-ВО

QUENCHING MEDIA

The quenching medium must provide for a cooling rate above the critical value to prevent austenite decomposition in the pearlite and intermediate regions. In the martensitic transformation temperature range, cooling should be slower to avoid high internal stresses, warping of the hardened part, and cracking.

The most widely employed quenching media are water, various aqueous solutions, oil, air, and molten salts.

Heat is extracted in three distinct stages of varying intensity, when water, oil, or other liquids, subject to boiling, are used for quenching:

1. At the first stage a thin vapour film or blanket surrounds the hot metal. Cooling proceeds by film boiling, the cooling rate is relatively slow and is determined by the radiation and conduction of the vapour.

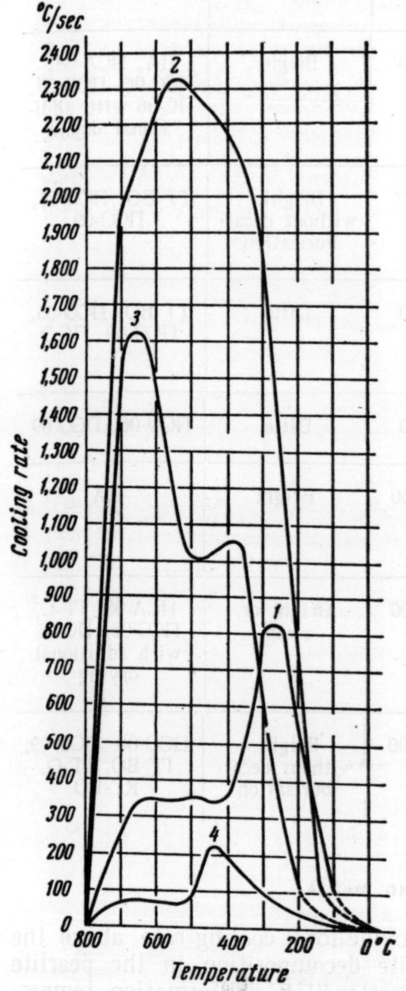

Fig. 157. Variation in the cooling rate with the temperature for various quenching media:
1—H_2O (at+20 °C), 2—H_2O+10 per cent NaCl, 3—H_2O+50 per cent NaOH, 4—oil

2. The vapour film breaks up and the liquid boils with bubbles on the surface of the metal being cooled. In this period, the liquid wets the metal surface in direct contact and cooling is accomplished by vapour generation on this surface. Since all quenching media have a high latent heat, this is the fastest stage of cooling.

3. At temperatures below the boiling point, cooling is much slower as heat is extracted mainly by convection. The cooling rate decreases as the temperature of the metal falls.

Water and aqueous solutions are most frequently used as quenching media in hardening carbon and certain low-alloy steels with a high critical cooling rate. Some features of water quenching are:

1) its pronounced vapour blanket stage (Fig. 157) extending through a wide temperature range. This reduces the cooling rate in the region of diffusional decomposition of austenite and makes it nonuniform;

2) the sharp reduction of cooling capacity at higher temperatures. An increase in the water temperature extends the

temperature range in which a stable vapour blanket may exist. This makes the cooling even less uniform;

3) the high cooling rate in the temperature range of martensite formation (Fig. 157). This leads to high structural stresses causing deformation, warping, and even quenching cracks.

It should be noted that an increase in the water temperature will not reduce the cooling rate in the martensite temperature range. On the opposite, the cooling rate will increase to a certain extent. This is the frequent cause of cracks when work is quenched in hot water.

An addition of sodium chloride, alkalis, soda and sulphuric acid to water substantially increases its cooling capacity, practically excludes the vapour blanket stage (Fig. 157) and provides more uniform cooling. Rapid destruction of the vapour blanket is due to the decrease in temperature of the surface of the metal to a point where the blanket loses its stability. The vapour blanket is broken up to some extent by scale and salt spalling off from the metal. The maximum cooling rate is attained when from 10 to 15 per cent salt is added.

Other advantages of salt and alkali solutions in comparison with pure water are the following:

1. They are less sensitive to heating.
2. They provide a lower cooling rate in the martensitic transformation temperature range, especially when concentrated solutions are applied, due to their higher boiling point.

In the last years, many Soviet plants use a 40 to 50 per cent solution of NaOH as a quenching liquid for carbon and alloy steels. It provides a high cooling rate in the region of pearlite and intermediate (bainite) transformations and slower cooling in the martensitic range (Fig. 157). Quenching in a 40 to 50 per cent solution of NaOH ensures minimum warping; it enables a clean surface to be obtained due to intensive descaling in the cooling process, as well as uniform hardness. Raising the temperature to 90° or 100°C does not reduce the quenching capacity of a 50 per cent solution of NaOH to an appreciable extent.

It must be noted, however, that the NaOH solution intensively absorbs carbon dioxide from the air and loses its properties after 20 or 30 hours. To prevent this, the bath is protected by a layer of mineral oil from 10 to 20 mm thick. In this case, the service life of the solution increases to a period from 6 to 12 months without any appreciable reduction in its cooling capacity. The oil layer also protects the workers, to a considerable extent, against the attack of harmful alkali vapours.

A 5 per cent solution of potassium permanganate and a 3 per cent solution of glycerine are used to reduce the cooling rate in the martensitic transformation range, particularly for the induction hardening of steels with a martensite point below 300°C. The cooling rate in the 5 per cent potassium permanganate solution will be from 60° to 35°C per sec in this range.

Mineral oils are suitable for quenching alloy steels in which austenite is highly stable and the critical cooling rate is therefore low.

Oil has a number of advantageous features as a quenching liquid:

1. Due to its relatively high boiling point (250° to 300°C), the cooling rate in the martensitic range for steel quenched in oil is comparatively low. This prevents quenching defects. At low temperatures (200°C), the cooling rate in oil is approximately 28 times less than in water (Fig. 157).

2. The quenching capacity of oil is subject to relatively small change in the temperature range from 20° to 150°C.

3. In comparison with water, oil cools the steel more uniformly over the whole temperature range. This also reduces quenching stresses.

Disadvantages of oil quenching include the comparatively low cooling rate in the range of pearlite and intermediate transformations, the high inflammability of the oil, and its tendency to thicken (become gummy) in the course of time. This last defect lowers its quenching capacity.

The most frequently used oils in the U.S.S.R. for quenching are the sulphuric-acid-refined distilled petroleum oils "Industrial 20" (Spindle oil-3), having a flash point of 170°C, and the cheaper "Industrial leached 20B" (according to GOST 2854-51) which provides for a higher cooling rate and has a satisfactory flash point (170°C).

"Industrial oil 12" (Spindle oil-2), with its lower cooling rate and lower flash point (165°C), is more seldom used.

Though vegetable oils provide for a higher cooling rate in the range of austenite decomposition by diffusion, they are not used for quenching because of their high cost and high oxidability.

A water and air mixture (moistened air), applied at a pressure of 3 atm, is sometimes employed for quenching in hardening heavy articles. Various cooling rates may be obtained by changing the amount of water in the mixture.

Air with a low moisture content cools steel at almost the same rate as oil. Pertinent factors, such as the optimum amount of water to use, the size of surface hardened at one time, and the distance of the sprayer from the quenched surface, should be established by experiment in each case.

No special difficulties are encountered in automating hardening facilities that use a water and air quenching system.

Proper immersion of the part being treated into the quenching medium is of prime importance. Disregard of this matter may lead to excessive warping and cracking due to nonuniform cooling of various portions of the part.

The following principal rules will eliminate most troubles in quenching. Long articles (both cylindrical and of other cross section) should be immersed with their main axis perpendicular to the bath surface. Thin and flat articles should be immersed on edge and recessed articles with the recess upward. The direction of movement of the article during cooling should coincide with the direction of immersion. Heavy massive articles should be held stationary in the bath and the liquid should be agitated.

HARDENABILITY OF STEEL

The hardenability of steel is that property which determines the depth of the hardened zone induced by quenching. The depth of hardening depends on the critical cooling rate: since this is not the same for the whole cross-section, full hardening may be achieved if the actual cooling rate, even at the core, exceeds the critical value.

Only the surface layer will be hardened if the actual cooling rate of this layer exceeds the critical rate and that of the core is less than the critical value. In this case, the structure of the core will consist of troostite, sorbite, or pearlite.

Fully hardened articles will have the same properties throughout their cross-section. The variation in structure in incomplete hardening will lead to a corresponding variation in properties (Fig. 158).

The depth of hardening is usually taken as the distance from the surface to the semi-martensite zone (50 per cent martensite + 50 per cent troostite).

The hardness of the semi-martensite zone, also called the 50 per cent martensite zone, depends upon the composition of the steel (Table 9).

Table 9

Relationship Between the Hardness of the Semi-Martensite Zone and the Carbon Content

Carbon content, per cent	Hardness of the semi-martensite zone, R_c	
	carbon steels	alloy steels
0.03 to 0.17	—	25
0.13 to 0.22	25	30
0.23 to 0.27	30	35
0.28 to 0.32	35	40
0.33 to 0.42	40	45
0.43 to 0.52	45	50
0.53 to 0.62	50	55

Full hardening of carbon steels is observed in articles of a diameter or thickness up to 20 mm.

Alloy steels harden to a considerably larger depth due to the high stability of the supercooled austenite and the correspondingly lower critical cooling rate.

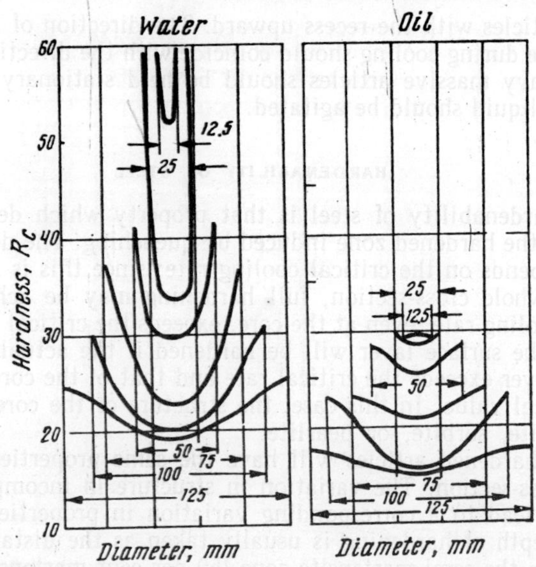

Fig. 158. Hardenability of steel containing 0.45 per cent C (hardening of specimens of different diameter)

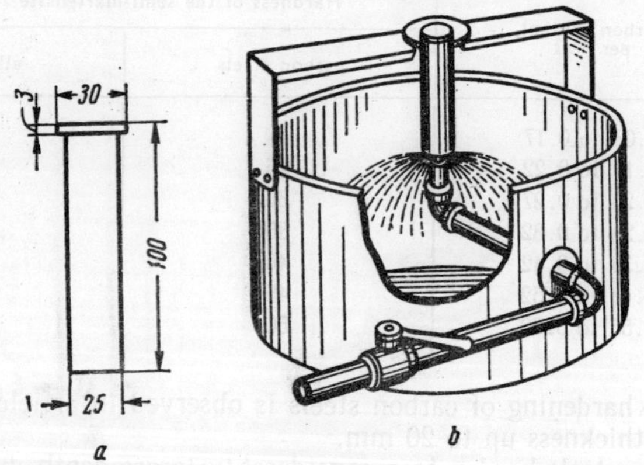

Fig. 159. End quench test for hardenability:
a—test bar; *b*—quenching fixture for the test

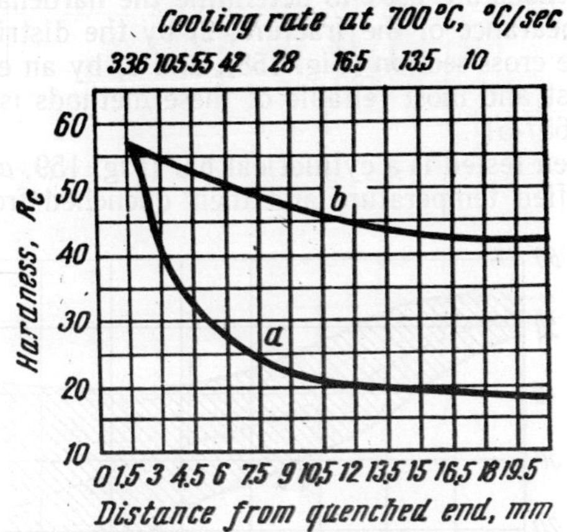

Fig. 160. Hardenability curves plotted from end quench test data (after A. P. Gulyaev):

a—for shallow, *b*—for deep hardening steel

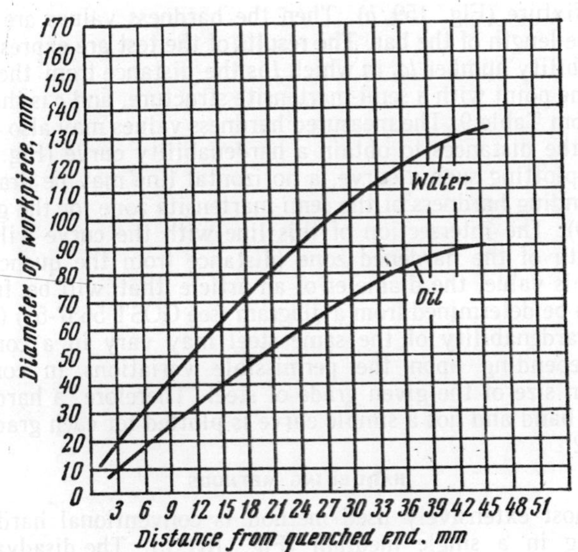

Fig. 161. Determining the diameter of fully hardened articles according to the distance from the quenched end (according to GOST 5657-51)

Several methods are used to determine the hardenability of steels: 1) by the appearance of the fracture, 2) by the distribution of hardness along the cross-section (Fig. 158), and 3) by an end quench test.

The simplest and most reliable of these methods is the *end quench test* (GOST 5657-51).

The specimen tested is a cylindrical bar (Fig. 159, *a*) which is heated to a specified temperature and then quenched from the end in a

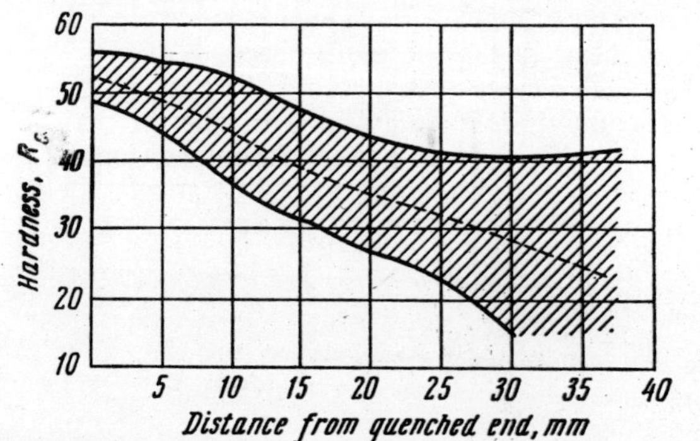

Fig. 162. Hardenability band of steel 30XГCA (after L. N. Davydova)

special fixture (Fig. 159, *b*). Then the hardness values are measured along the length of the bar. The results of the test are expressed by the hardenability number *lc*, in which *l* is the distance from the quenched end to the point with a semi-martensite structure, and *c* is the hardness value from Table 9. The measured hardness values may also be plotted against the distances to obtain a hardenability curve (Fig. 160).

After plotting such a curve, a horizontal line may be drawn at the corresponding hardness of the semi-martensite zone for the given steel (Table 9). The intersection of this line with the curve will indicate the length of the hardened zone (distance from the quenched end). Using this value, the diameter of an article that will be fully hardened can be determined from a diagram (see GOST 5657-51) (Fig. 161).

The hardenability of the same steel may vary in a considerable range depending upon the permissible variations in composition and grain size of the given grade of steel. Therefore, a hardenability range or band and not a simple curve is plotted for each grade of steel (Fig. 162).

HARDENING METHODS

The most extensively used method is conventional hardening by quenching in a single medium (Fig. 163, *a*). The disadvantage of this method, however, is that the cooling rate in the martensitic transformation range will be very high. It will differ only slightly from the rate in the upper zone of supercooled austenite of low stability and, therefore, cracks, distortion, and other defects may occur in this method. Other hardening methods, which shall be briefly described, are extensively employed to avoid these defects and to obtain the required properties.

Quenching in two media. Articles hardened by this method are first quenched in water to a temperature from 300° to 400°C and then quickly transferred to a less intensive quenching medium (for example, oil or air) where they are held until they are completely cooled.

The purpose of the transfer to the second quenching medium is to reduce internal stresses associated with the austenite-to-martensite transformation. It is not advisable to quench first in water and then in oil as this may lead to partial decomposition of the austenite in its zone of least stability (500° to 600°C) and to the development of high residual stresses due to rapid cooling in the martensitic transformation range.

Quenching in two media is widely employed in the heat treatment of carbon steel tools (taps, dies, milling cutters, etc.) of a shape unfavourable as regards cracking and warping.

Hardening with self-tempering. Here the article is not held in the quenching medium until it is completely cooled but is withdrawn to retain a certain amount of heat in the core which accounts for the tempering (self-tempering). The moment when quenching must be interrupted may be established by experiment. Frequently, more heat is retained in the core than is required for tempering and, when the tempering temperature is reached, the article is reimmersed in the quenching liquid.

This hardening method is applied for chisels, sledge hammers, hand hammers, centre punches, and other tools that require a high surface hardness in conjunction with a tough core.

This method has found wide application recently for induction hardening operations.

Stepped quenching or martempering (Fig. 163, *b*). After heating the steel to the hardening temperature, it is quenched in a medium having a temperature slightly above (or below) point *M* (usually from 150° to 300°C). The article is held until it reaches the temperature of the medium and then it is cooled further to room temperature in air and, sometimes, in oil. The holding time in the quenching bath should be sufficient to enable a uniform temperature to be reached throughout the cross-section but not long enough to cause austenite decomposition. Austenite is transformed into martensite during the subsequent period of cooling to room temperature. This treatment will provide a structure of martensite and retained austenite in the hardened steel.

Martempering has the following advantages over conventional quenching: 1) less volume changes occur due to the presence of a large amount of retained austenite and the possibility of self-tempering of the martensite; 2) less warping since the transformations occur almost simultaneously in all parts of the article, and 3) less danger of quenching cracks appearing in the article.

On the other hand, the extremely low stability of austenite in the range from 500° to 600°C requires a cooling rate of 200° to 500° per second in this range to obtain supercooling. At the same time,

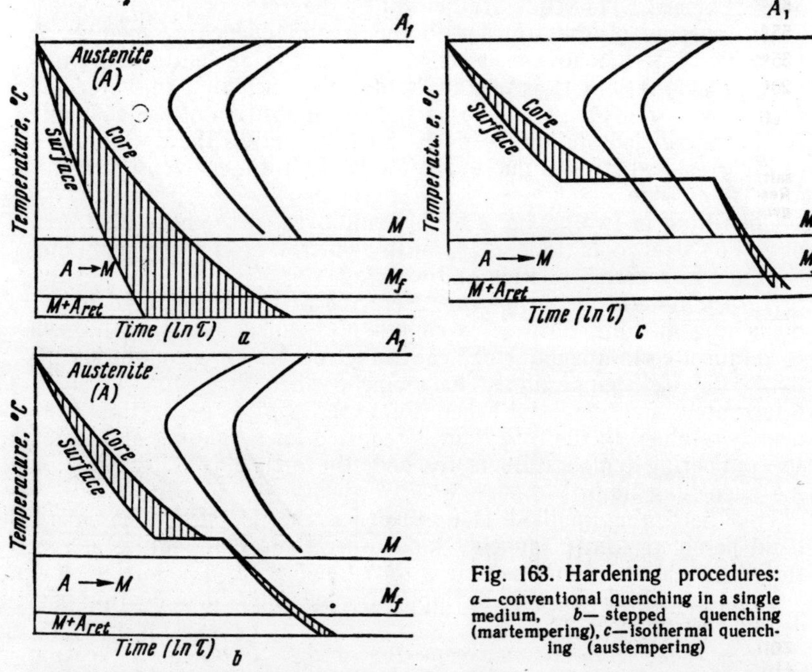

Fig. 163. Hardening procedures: *a*—conventional quenching in a single medium, *b*—stepped quenching (martempering), *c*—isothermal quenching (austempering)

cooling in hot media is much slower than in water or oil at room temperature (Tables 10 and 11). Therefore, austenite in carbon steel can be cooled through the zone from 600° to 500°C, without decomposition, only in thin articles (up to 5-8 mm in thickness). Such articles are expediently hardened by this method. Alloy steel articles, hardened by this method, may be considerably thicker.

Isothermal quenching or austempering (Fig. 163, *c*) is performed in the same manner, principally, as martempering but with a longer holding time at the hot bath temperature (above the martensite point) to ensure a sufficiently complete austenite decomposition (usually into acicular troostite or bainite).

Structural steels (0.3 to 0.5 per cent C) attain their optimum mechanical properties if they are held in the lower part of the intermediate

Table 10

Media for Stepped and Isothermal Quenching (Martempering and Austempering)

Composition	Melting point, °C	Range of application, °C
55% KNO_3+45% $NaNO_2$	137	160 to 500
55% KNO_3+45% $NaNO_3$	221	230 to 550
55% $NaNO_3$+45% KNO_3	218	230 to 600
35% NaOH+65% KOH	155	180 to 500
20% NaOH+80% KOH+6% H_2O (of weight of salts)	130	150 to 500

Note: At temperatures above 550-600°C the saltpetre decomposes. Overheated saltpetre reacts chemically with iron and cast iron and may cause an explosion. Reaction of saltpetre with water (moist articles), soot, or carbon is especially dangerous.

Table 11

Relative Cooling Rates in Agitated Molten Salts (after I. I. Sidorin and B. N. Arzamasov)

Quenching medium	Temperature, °C	Relative cooling rate in the range 650-550°C
Water	20	1.00
KNO_3-$NaNO_2$	160	0.390
KNO_3-$NaNO_2$ or KOH-NaOH	200	0.334
Ditto	300	0.300
Ditto	350	0.260
Ditto	400	0.210

zone on the isothermal austenite decomposition (TTT) diagram (slightly above point *M*). Increasing the temperature of austenite decomposition in the intermediate zone reduces the ductility and toughness of steel. The holding time in the quenching medium depends upon the austenite stability at the given temperature, determined from the TTT diagram for steel of the given composition.

Acicular troostite (bainite) is formed as a result of austempering of carbon steels in which austenite decomposition is fully completed in the intermediate zone. This decomposition ceases without being completed in many alloy steels. In such cases, the structure, after properly conducted austempering, consists of acicular troostite and 10 to 20 per cent undecomposed (retained) austenite,* enriched with

* If the undecomposed austenite undergoes martensitic transformation in subsequent cooling, the properties of the steel will become sharply deteriorated.

carbon. Such a structure ensures a very high strength in conjunction with sufficient toughness. This is especially essential for heavy duty structures. Experience shows that austempering of many grades of steel provides a substantial increase in structural strength, i.e., strength of complex specimens. In comparison with conventional hardening followed by tempering at 250° to 400°C, austempering reduces notch sensitivity and sensitivity to eccentric loading and increases the ductility in the notch by 1.5 to 2 times.

It must be noted, however, that hardening with quenching in a hot medium is not suitable for all grades of steel and for articles of all sizes.* Improper procedure may substantially reduce the mechanical properties.

Molten salts (Table 10) are usually used as a medium in martempering and austempering. The lower the temperature of the salt bath, the higher the cooling rate it provides (Table 11). Since cooling in molten salts is achieved only by conduction, their cooling capacity is increased to a great extent by agitation. The addition of water (6 to 10 per cent) to a melt of caustic alkalis increases the cooling rate to 100-200°C per second. Quenching in molten caustic alkalis, in cases when the heating was conducted in molten chlorides (see Table 5), will enable a clean light grey surface to be obtained (bright hardening).

Steel is not oxidised when it is heated in chlorides. The thin film of chlorides**, covering the article, protects it against oxidation while it is being transferred to the quenching bath. At the moment of immersion into the molten caustic alkali, the film breaks off (or is dissolved) and bares the metal surface. Contact with the caustic alkali, however, does not oxidise steel parts to any appreciable extent.

EFFECT OF HARDENING ON THE PROPERTIES OF STEEL

The effect of various quenching conditions on the hardness of steels of different carbon content is shown in Fig. 164. The diagram indicates that the higher its carbon content, the harder a steel will be after hardening to a martensitic structure. An increase in the amount of retained austenite in the hardened steel will noticeably reduce its hardness.

Notwithstanding the high hardness, hardened steel has a low cohesive strength, a lower tensile strength and, particularly, a low elastic limit. This is due to the stress conditions after hardening. Tension tests, conducted on hardened steels with a carbon content exceeding 0.4 per cent, result in brittle fracture by separation.

The impact strength, relative elongation and reduction of area are also considerably reduced by hardening.

---

* L. M. Pevzner, "Structure and Properties of Steel Hardened in Hot Quenching Media", in the symposium *Physical Metallurgy and Heat Treatment*, VNITOMASH, Mashgiz, Moscow, 1955.

** An increase in the chloride content of an alkali bath reduces its fluidity.

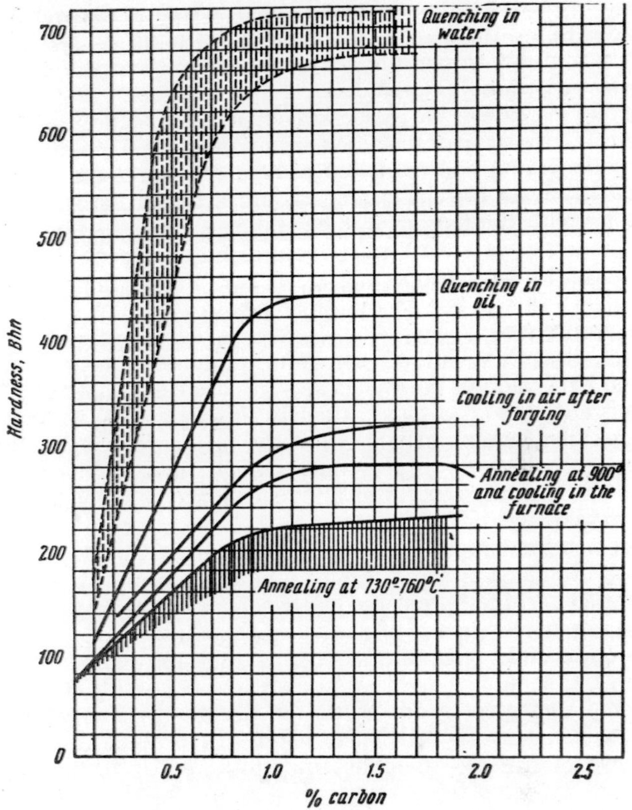

Fig. 164. Effect of various hardening procedures on the hardness of steels of different carbon content (after I. S. Gayev)

Hardening increases the electrical resistivity of steel. This is due to the fact that carbon and other admixtures pass into the solid solution. The specific volume and coercive force of steel increase after hardening while the residual induction and magnetic permeability are reduced.

9-4. Tempering of Steel

Tempering, as mentioned above, leads to the decomposition of martensite into a ferrite-cementite mixture and strongly affects all properties of steel.

At low tempering temperatures (up to 200° or 250°C), the hardness changes only to a small extent and the true tensile strength and bending strength are increased. This may be explained by the separation of carbon from the martensite lattice and the corresponding reduction in its stressed state.

A further increase in the tempering temperature reduces the hardness, true tensile strength, proportional limit, and yield point, while the relative elongation and reduction of area are increased. The data in Fig. 165 illustrate the changes in the properties of hardened steels in accordance with the tempering temperature.

The properties of steel after structural improvement, i.e., hardening followed by high tempering, are always higher than those of annealed steel. This is due to the difference in structure of the ferrite-cementite mixture.

Impact strength varies in a somewhat different manner. Tempering at 250° to 400°C reduces the impact strength of steel (temper brittleness). All steels are subject to some extent to this tempering embrittlement which is called temper brittleness of the first order. Some steels, for example, carbon steels display only a slight loss in toughness (Fig. 165, *a*). The impact strength of alloy steels may be reduced by 50 or 60 per cent (Fig. 165, *b*).

The reason for temper brittleness is not yet sufficiently clear. This reduction in impact strength is most probably associated with the precipitation of carbides* from the martensite in tempering. It may also be due to decomposition of the retained austenite.

It is quite clear from the aforegoing that the temperature range 250° to 400°C should be avoided in assigning tempering temperatures.

Austempering gives good results for steels with a marked susceptibility to nonrecurring brittleness. This will provide a considerably higher impact strength than hardening followed by tempering in the range from 250° to 400°C.

Fig. 166 gives data showing the change in the physical properties of steel, after tempering at various temperatures. Not only the tempering temperature, but its duration, affects the properties.

In assigning tempering procedures it must be noted that longer tempering times are required for heavy parts than for small ones to obtain the same results.

Effect of tempering on the magnitude of residual (internal) stresses. The volume increase and the fact that the martensitic transformation does not take place at the same time, throughout the cross-section of the part being hardened, inevitably lead to high internal stresses. Purely thermal stresses, due to nonuniform cooling of surface and internal layers, also appear.

---

* Y. N. Sokolkov, and V. D. Sadovsky, "Nonrecurring Temper Brittleness of Structural Alloy Steels", *Metal Physics and Physical Metallurgy*, Vol. I, Issue 2, U.S.S.R. Acad. of Sciences, Moscow, 1955.

The simultaneous action of thermal compression and phase transformations results in a very complex distribution of residual stresses in a hardened article. Residual stresses may reach high values. They cause distortion (warping) and cracking in quenching.

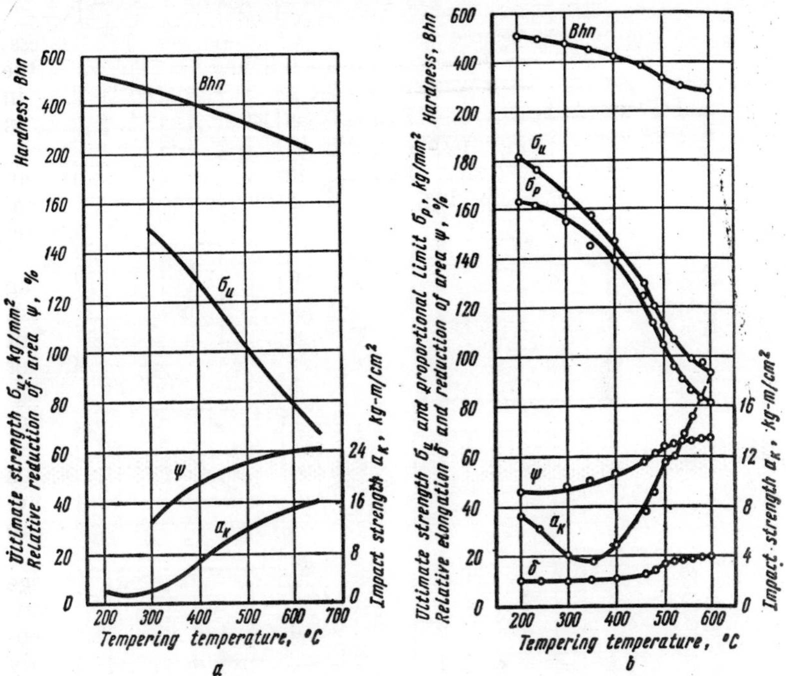

Fig. 165. Effect of tempering temperature on mechanical properties (after V. M. Doronin):

a—carbon steel (0.4 per cent C), *b*—alloy steel (0.35 per cent C, 1 per cent S, 0.9 per cent Mn, and 1 per cent Cr) (Grade 35ХГСА)

In the most favourable cases, residual compressive stresses appear at the surface of the article and tensile stresses of a somewhat less magnitude—in the core.

A less favourable case may occur, however, in which tensile stresses appear at the surface. This may lead to quenching cracks.

The higher the cooling rate (especially in the martensitic range), the heavier the cross-section of the article, and the coarser the actual austenite grains, the larger the residual stresses will be after hardening.

Tempering is the principal method for relieving residual stresses in hardened steel. The higher the tempering temperature, the more completely the internal stresses, caused by quenching, will be relieved.

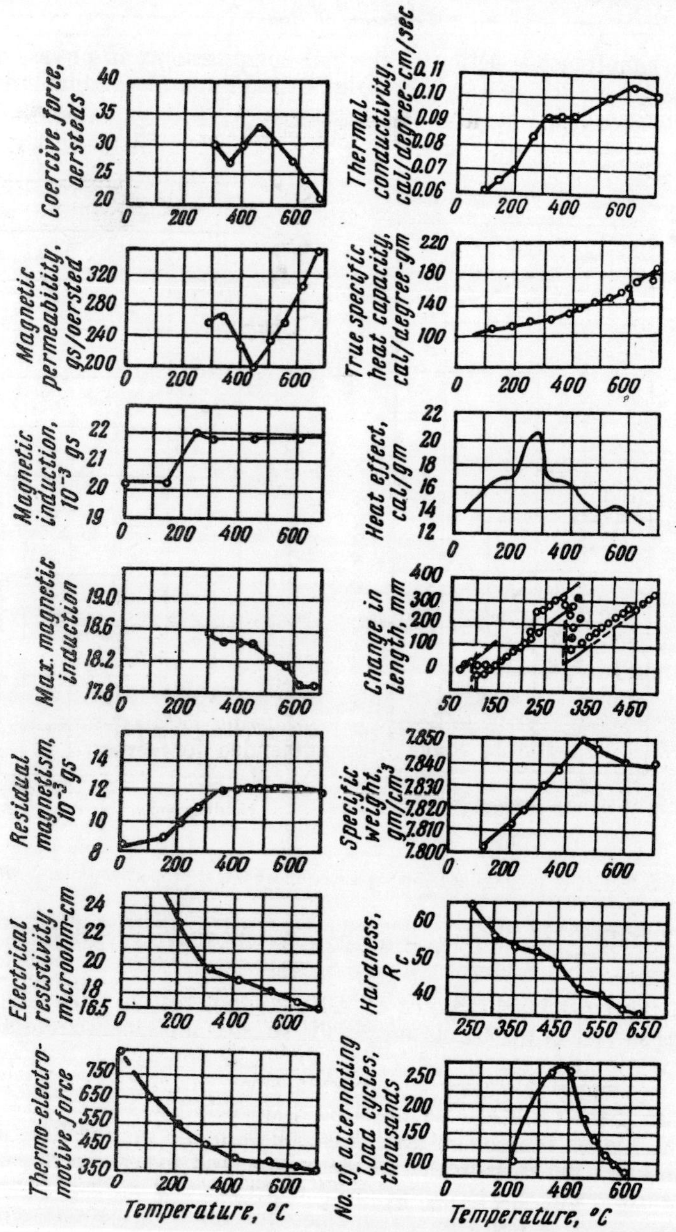

Fig. 166. Effect of tempering temperature on the physical properties of steel (after K. F. Starodubov)

After tempering a cylindrical specimen at 550°C, the maximum axial stresses were reduced from 60 to 8 kg per sq mm.

Shear and radial stresses were reduced by approximately the same amount. Residual stresses are most intensively relieved in the first 15 to 30 minutes of the tempering operation. After a holding time of 1.5 hours, the stresses reach the minimum value for the given temperature.

The cooling rate after tempering has a great effect on the residual stresses. The slower an article is cooled, the less the residual stresses will be. Rapid cooling in water from 600°C will produce stresses of a value near to those produced by hardening. Cooling in air after tempering will result in stresses 7 times less than those obtained by water cooling; cooling in oil will give stresses 2.5 times less than in water.

Thus, articles of complex form, of carbon and certain alloy steels, should be cooled slowly, after tempering at a high temperature, to prevent the appearance of new stresses.*

Tempering procedures are classified in accordance with the heating conditions.

Low-temperature tempering is performed in the range from 150° to 250°C and its purpose is to reduce internal stresses and to increase the toughness without any appreciable loss in hardness. This type of tempering is usually used for measuring and cutting tools of carbon and low-alloy steels, as well as for parts that have been surface hardened, case carburised, etc. The following holding times at the tempering temperature may be recommended for carbon and alloy steel tools:

Diameter (thickness) of tool, mm	Holding time, hrs
up to 20	1.0
21 to 40	1.5
41 to 60	2.0
over 60	2.5

After low-temperature tempering, the martensite produced by quenching is transformed into tempered martensite.

Temperature control in low tempering is achieved by watching the so-called temper colours which appear on the ground surface of steel at temperatures from 200° to 300°C.

At temperatures from 220° to 240° C, the oxide layer which forms on carbon steel will have a straw-yellow colour (thickness of the oxide layer 450 Å); in the range 240-260° C, it will be of an orange colour (oxide layer thickness 500 Å); in the range 260-280° C, it will be violet in colour (layer 650 Å), and from 280° to 300° C the colour will be blue (layer approx. 700 Å).

---

* Articles made of certain grades of alloy steels should be cooled rapidly after tempering at 500-650° C.

The temper colour is the colour of the thin layer of oxides that forms on the surface of heated steel. The colour depends on the thickness of the oxide layer.

Medium-temperature tempering at 350° to 450°C is employed for coil and laminated springs and provides the highest attainable elastic limit in conjunction with ample toughness. Steel has a troostite structure after this tempering procedure.

High-temperature tempering is performed in the range from 500° to 650°C. It almost completely eliminates internal stresses and provides the most favourable ratio of strength to toughness for structural steels. The tempered steel has a sorbite structure after this treatment

9-5. Sub-Zero Treatment of Steel

A certain amount of retained austenite may always be found in hardened steel. Retained austenite reduces the hardness, wear-resistance, and thermal conductivity of steel and makes its dimensions unstable.

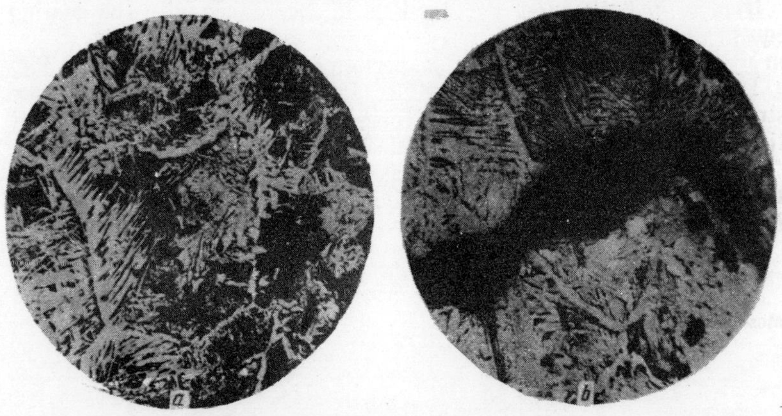

Fig. 167. Microstructure of steel, × 300:
a—overheated steel, *b*—burnt steel

A sub-zero treatment has been devised to reduce the retained austenite in hardened steel. It consists in cooling the metal being treated to sub-zero temperatures. Such treatment is suitable only when the temperature, at which the martensitic transformation is completed (M_f), is below zero.

Cooling to M_f transforms the retained austenite into martensite. This increases the hardness of the part and its dimensions will become more stable. There is no purpose in cooling below M_f, since no additional transformation of retained austenite occurs below this temperature. Sub-zero treatment is usually conducted in the temperature range from —30° to —120°C. The holding time at this temperature is from 1 to 1.5 hours.

Prolonged holding at room temperature after hardening will stabilise the austenite of many grades of steel and reduce the effect of sub-zero treatment. Therefore, it is advisable to perform sub-zero treatment directly following the hardening operation.

Sub-zero treatment* is most frequently used for high-speed steel tools (see page 336), measuring tools, and also for carburised gears and other machine components of alloy steels.

9-6. Defects Due to Heat Treatment of Steels

The main types of rejects in annealing and normalising are due to faulty regulation of heating temperature, including overheating, burning, underheating, etc.

Due to the application of perfected pyrometry and automatic equipment, the above types of rejects have been practically eliminated in up-to-date heat-treating departments.

The chief cause of quenching defects is high residual (internal) stresses occurring in hardened articles. These stresses may cause distortion, warping, and even cracking.

The types of defects encountered in the heat treatment of steel, their causes and measures for their prevention or correction are listed in Table 12.

---

* For more detail see V. G. Vorobyov, *Sub-Zero Treatment of Steel*, Oborongiz, Moscow, 1954.

Table 12

Principal Defects in Steel Due to Faulty Heat Treatment and Measures for Their Elimination

Defect and its characteristics	Causes	Measures for avoiding or correcting the defect
Overheating—coarse-grained microstructure and fracture, Widmanstätten structure in the annealed steel, coarse crystallic martensite in the hardened steel. Reduced ductility and, especially, impact strength (Fig. 167, *a*)	Heating for long periods at temperatures considerably exceeding the normal values	Elimination: 1) for slight overheating—apply normal annealing and normalising, 2) for considerable overheating—apply double annealing or normalising (first annealing or normalising at a temperature of $Ac_3 + 100°$ to 150°, second at normal temperature)
Burning—grain boundaries have. 1) regions enriched in carbon—1st stage of burning, 2) nonoxidised cavities and blowholes—2nd stage, 3) iron oxide inclusions—3rd stage (Fig. 167, *b*). Stone-like fracture Low ductility	Heating for a long period at a high temperature in an oxidising atmosphere or at temperatures near the melting point	1) 1st stage—use homogenising followed by double annealing (see overheating) 2) 2nd stage—use forging followed by annealing, 3) 3rd stage—defect cannot be remedied
Oxidation. Thick layer of scale on the surface of the steel article	Oxidising atmosphere in the heating furnace	Prevention of the defect: 1) heat in furnaces with reducing, neutral, or protective atmospheres, 2) heat in boxes with used carburising agent or cast iron chips, 3) heat in molten salt baths
Decarburisation—loss of carbon in the surface layers of the article. Lower hardness after quenching, lower fatigue limit	Ditto	Ditto Correction of the defect—machine off the decarburised layer if the machining allowance is sufficiently large

Continued

Defect and its characteristics	Causes	Measures for avoiding or correcting the defect
Excessive hardness of hot-worked annealed steel	Excessive cooling rate for ordinary annealing or insufficient holding time for isothermal annealing	Repeat annealing with cooling at specified rate
Black fracture—free graphite (annealing carbon) inclusions in the steel	Excessive heating time and slow cooling after annealing	Heat the steel to a high temperature and forge it thoroughly
Quenching cracks. Usually appear at the grain boundaries and are zigzag in form. They may be external and internal. Cracks may be longitudinal, arc-like, etc.	1) Internal stresses due to volume changes at low temperatures, caused by the martensitic transformation, 2) the martensitic transformation is not simultaneous throughout the article being hardened	This defect cannot be corrected. It may be prevented by: 1) avoiding sharp projections, sharp corners, and sudden transitions from thick to thin sections in designing the article, 2) the article should be free of stresses before being hardened; stresses should be relieved by previous annealing, 3) heat to the minimum suitable temperature for hardening, 4) cool slowly in the martensitic range (apply interrupted quenching, oil quenching, martempering, etc.), 5) apply austempering, 6) temper immediately after quenching
Deformation and volume (dimension) changes after hardening	The volume of the steel increases due to the martensitic transformation. The higher the hardenability of the steel, the more severe the deformation in hardening will be	1) Use alloy steels that are only slightly deformed by quenching 2) cool slowly in the martensitic range 3) apply surface hardening (when feasible)

Continued

Defect and its characteristics	Causes	Measures for avoiding or correcting the defect
Warping—asymmetrical deformation of the article in quenching	1) Volume changes in cooling (or heating), 2) nonuniform heating (or cooling) of the article, 3) internal stresses in the article before heating, 4) the article is lowered into the quenching bath in an inclined position	Ditto, and also: 1) anneal, normalise, or temper at a high temperature before hardening, 2) heat uniformly for hardening, 3) quench as uniformly as possible in hardening, 4) hold the article in the proper position when lowering it into the quenching bath, 5) use special quenching jigs. The defect may be corrected by straightening or grinding (if the grinding allowance is sufficiently large)
Insufficient hardness after guenching	1) Hardening temperature is too low, 2) holding time is insufficient at the hardening temperature, 3) cooling rate is too low	The defect may be corrected by normalising or annealing followed by hardening with the proper specified procedure
Soft spots—zones on the surface of hardened articles with lower hardness	1) Vapour accumulates (forming bubbles) on the surface of the quenched articles reducing the cooling rate at the given places, 2) localised decarburisation, 3) inhomogeneity of initial structure in solidification of the steel	The defect may be prevented by: 1) using a more effective quenching medium, 2) obtaining a more homogeneous structure, employing annealing or normalisation before hardening, 3) protecting against decarburisation in heating. The defect may be corrected by normalising followed by quenching at a higher cooling rate

Continued

Defect and its characteristics	Causes	Measures for avoiding or correcting the defect
Excessive hardness after tempering	Low temperature or insufficient holding time in tempering	Corrected by a second tempering with proper temperature and holding time
Insufficient hardness after tempering	Tempering temperature was too high	Anneal, reharden, and temper at normal temperature
Erosion—reduction in size of the article or distortion of its form due to loss of metal from its surface	Chemical reaction of chlorous salts and oxidation of metals heated in molten salt baths	Preventive measures: 1) deoxidise salt bath with borax or ferro-silicon, 2) properly position the article in reference to the electrodes of electrical salt baths
Corrosion—pitting or groove type destruction of the surface of an article	In heating in molten salt baths: a) high content of sulphuric salts (over 0.7-0.8 per cent), b) bath has become rich in oxygen or iron oxides. In heating in flame furnaces—nonuniform scale formation	Preventive measures: 1) carefully control salt composition; 2) deoxidise the bath; 3) eliminate the oxidising atmosphere when heating in flame furnaces

Notes: 1. Factors conducive to cracking are: sharp corners, large volume changes at low temperatures, and low cohesive strength. All factors that favour deeper hardenability are also conducive to cracking.

2. Deformation in hardening may be substantially reduced by employing special heat treating procedures which enable tempered martensite and an increased amount of retained austenite to be obtained after hardening. This is achieved by increasing the hardening temperature to a certain extent and by slow cooling in the temperature range of martensitic transformation.

Chapter 10

SURFACE HARDENING OF STEEL

Surface hardening is a selective heat treatment in which the surface layers of a metal are hardened to a certain depth while a relatively soft core is maintained. The chief types of surface hardening are:

1) hardening with high frequency induction heating,
2) hardening with electrical contact resistance heating,
3) hardening with electrolytic heating, and
4) oxyacetylene flame hardening.

The principal purpose of surface hardening is to increase the hardness and wear resistance of the surfaces of metal articles. At the same time, the reliability in operation of a machine component is increased, as well as its fatigue limit.

10-1. High-Frequency Induction Hardening

High-frequency currents were first used in heating metal for hardening by V. P. Vologdin in 1923. The first industrial applications of induction heating, however, date from 1935.

Induction heating has the following important advantages: 1) it enables the time required for heat treatment to be sharply reduced, thereby increasing the labour productivity, 2) articles may be heated with practically no scaling so that the metal allowance for subsequent machining may be reduced, 3) deformation, due to heat treatment, is considerably reduced, etc.

Induction hardening also facilitates automation of heat treating processes and enables heat treatment to be performed directly in the machining production line without interrupting the technological sequence of operations.

Induction hardening is at present extensively applied in Soviet plants. For example, crankshafts, camshafts, gears, and many other automobile and tractor parts are induction hardened directly in the main production lines where these parts are machined.

Heating by high frequency current (induction heating) is accomplished by the thermal effect of current induced in the article being heated. The latter is placed in an alternating magnetic field set up by the high frequency current.

The part to be heated is placed in the so-called inductor or inductor coil (Fig. 168) which comprises one or several turns of copper tube or busbar. When alternating current is passed through the inductor, it sets up a magnetic field the intensity of which varies periodically in magnitude and direction.

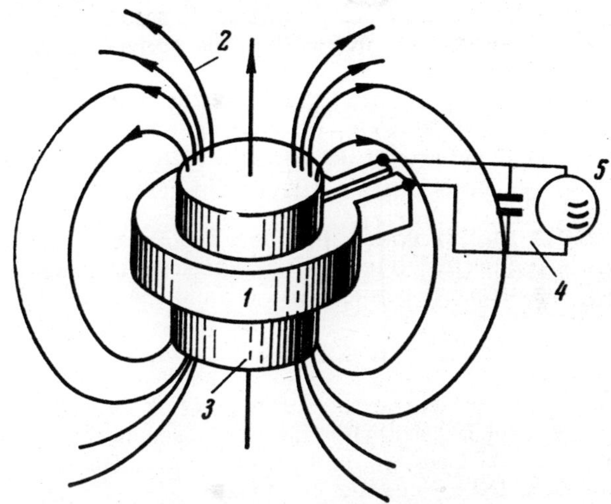

Fig. 168. Principle of high-frequency induction heating (after V. P. Vologdin): 1—inductor, 2—magnetic field, 3—article being heated, 4—capacitor battery, 5—high-frequency generator

The alternating magnetic lines thread through the surface of the article being heated in the inductor and induce in this surface an alternating current of the same frequency but reversed in direction. The density of the induced alternating current is not uniform throughout the cross-section of the conductor (the heated article). The current passes chiefly along the surface of the conductor. This is known as skin effect.

The depth to which the current penetrates may be determined from the formula:

$$y = 5{,}000 \sqrt{\frac{\rho}{\mu f}} \text{ cm}$$

in which y is the depth of current penetration, cm;
ρ is the electrical resistivity, microohm-cm;
μ is the magnetic permeability of the steel, gs/oersted;
f is the current frequency, cps.

The depth of penetration y of the current increases with the temperature. It increases most sharply at temperatures above the Curie point (768°C) where steel is transformed from the ferromagnetic to the paramagnetic state. For example, the depth of penetration of current in a steel containing 0.45 per cent C is increased 17 times as the temperature rises from 20° to 850°C.

If ample power is available (0.3 to 1.5 kW per sq cm), the surface layer, through which the induced current passes, will be rapidly heated to the hardening temperature (from 2 to 10 seconds).

If heating is followed by quenching, the surface layer will be hardened.

From the energy point of view, induction hardening is adequately characterised by two parameters: the heating time and the specific power. The latter is defined as the energy transformed into heat per unit volume of the surface layer. The specific power determines the heating rate.

The higher the specific power, the more rapidly the steel will be heated. At a given current frequency (f), the specific power depends mainly upon the magnetic permeability (μ) and the electrical resistivity (ρ):

$$\Delta W = \alpha \sqrt{\mu\rho}$$

When steel is heated above the Curie point, where the magnetic permeability of steel is practically the same as that of air, the specific power and, consequently, the heating rate are sharply reduced. This greatly affects the character of induction heating in steels. In the heating process, current is first concentrated in a thin surface layer (similar to its penetration at low temperatures—so-called *cold* depth of penetration) which is rapidly heated to the Curie point. Above this point, heating is retarded in this surface layer while in the consecutive layer heating is accelerated since maximum current density is now concentrated in this second layer. When the second layer reaches the Curie point, its heating is retarded and maximum current density is shifted to the third layer, and so forth.

Since each successive layer is heated more rapidly than the preceding layer, their temperature difference is quickly reduced and the layers reach the hardening temperature practically at the same time. Such heating ensures a uniform temperature throughout the heated layer and usually prevents overheating.*

The higher the current frequency, the more shallow the hardened layer will be. Table 13 indicates recommended current frequencies to obtain specified depths of the hardened layer in induction hardening.

---

* It must be noted, however, that induction heating does not eliminate the possibility of overheating steel.

Table 13

Recommended Frequencies for Induction Hardening to a Specified Depth

Depth of hardened layer, mm		1.0	1.5	2.0
Maximum frequency, cps		250,000	100,000	60,000
Minimum frequency, cps		15,000	7,000	4,000
Optimum frequency, cps		60,000	25,000	15,000
Recommended generator		Vacuum-tube type	Vacuum-tube or machine type, 8,000 cps	Vacuum-tube or machine type, 8,000 cps
Depth of hardened layer, mm	3.0	4.0	6.0	10.0
Maximum frequency, cps	30,000	15,000	8,000	2,500
Minimum frequency, cps	1,500	1,000	500	150
Optimum frequency, cps	7,000	4,000	1,500	500
Recommended generator	Machine type, 8,000 cps	Machine type, 2,500 cps	Machine type, 2,500-1,000 cps	Machine type, 500 cps

Fig. 169 illustrates various types of inductors employed in induction heating.

High frequency current is supplied from either vacuum-tube or machine-type generators.

At the present time machine-type generators are available for frequencies from 1,000 to 10,000 cps, and have ratings from 60 to 1,000 kW. Standard vacuum-tube generators have frequencies from 100,000 to 10,000,000 cps and a rating from 5 to 220 kW. Machine-type generators are more widely used.

Special appliances are extensively made use of at present for induction heating. They are automated to provide for more uniform results in hardening, higher production capacity, and more effective utilisation of the generators. The heating time, power consumption, quenching procedure, heating temperature, and water pressure are automatically controlled in such machines. Handling operations are frequently mechanised or automated as well. These may include: transferring the article to the appliance, feeding it to the heating and quenching devices, etc. Such equipment is designed to suit the accepted heat treatment and the article produced.

Induction hardening procedures that find application at the present time may be classified into the following groups:

1. The whole surface is simultaneously heated and quenched (Fig. 170, *a*). This method is applicable for hardening small surfaces, as for pins, small shafts, drills, etc.

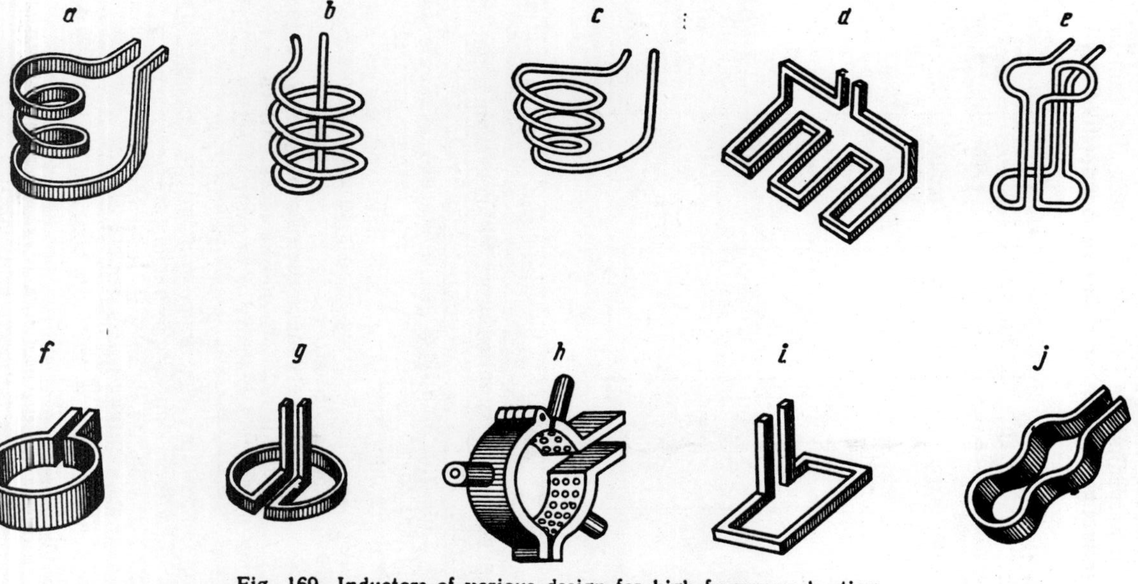

Fig. 169. Inductors of various design for high frequency heating

Multiple-turn inductors for heating: *a*—external surfaces of cylindrical parts, *b*—internal surfaces of hollow cylindrical parts, *c*—external surfaces of conical parts, *d*—plates and flat surfaces by progressive movement, *e*—side and end surfaces of cylindrical parts simultaneously with rotation of the part about a vertical axis. Single-turn inductors for heating: *f*—external surfaces of cylindrical parts, *g*—internal surfaces of hollow cylindrical parts, *h*—crankshaft journals. Split-type inductor with quenching jets on the inner surface for heating: *i*—clamps of complex form and plates by progressive movement, *j*—one gear tooth in consecutive "tooth-by-tooth" hardening

2. Consecutive heating and quenching (Fig. 170, *b*) of sections of the article is used for hardening the journals of crankshafts, cam surfaces of camshafts, teeth of gears with a module over 6 mm ("tooth-by-tooth" hardening), etc.

3. Continuous consecutive or progressive heating and quenching (Fig. 170, *c*) is most suitable for hardening long shafts, axles, and other similar articles. The speed of inductor (or workpiece) travel,

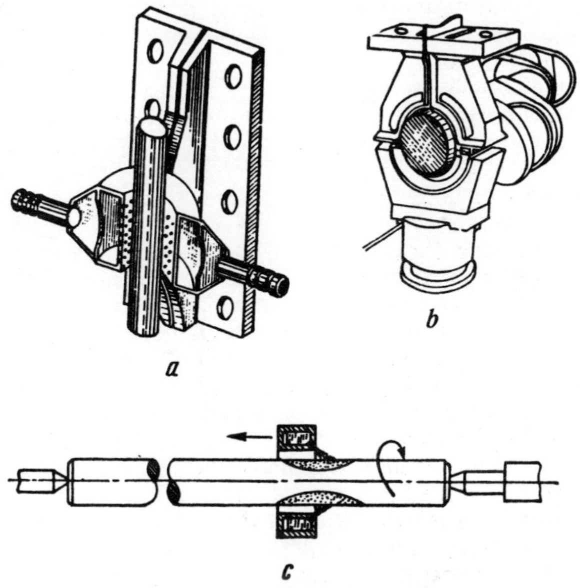

Fig. 170. Induction heating methods:
a—simultaneous heating and quenching of the whole surface to be hardened, *b*—consecutive heating and quenching, *c*—continuous consecutive (progressive) heating and quenching

required to obtain a hardened layer from 1 to 10 mm thick at a frequency *f* from 2,000 to 200,000 cps, will be from 0.3 to 3 cm per sec.

In induction heating, the proximity effect should be taken into consideration. The essence of this phenomenon is that in a system of two conductors, through which alternating current flows in different directions, the maximum current density will be in the regions of the conductors nearest to each other. Therefore, the distance from the inductor to the article being hardened must be the same all around to obtain a uniform depth of the hardened layer. Rotation of the article within the inductor gives good results in this respect.

The quenching liquid is usually delivered through a spraying device (Fig. 170, *a*). Induction hardened articles are subject to subsequent low tempering (160° to 200°C). Frequently, self-tempering is resorted to. In this case, the article is not completely cooled in quenching and retains sufficient heat for tempering.

Steels containing from 0.35 per cent to 0.55 per cent C are most frequently induction hardened. They are first normalised or are subject to structural improvement (high tempering).

Due to the high heating rates in induction hardening, in the order of hundreds of degrees per second, the transformation of pearlite into austenite is shifted into a range of higher temperatures. Therefore, the heating temperature for this process considerably exceeds the temperature to which steel is usually heated in ordinary furnaces for hardening.

Investigations, conducted by I. N. Kidin, show that the hardening temperature depends on the heating rate and may vary in a wide range. At low heating rates, best results are obtained at comparatively low hardening temperatures; at high rates, optimum properties are obtained at higher temperatures. Diagrams of permissible heating temperatures for induction hardening are used at present to assign proper hardening procedures. Several such diagrams are given in Fig. 171. These diagrams show that the more dispersed the initial structure of a steel, the lower its hardening temperature should be. It must be noted that at a constant heating temperature, variation in the heating rate will result in considerable changes in hardness of the hardened layer.

The heating temperature and rate also affect the depth of the hardened layer. The higher the temperature, the deeper the hardened layer will be at a given heating rate. At a given heating temperature, the depth of the layer decreases with an increase in the heating rate. Thus, the chief parameters of induction hardening, temperature and heating rate, cannot be assigned independently of each other.

Fine acicular martensite forms on the surface of properly hardened hypoeutectoid steels. In passing gradually from the surface to the core, zones of ferrite appear in addition to the martensite. The amount of ferrite increases toward the core. Finally, martensite disappears, and only the initial structure, pearlite+ferrite or sorbite, remains.

Table 14

Relationship Between Mechanical Properties of Steel and Its Heat Treatment (after K. Z. Shepelyakovsky)

Heat Treatment	Hardness, R_c	Fatigue limit σ_w, kg/mm²
Normalising	10	9.7
Hardening followed by tempering at 500° C	28	13.7
Hardening followed by tempering at 180° C	52	31
luduction surface hardening followed by tempering at 180°	57	70

Note: Fatigue tests were conducted on specimens 20 mm in diameter with a notch on the gauge length, 0.6 mm wide, 0.8 mm deep, and with a radius of 0.3 mm The hardened case was 2.8 mm deep.

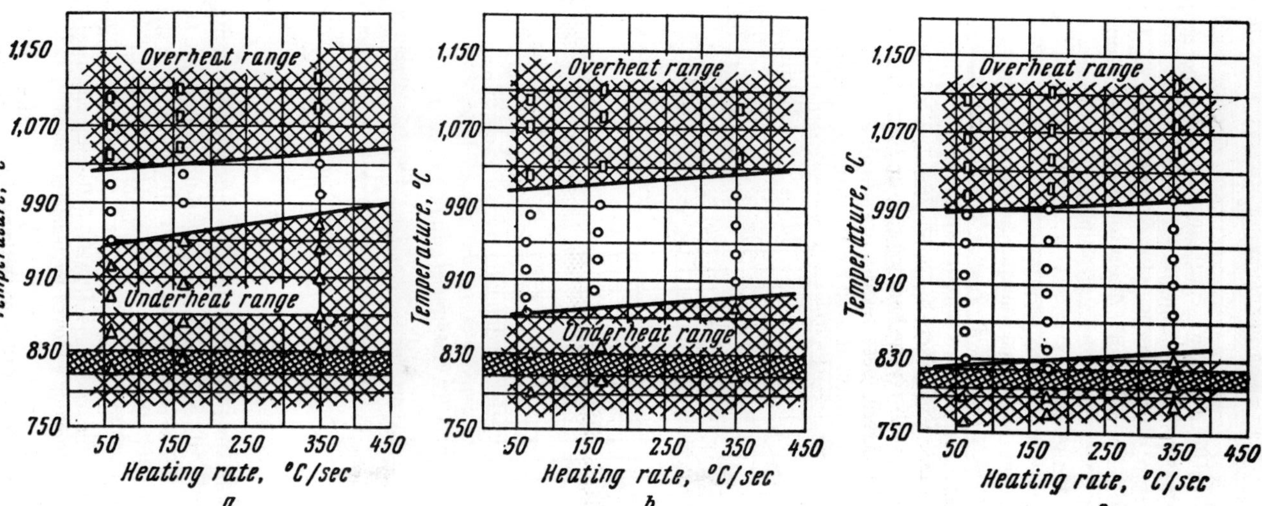

Fig. 171. Permissible heating temperature diagrams for induction hardening of steel Grade 50 after:
a—annealing, *b*—normalising, *c*—structural improvement (high-temperature tempering) (after A. P. Gulyaev and V. V. Kukolev)

The high heating rate and, consequently, the large degree of superheating which results in the formation of fine initial austenite grains, in conjunction with the absence of holding time at the hardening temperature, enable a finer actual grain to be obtained in induction hardening as compared to ordinary hardening with furnace heating.

In comparison with ordinary hardened steels, induction hardened steels have: 1) higher hardness, 2) higher wear resistance, 3) higher impact strength at the same tensile strength, and 4) a higher fatigue limit even if the components operate in a corroding medium (Table 14).

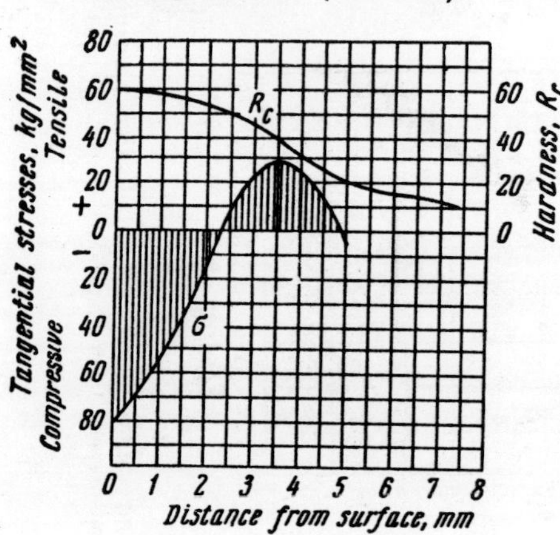

Fig. 172. Residual stress distribution diagram for surface hardened specimens (after M. S. Kosoi)

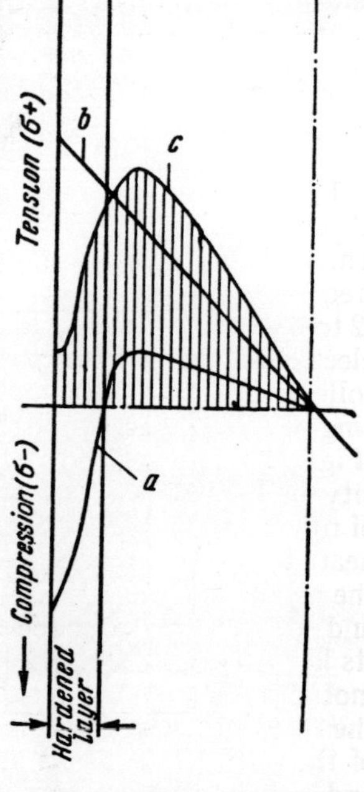

Fig. 173. Variation in the stress condition of the external layer after induction hardening:

a—distribution of residual stresses after surface hardening, *b*— distribution of stresses caused by the bending moment, *c*—resultant stress distribution diagram

The increase in the fatigue limit after induction hardening (Fig. 172) is associated with the appearance of residual compressive stresses in the hardened layer. These stresses reduce the effect of tensile stresses arising from the application of external forces. This is quite evident from Fig. 173 which schematically represents the stress distribution diagram for a bending fatigue test on a specimen having residual compressive stresses.

This reduction of tensile stress at the surface, where the main design and processing stress raisers are concentrated, is the principal reason why the fatigue limit is increased.

In planning a surface hardening procedure, provision must be made for hardening the whole working part of the component and, particularly, places where there is a sharp change in cross-section. Places, where the hardened layer is interrupted, are weakened since here residual tensile stresses appear instead of compressive stresses, and they lower the fatigue limit.*

10-2. Hardening with Electric Contact Resistance Heating

This surface hardening method was invented and developed by one of the outstanding Soviet physical metallurgists, N. V. Geveling. This method consists in supplying alternating current of industrial frequency (50 cps) and at low voltage (2 to 6 V) to the workpiece through electrodes which have the form of rollers. The power is supplied from single-phase step-down transformers (Fig. 174). The current has a density of 400 to 700 A per millimetre of roller width. The steel is rapidly heated at the point of contact with the roller. The workpiece is rotated and the roller is fed gradually along its length (Fig. 174). Due to these motions, the point of contact with the roller travels along the surface of the shaft (workpiece) in a helix and consecutively heats the shaft (band hardening). Cooling is done, either by heat conduction into the metal or by quenching in the ordinary manner.

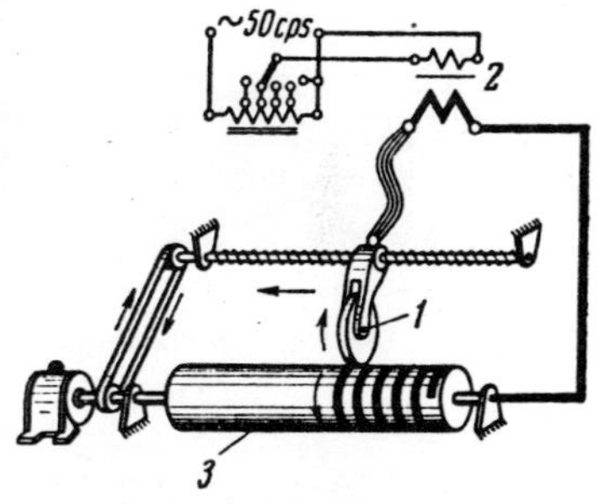

Fig. 174. Hardening with electric contact heating (after N. V. Geveling):
1—contact roller, 2—transformer, 3—workpiece

The feed of the roller along the work is varied in a range from 2 to 15 mm per second in accordance with the depth of hardened layer required.

The roller is pressed against the surface of the workpiece with a pressure of 10 to 15 kg per millimetre of roller width. The chief disadvantage of this method is the helical "tempered band" of lower hardness on the surface after hardening. The tempered band is obtained because, in heating each new turn of the helix, the roller partially heats and tempers the preceding, already-hardened turn. This hardening method is seldom employed at the present time.

* In cases when the complete working section of the component cannot be surface hardened, the places, where the hardened layer ends, are shotpeened.

The surface of steel hardened by this method usually has a lower hardness due to excessive cooling by the water-cooled copper rollers. The maximum hardness will lie at a depth of 0.05 to 0.1 mm below the surface.

10-3. Hardening with Electrolytic Heating

This method was developed by I. Z. Yasnogorodsky. The workpiece, which serves as the cathode, is immersed in an electrolyte (a 5 to 10 per cent solution of soda or potash, Fig. 175). The shell of the bath is the anode. When direct current at a high voltage (250 to 300 V) is passed through the electrolyte, a hydrogen blanket is formed surrounding the cathode (i.e., workpiece).

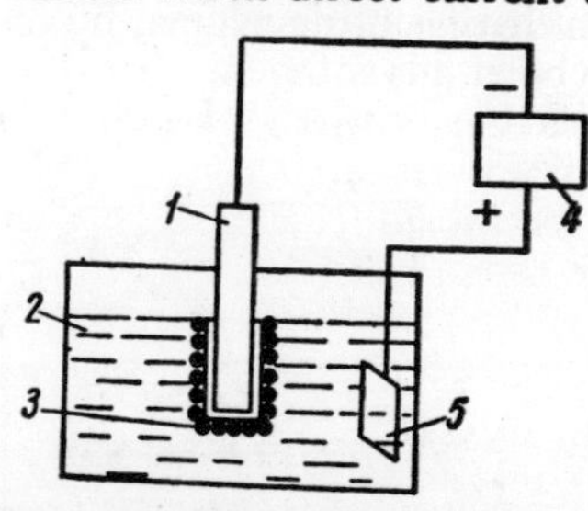

Fig. 175. Electrolytic heating (schematic diagram):

1—workpiece (cathode), 2—electrolyte, 3—hydrogen blanket, 4—current supply, 5—anode

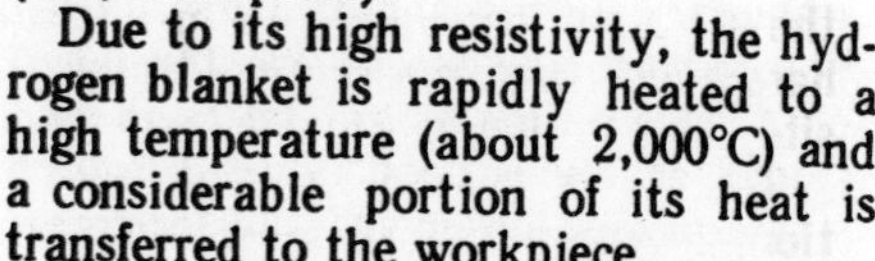

Due to its high resistivity, the hydrogen blanket is rapidly heated to a high temperature (about 2,000°C) and a considerable portion of its heat is transferred to the workpiece.

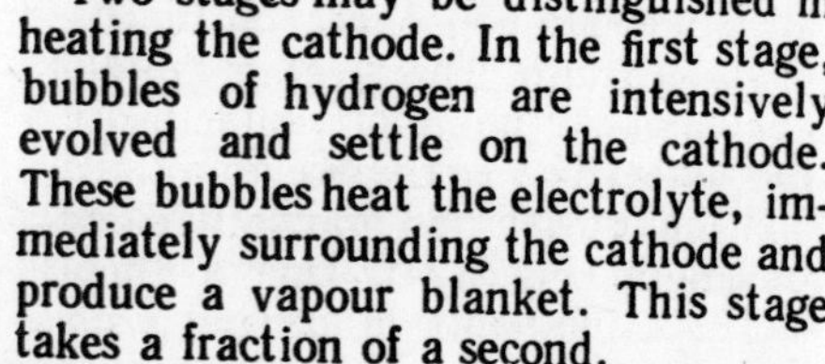

Two stages may be distinguished in heating the cathode. In the first stage, bubbles of hydrogen are intensively evolved and settle on the cathode. These bubbles heat the electrolyte, immediately surrounding the cathode and produce a vapour blanket. This stage takes a fraction of a second.

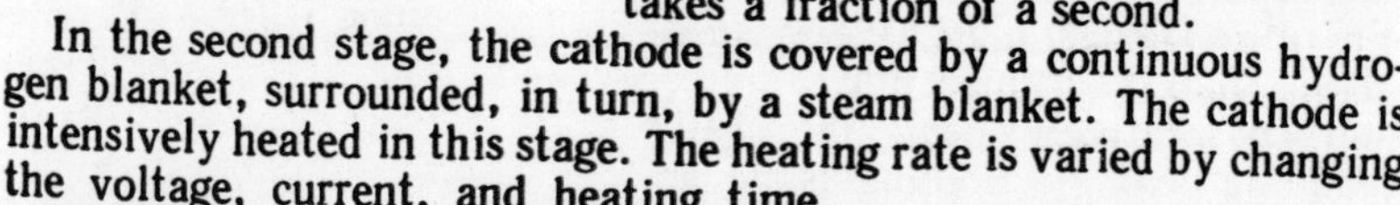

In the second stage, the cathode is covered by a continuous hydrogen blanket, surrounded, in turn, by a steam blanket. The cathode is intensively heated in this stage. The heating rate is varied by changing the voltage, current, and heating time.

Heating time is varied in a range from 10 to 40 seconds to suit the type of workpiece being heated. The current density varies from 3 to 6 A per sq cm.

The principal disadvantage of this method is that the surface is inevitably overheated if a comparatively shallow depth of hardening is required. This shortcoming, in conjunction with the high power consumption and difficulties encountered in mechanising the process, has limited the application of this method.

10-4. Oxyacetylene Flame Hardening

The surface of workpieces, treated by this method, is heated by an oxyacetylene flame having a temperature of 3,000° to 3,200°C. The large amount of heat transferred to the surface of the workpiece, notwithstanding heat removal into the metal due to high thermal conductivity, rapidly heats it to the hardening temperature before the core is appreciably heated. Subsequent quenching hardens the surface layer.

This method is applied in hardening large cast gears, worms, shafts, steel guideways, mill rolls, etc.

Ordinary welding torches are employed for heating; they may have single-flame, slit-type, or multiple-flame tips. The heating time is $7y^2$ seconds (in which y is the depth of the hardened layer in mm) while the torch is traversed along the workpiece (or vice versa) at a speed of $\frac{72}{y}$ mm per sec.

A flame hardening unit comprises the source of acetylene supply, an oxygen plant, quenching devices, hardening control desk, instruments, and a set of torches and tips. The chief advantage of this method is its simplicity. A disadvantage is the danger of overheating the surface. Coarse acicular martensite is formed at the surface after hardening. In deeper layers, this goes over into fine acicular martensite, then troosto-martensite and, finally, the initial structure is found.

The coarse acicular martensite at the surface and the smaller transition zone (as compared with other hardening methods) lower the quality of the hardened layer.

Chapter 11

CHEMICAL HEAT TREATMENT OF STEEL (CASE-HARDENING)

11-1. Physical Principles Involved in Chemical Heat Treatment

The saturation of the surface of steel with a certain element by diffusion of this element from the surrounding medium at a high temperature is called chemical heat treatment or, more simply, case-hardening.

Diffusion in a crystalline solid is based on the fact that atoms can change their equilibrium positions within the crystal lattice.

Each temperature of a solid corresponds to a definite energy level associated with its thermal vibrations. Energy is not uniformly distributed among the individual atoms of a crystal lattice. The great majority of atoms possess a certain mean energy corresponding to the temperature to which the solid is heated. On the other hand, due to energy fluctuations, there is always a sufficient number of atoms possessing either more or less energy than the mean value.

A theory developed by the Soviet scientist Y. I. Frenkel states that atoms possessing a sufficiently high energy level may leave their proper positions at the points of the crystal lattice (see Fig. 11) and jump to interstitial positions. This will create vacancies in the crystal lattice ("holes") which may be occupied by any other atom.

As the temperature is raised, the number of interstitial atoms and vacancies in the crystal lattice will increase, thus increasing the rate of diffusion.

This brief description of the elementary diffusion process shows that diffusion may occur in alloys only if the diffused element forms a solid solution with the matrix metal. The principal reason for diffusion in solid solutions, and particularly in chemical heat treatment, is the tendency to homogenise the concentration throughout the metal.

The rate of diffusion of atoms entering the iron crystal lattice will vary in accordance with the type of solid solution that is formed.

When iron is saturated by carbon or nitrogen which form interstitial solid solutions, diffusion will proceed much more easily than in saturation by metals that form substitutional solid solutions.

In interstitial solid solutions, the carbon and nitrogen atoms move about easily between interstitial positions; they jump easily from one interatomic position to another.

Diffusion in substitutional solid solutions requires that a part of the atoms leave the lattice points (they jump into interstitial positions) and that the resulting vacancies be occupied by the diffusing atoms. Naturally, such movement is much slower.

Due to the low rate of diffusion when iron (or steel) is saturated by metals which form substitutional solid solutions, such a process must be carried out at high temperatures and long holding times to obtain a diffused layer of sufficient depth.

Any case-hardening operation comprises three elementary processes: 1) processes that take place in the external medium and result in the liberation of the diffusing element in an atomic (ionic) state, for example, dissociation of ammonia with the liberation of atomic hydrogen ($2NH_3 \rightarrow 2N+6H$) or the decomposition of carbon monoxide to produce atomic carbon ($2CO \rightarrow CO_2+C$); 2) contact of the diffusing atoms with the surface being treated and the forming of chemical bonds with the atoms of the matrix metal (adsorption); and 3) penetration of the saturating element into the matrix metal, i.e., diffusion.

The concentration of the diffused element on the surface depends upon the potential of the surrounding medium, which ensures an ample supply of the atoms of this element to the surface, and upon the rate of diffusion which enables these atoms to penetrate into the metal.

The depth of penetration is determined by the rate which, for a given definite case, depends chiefly upon the temperature of the process and the concentration of the diffused element at the surface.

The total depth of the diffused layer (y) varies with the time (τ) the process continues, at a given temperature, in a parabolic relationship:

$$y=k\sqrt{\tau}$$

This equation shows that in the course of time, the rate, at which the layer increases, is reduced. This is associated with the reduction in the concentration gradient between any adjacent zones as time elapses. Concentration gradient is one of the main factors that determine the rate of diffusion.

The higher the concentration of the diffused element on the surface of the metal, the deeper the diffused layer may be, all other conditions being equal.

Higher temperatures naturally increase the rate of diffusion. The higher the temperature, the deeper the diffused layer obtained in a given interval of time.

The amount of substance (in g) diffusing in unit time (second) through unit area (sq cm) at a unit concentration gradient is called the diffusion coefficient (D, cm^2/sec).

The variation of the diffusion coefficient with temperature is expressed by the equation

$$D = D_0 e^{-\frac{Q}{RT}}$$

in which: D_0 is the temperature-independent multiplicator in cm^2/sec;
Q is the heat of diffusion (activation energy, i.e., the energy required for the atom to jump to a new position in the lattice) in cal/gramme-atom;
R is the gas constant (1.98);
T is the absolute temperature in degrees Kelvin.

The values of the constants for this equation for several specific substances are given in Table 15.

Table 15

Values of the Constants in the Diffusion Coefficient Equation

Matrix metal	Diffusing element	D_0, cm^2/sec	Q, calories per gramme-atom
γ-iron	Carbon	0.04+0.08 (per cent C*)	31,400±800
	Nitrogen	0.335×10^{-2}	34,660
	Aluminium	—	44,000
	Chromium	—	80,000
α-iron	Carbon	0.02	20,000
	Nitrogen	4.67×10^{-4}	17,950

* Per cent C is the carbon content of the austenite.

The greater the value of Q, the smaller the value of D at a given temperature, the slower the rate of diffusion and the higher the temperature and the longer the time of a chemical heat treatment must be to obtain a given depth of the diffused layer.

The nature of primary formations and the structure of the diffused layer may be described by an equilibrium diagram of the system, "diffusing element-solvent metal (metal being treated)".

Let us suppose that component A is to be saturated with component B. These two metals have the equilibrium diagram shown in Fig. 176. At the beginning of diffusion, at temperature t_1, the atoms of component B penetrate the lattice of component A and form the solid solution α (Fig. 176). As the atoms of component B continue to enter from the external medium, the depth of the α solution increases and the component B concentration gradually decreases in a direction away from the surface and into the metal.

When a certain time has elapsed, the component B concentration at the surface of component A reaches its maximum value for the given temperature (point a).

As the maximum solubility for the given temperature is reached, conditions arise for the nucleation of compound γ at the surface. The appearance of γ-phase nuclei is associated with concentration fluctuations in the saturated α solution. The growth of γ-phase nuclei

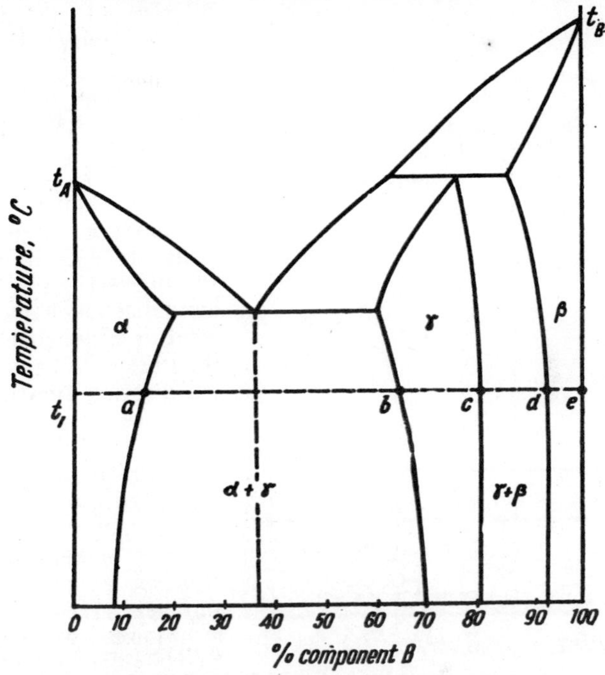

Fig. 176. Equilibrium diagram of the system *A-B*

along the surface is much more rapid than into the depth of the specimen and soon the surface is covered with a thin continuous layer of the γ-phase.

At the interface between the α- and γ-phases, the concentration of component *B* in the α solution corresponds to the maximum saturation value (point *a*) at all times. This provides for a continuous transformation of α into γ and growth of the γ-phase layer. The atoms of metal *B* penetrate into the α solution by passing through the γ-phase layer.

The *B* metal concentration in the γ-phase, after it is formed, is near to point *b* (Fig. 176). In the course of time, the *B* concentration in the γ-phase increases. When it reaches maximum saturation (point *c*) crystals of the β-phase nucleate on the surface. This phase soon forms a continuous layer over the surface of the specimen. As diffusion proceeds, the β-phase layer gradually moves inward, into the specimen.

The structure of the diffusion layer and the concentration gradient from the surface into the metal are shown schematically in Fig. 177 for the moment when the concentration of the solid solution at the surface reaches point *e* (Fig. 176).

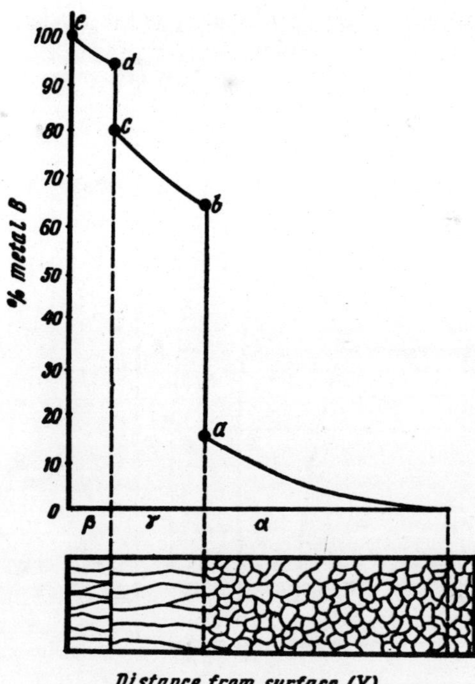

Fig. 177. Structure and concentration gradient of component *B* along the depth of the diffusion layer, for the system of Fig. 176

Fig. 177 shows that a sharp change in concentration is established at the interface between phases β and γ, as well as between γ and α.

Boundary concentrations are determined by the intersections of the isohermal, at which the treatment is conducted, with the lines of the equilibrium diagram limiting the existence of the various phases (Fig. 176). The concentration varies gradually within the limits of each phase.

Nuclei of the new γ- and β-phases appear on the surface. These nuclei grow inward in the direction of diffusion and form characteristic columnar crystals.

The columnar character of the diffusion layer always indicates that phase recrystallisation occurs in the process of saturation.

The mechanism here described for the formation of the diffusion layer in the chemical heat treatment of iron shows that no phase mixtures, for example $\alpha+\gamma$ or $\beta+\gamma$ may exist at the diffusion temperature. Phase mixtures form during subsequent cooling as the result of the decomposition of solid solutions.*

* In the treatment of alloy steels, two-phase states may be formed directly at the temperature of the diffusion process.

11-2. Carburising of Steel

Carburisation is the process of saturating the surface layer of steel with carbon. In accordance with the carbon source, carburisation is classified as: 1) pack carburising with solid carbonaceous mixtures (carburisers), 2) gas carburising, and 3) liquid carburising.

The main purpose of the carburising process is to obtain a hard and wear-resistant surface on machine parts by enrichment of the surface layer with carbon to a concentration from 0.75 to 1.2 per cent and subsequent quenching. Steel which has been carburised and quenched (case-hardened) has a higher fatigue limit.

Low carbon steels, containing from 0.1 to 0.18 per cent C, may be subjected to carburising. Steels with a higher carbon content (0.2 to 0.3 per cent C) may be employed for large components. Recently, there has been a tendency to use carburising steels of higher carbon content for medium and small machine components as well. Steel with an increased carbon content (0.25 to 0.35 per cent C) has a stronger core. This enables the depth of the carburised case to be reduced thus simplifying subsequent heat treatment.

In the Soviet Union, low-carbon steels, Grades 15, 20, and 25 (see Sec. 13-3) are extensively used as carburising steels. Alloy steels are used for highly stressed components, especially of large size, in applications when the core must possess high mechanical properties and the case must be highly wear resistant. Such alloy steels include chromium steels (15X and 20X), chromium-manganese-titanium steels (18XΓT and 30XΓT), chromium-nickel steels (12X2H4A and 12XH3), and others (see Sec. 13-7).

The use of alloy steels, as mentioned above, reduces distortion in quenching since oil is used instead of water as a quenching medium. This enables complex machine parts to be successfully carburised and subsequently heat treated.

Machine parts are carburised directly after machining or they may have an allowance of 0.05 to 0.1 mm for finish grinding. Selective carburising is frequently applied. Here, only certain portions of the surface of the work are carburised. The remaining surface is protected against the carburising effects by a thin layer of copper (0.03-0.04 mm) applied by electrolytic plating, or by coating with special pastes. Sometimes, an allowance is provided, and these already-carburised sections of the surface are machined off before hardening.

An anti-carburising paste called linite (talc and fine ground white clay mixed in water glass) is often employed. It must be noted that in drying and carburising, the paste may crack easily and so does not provide reliable protection. Copper plating is the most reliable procedure.

THE MECHANISM BY WHICH THE CARBURISED LAYER IS FORMED

Carbon may diffuse into steel only if it is in the atomic state, i.e., as elemental carbon, for example, as the result of dissociation of carbon-containing gases (CO, CH_4, etc.).

Carburisation comprises three consecutive stages: 1) dissociation of the carbonaceous gases with the evolution of atomic carbon:

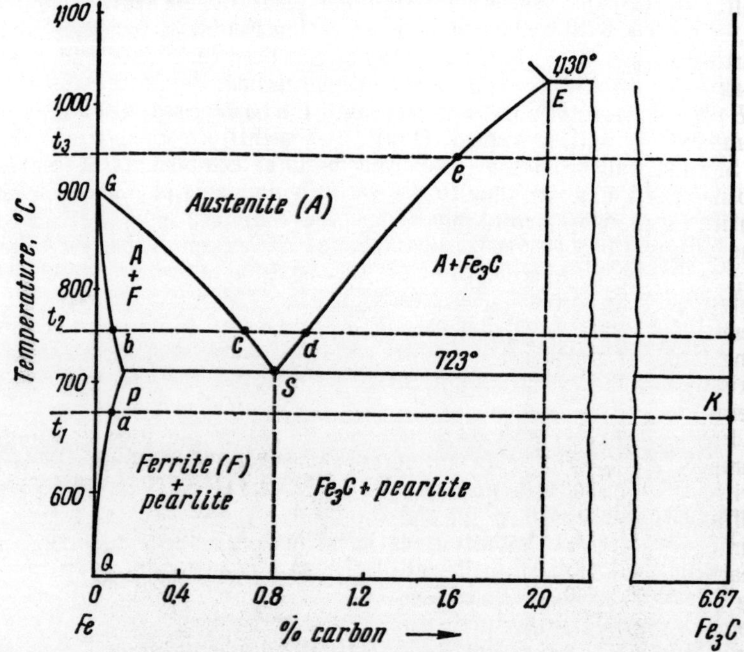

Fig. 178. A section of the Fe-C equilibrium diagram

$2CO \rightarrow CO_2 + C_{at}$ or $CH_4 \rightarrow 2H_2 + C_{at}$; 2) enrichment of the surface layer with carbon (the degree of saturation in this layer depends upon the carburising temperature and the composition of the carburiser); and 3) diffusion of carbon, adsorbed by the surface, deep into the metal. The diffusion rate depends on the carburising temperature. The higher the temperature, the more rapidly carbon will be diffused.

The structure of the carburised case may be described with the aid of the iron-carbon equilibrium diagram.

If carburisation is carried out at temperature t_1 (Fig. 178), which is below the eutectoid temperature (point A_1), then a solid solution of carbon in α-iron is formed first. When the solid solution is saturated with carbon, as determined by the line PQ on the equilibrium diagram (point *a* in our case) cementite begins to form on the surface. Thus, in this case, we obtain a layer of only cementite at the surface of the steel in the diffusion zone. Under the cementite will be the solid solution of carbon in α-iron. Upon subsequent slow cooling, excess carbon will separate from the α solution as tertiary cementite.

Saturation with carbon at the temperature t_2, between critical points A_1 and A_3, will proceed in the following manner. First carbon diffuses in the α-iron. When maximum saturation is reached (point *b*), a solid solution of carbon in γ-iron (austenite) is formed. Upon further supply of carbon from the external medium and when the maximum saturation of the austenite is reached, determined by line *SE* on the diagram (point *d* in our case), the next phase, stable at this temperature, will be formed. It will be cementite.

Slow cooling from the carburising temperature will cause decomposition of the austenite with the precipitation of either surplus cementite or ferrite, depending on the composition. At critical point A_1, pearlite is formed by the eutectoid transformation. Excess cementite is precipitated in cooling from the alpha solid solution existing at the carburising temperature.

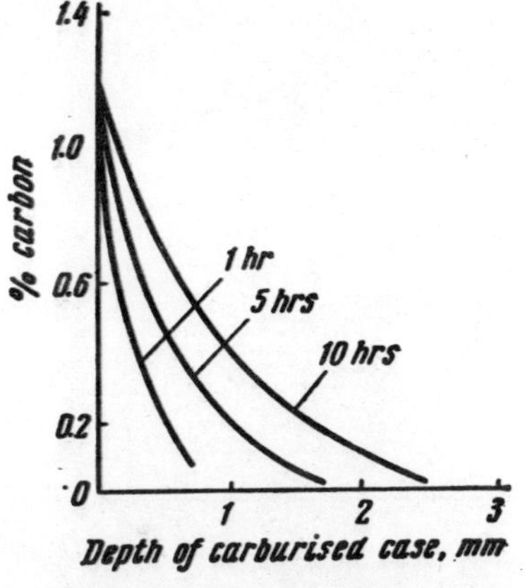

Fig. 179. Variation in carbon content along the depth of the carburised case; carburising temperature 930°C

Finally, in carburising at the temperature t_3 (Fig. 178), above critical point A_3, carbon will first diffuse in γ-iron. When the maximum saturation (point *e*) is reached, a layer of cementite will be formed. The austenite will decompose in accordance with the equilibrium diagram upon cooling from the carburising temperature.

In practice, steel is carburised in the range from 900° to 950°C, where carbon is diffused in γ-iron.

The above discussion on the formation of the carburised layer referred to extreme cases. Under actual conditions, the formation of a continuous cementite layer on the surface is rarely observed. To obtain a continuous layer of cementite in the diffusion zone, steel must be carburised for a long time in a sufficiently active carburiser containing a considerable amount of CO or CH_4 in the gaseous phase.

Carburising normally takes several hours. More time (10 or 20 hours) is seldom required. The carbon concentration in the surface layer does not usually exceed 0.8 to 1.2 per cent for a holding time of several hours. This means that only austenite will be present at

Fig. 180. Structure of steel after carburising at 920° C, × 200

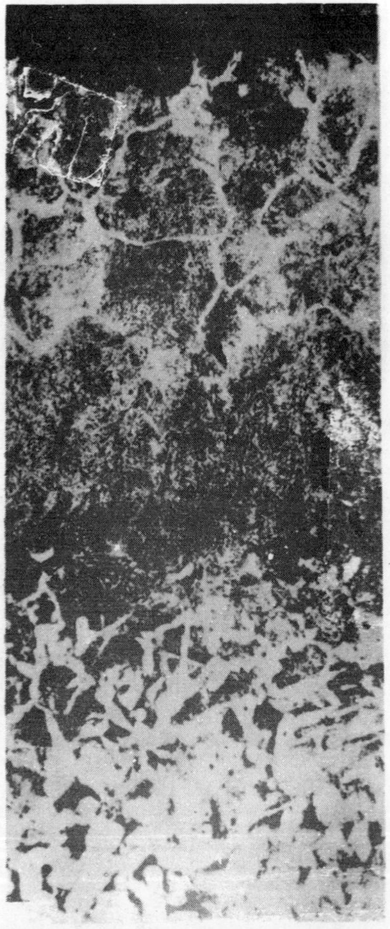

Fig. 181. Coarse cementite network in the structure of the carburised case, × 200

temperatures above Ac_3 and its products of decomposition will be obtained after cooling.

The carburised case has a variable carbon concentration along its depth (Fig. 179), decreasing from the surface toward the core. Consequently, the structure of the case will change gradually in depth from one inherent to high-carbon steels to the initial structure of the low-carbon steel.

A typical structure of carburised steel is illustrated in Fig. 180. Here, three structural zones may be distinguished: 1) hypereutectoid

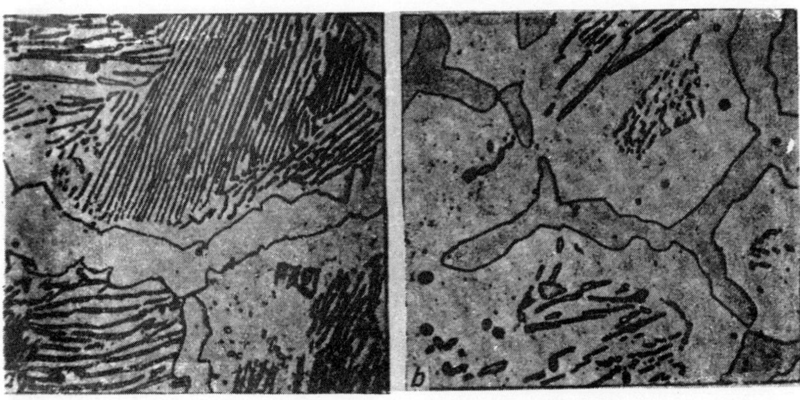

Fig. 182. Microstructure after carburising:
a—partly abnormal, *b*—abnormal steel

zone, consisting of pearlite and proeutectoid cementite; 2) eutectoid zone, consisting of pearlite only, and 3) hypoeutectoid zone consisting of pearlite and ferrite. The amount of ferrite, in the last zone, continuously increases toward the core.

Carbon content at the surface depends on the temperature of the process, the holding time, the steel composition, and the activity of the surrounding medium which supplies carbon atoms to the surface. The carbon content in the surface layer should be from 0.8 to 1.0 per cent. A higher concentration will lead to the formation of a continuous coarse carbide (cementite) network on the surface of the carburised case (Fig. 181). This network cannot be eliminated by subsequent heat treatment and it deteriorates the quality of the layer. The depth of the carburised case is usually accepted to be the sum of the hypereutectoid, eutectoid*, and one half of the transition (hypoeutectoid) zones.

* In some cases, only the extent of the hypereutectoid and eutectoid zones is taken as the depth of carburisation.

The normal structure for the hypereutectoid zone of the case is lamellar pearlite, surrounded at certain regions by a cementite network. Sometimes, however, the cementite precipitates as massive inclusions, surrounded by free ferrite (Fig. 182). This is called an abnormal structure. In the heat treatment of steels with an abnormal structure cementite passes over into the solid solution with difficulty. This results in the formation of "soft spots" on the surface of a part after quenching. This phenomenon may be eliminated by heating to higher hardening temperatures and by using a highly effective quenching medium.

PACK CARBURISING

In this process the carbonaceous medium is a so-called solid carburiser. The chief carburisers for pack carburising are activated charcoal (oak or birch) in grains from 3.5 to 10 mm in diameter, coal semi-coke, and peat coke.

Carbonates are added to the charcoal to accelerate the carburising process. They include barium carbonate ($BaCO_3$) and soda ash (Na_2CO_3) which are added in an amount from 10 to 40 per cent of the weight of the charcoal.

Table 16 indicates the compositions of carburisers that are extensively employed in Soviet industry (this table is an excerpt from GOST 2407-51).

Table 16

Compositions of Solid Carburisers

Grade	Content, per cent						
	Barium carbonate	Calcium carbonate	Sulphur	Silica	Moisture	Volatile matter	Charcoal
			maximum			minimum	
1	20 to 25	3.5	0.06	0.5	5.0	10.0	Remainder
2	20 to 25	5.0	0.1	1.5	5.0	10.0	Remainder

A working mixture consists of 25 to 35 per cent fresh carburiser and 65 to 75 per cent used material.

A semi-coke carburiser has been used to some extent recently. Its specifications are stipulated by GOST 5535-50 and it consists of granules of activated coal semi-coke coated with a film of barium carbonate. The carburiser contains 10 to 15 per cent $BaCO_3$, up to 3.5 per cent $CaCO_3$, not over 0.35 per cent S and not over 6 per cent moisture. Limestone ($CaCO_3$) is added to prevent caking. Sometimes mazout or oil is added to improve the bond between the coal and the carbonates.

Workpieces, which are to be pack carburised, are first cleaned of dirt, scale, and rust. They are then placed in a box. Welded steel boxes are employed, as a rule; cast steel or cast iron boxes are more seldom used. Welded boxes of cast heat-resistant alloys, for example of nichrome, are frequently used in large mass-production plants. Their service life (3,000 to 6,000 hrs) considerably exceeds that of steel boxes (250 to 300 hrs). The boxes are usually rectangular or cylindrical in form. It is advisable to employ cylindrical boxes or pots (better with an internal pipe) or boxes, coinciding in form to a certain extent with the work, to reduce the carburising time. Packing the work is accomplished by first covering the bottom of the box with a 40 to 45 mm layer of carburiser (Fig. 183). The workpieces to be carburised are placed on this layer with spaces of 20 to 25 mm between them and the box walls. Then they are covered with a dense layer of carburiser, 20 to 25 mm thick, which is rammed before laying the next row of workpieces. The upper row is covered with a layer of carburiser 40 to 50 mm thick. The box is closed with a cover whose edges are luted with fire clay or with a mixture of clay and talc with water glass. The packed box is placed in the furnace.

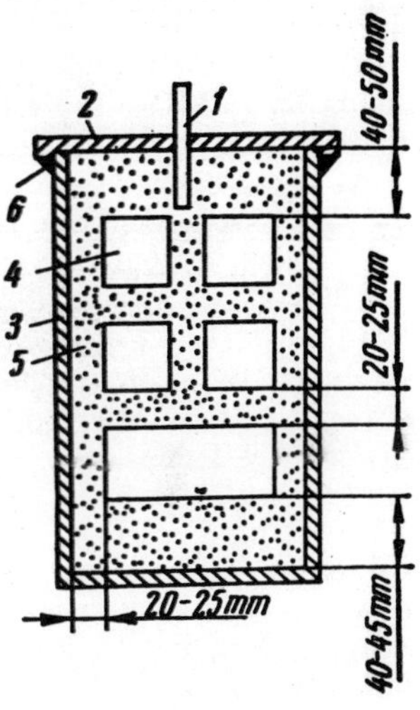

Fig. 183. Packing workpieces in a box for carburising:
1—test bar, 2—cover, 3—box, 4—workpieces, 5—carburiser, 6—lute

The carburising temperature ranges from 900° to 930°C. The total time required depends on the specified depth of the carburised case. The deeper the case required, the longer the holding time should be at the carburising temperature.

The required holding times for various depths of the carburised case are given in Table 17.

Table 17

Relationship Between Case Depth and Holding Time (Carburising temperature 930° C)

Case depth, mm	Holding time, hrs	Case depth, mm	Holding time, hrs
0.4 to 0.7	4.5 to 5.5	1.4 to 1.8	11.5 to 16
0:6 to 0.9	5.5 to 6.5	1.5 to 1.9	13 to 18
0.8 to 1.2	6.2 to 10.0	1.6 to 2.0	14 to 19
1.0 to 1.4	8.0 to 11.5	1.8 to 2.2	16 to 22

Frequently, the holding time is assigned on the basis of 0.1-0.12 mm per hour for cases up to 1.5 mm deep.

The majority of small machine parts require a carburised case from 0.8 to 1.5 mm deep.

Raising the carburising temperature to 950-980°C and using more active carburisers enable the carburising process to be substantially accelerated without lowering the quality. This procedure is feasible, however, only if inherently fine-grained steels are used and if carburising is followed by reheating for hardening. The whole process, in this case, must be carefully controlled.

As noted before, steel is saturated in carburising by atomic carbon. In pack carburising, the atomic carbon is liberated by the decomposition of carbon monoxide. This probably occurs in the following manner.

There is always air present in the carburising box, even when it is filled with carburiser. At a high temperature, the oxygen in the air reacts with the carbon in the carburiser to produce carbon monoxide. In the presence of iron, the carbon monoxide dissociates according to the equation

$$2CO \rightarrow CO_2 + C_{at}$$

The atomic carbon, evolved in this reaction (at the moment of nascence), diffuses into the steel. The addition of carbonates to the charcoal greatly activates the carburiser by enriching the atmosphere in the box with carbon monoxide. This is accomplished by a reaction, for example, of the following type:

$$BaCO_3 \rightarrow BaO + CO_2;$$
$$CO_2 + C \rightarrow 2CO$$

GAS CARBURISING

Gas carburising has many advantages over pack carburising and is applied in mass-production plants. These advantages are:

1) the possibility of better regulation of the process and of obtaining more accurate case depth since the composition of the gaseous phase may be controlled,

2) less time is required since there is no need to heat boxes filled with low heat-conductive carburiser,

3) only about one half as much floor space is required for the process since such operations are eliminated as preparation of the carburiser, packing the workpieces in the boxes, cooling and unpacking the boxes, and others,

4) many laborious, time-consuming operations, injurious to health, are excluded,

5) the operation may be feasibly mechanised; carburising may be combined with subsequent hardening.

The gas carburising process is considerably simpler than pack carburising. It is accomplished by heating the work in a gaseous medium containing carbon. Such include both natural and producer gases (town gas, gas obtained in the pyrolysis or cracking of kerosene, solar oil, benzol, and others).

Natural gases (for example, Saratov gas in the U.S.S.R.) are almost pure methane (CH_4) while producer gases consist of a mixture of saturated hydrocarbons (C_nH_{2n+2}), chiefly CH_4, and unsaturated hydrocarbons (C_nH_{2n}) with CO, CO_2, H_2, and N_2. These gases are used either in the pure state or mixed with diluent gases.

Natural gases are widely used in the U.S.S.R. for gas carburising, as well as the gases obtained from benzol, kerosene, and other liquid petroleum products supplied to the carburising furnace.

Recently Soviet plants have been using syntin for gas carburising. Syntin is a mixture of hydrocarbons, obtained in a catalytic reaction by synthesis from carbon monoxide, hydrogen, and triethanolamine $(C_2H_5O)_3N$, which saturates steel with both carbon and nitrogen.*

The part to be carburised is introduced by means of a special fixture into a preheated retort in a furnace. Then gas is admitted to the retort. The carburising temperature is usually between 930° and 950°C. The holding time is selected to suit the specified depth of the case (Fig. 184).

There is no difference in principle between gas and pack carburising since the processes are carried out by gases in both cases. The chief reaction which supplies active carbon atoms to the surface in pack carburising is:

$$2CO \rightarrow C_{at} + CO_2;$$

in gas carburising it is the dissociation of methane (CH_4):

$$CH_4 \rightarrow C_{at} + 2H_2$$

The latter is frequently accompanied by the decomposition of unsaturated hydrocarbons (ethylene C_2H_4, propylene C_3H_6, and others), present in the gas, which produce a great deal of carbon.

The unsaturated hydrocarbons are decomposed by the reaction

$$C_nH_{2n} \rightarrow C + nH_2$$

In decomposition, unsaturated hydrocarbons may form intermediate, more stable compounds and coking pitches. Excess carbon and coking pitch cover the surface of the work with soot and coke which impede carburisation. Therefore, a large amount of unsaturated hydrocarbons is undesirable in the carburising gas as they will affect the uniformity of carbon concentration in the carburised case. Gas carburising is accomplished in either batch type or continuous muffle furnaces, as well as in retort and shaft furnaces, designed so as to operate both on gas produced by special units and on gas obtained directly in the working chamber of the furnace by vaporising syntin, kerosene, etc.

* The use of syntin as a carburiser was proposed by A. T. Kalinin, M. N. Kunyavsky and A. Y. Zaitseva. A special feature of syntin and triethanolamine is that they produce very little soot and coke in carburising.

The high thermal stability and evaporativity of kerosene and other liquid hydrocarbons enable the generation of the carburising gas and the carburising process itself to be combined within the furnace. To increase the vaporising surface, liquid hydrocarbons are atomised by fans mounted in shaft furnaces and by high-pressure nozzles, arranged at the loading end of the retort, in continuous furnaces.

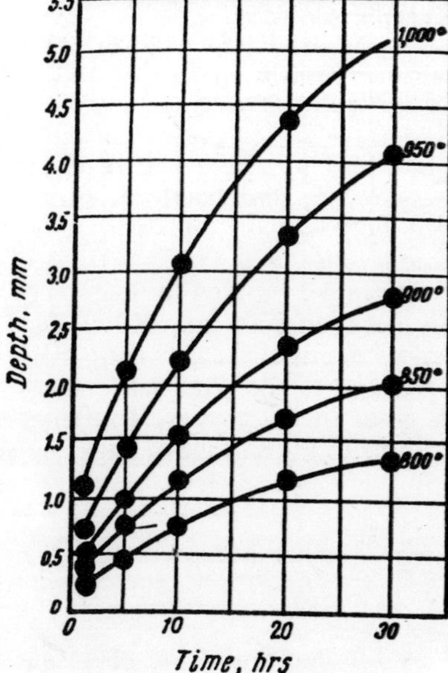

Fig. 184. Effect of carburising temperature on case depth obtained

Direct (drop) feed of liquid hydrocarbons into a muffle furnace is poor practice since much soot will be produced, it will be difficult to obtain large volumes of gas of uniform composition, stable results cannot be obtained in carburising, and the furnace output will be reduced.

It is advisable to dilute the carburising gas with ammonia when muffle furnaces, operating on liquid hydrocarbons, are used. This will prevent the formation of a large amount of soot while the volume of gas is increased. Ammonia somewhat reduces the time required for carburising.

Special controlled endothermic atmospheres (endogas)*, whose carburising potential is regulated to the dew point of the atmosphere, are used to obtain carburised layers of specified carbon content and to retain a clean (bright) surface on the treated workpieces.

Endogas is obtained from natural gas, butane, propane, or some other gaseous hydrocarbon by partial combustion in a special endothermic generator in the presence of a special catalyst at a temperature of 950-1,000°C. The air-to-gas ratio is such that no CO_2 and H_2O are formed. Endogas should have the following composition: 20 per cent CO, 40 per cent H_2 and 40 per cent N_2.

Endogas has a low carburising potential. A certain amount of untreated gas (natural gas, propane, butane, etc.) is added to the endogas at the beginning of the process to obtain carburisation.

The amount of untreated gas introduced into the furnace, in proportion to the endogas, should be: 8 to 10 per cent natural gas or 3 to

* Endothermic atmosphere, obtained under conditions when heat is absorbed in the reaction.

5 per cent propane or 1 to 3 per cent butane, in accordance with the extent of the surface on the work, case depth, and other factors.

At the end of the process, no more untreated gas is supplied to the furnace and the work is held in the endothermic atmosphere (diffusion period). The carbon potential of the atmosphere (regulated to the dew point) provides for equilibrium conditions with the specified carbon concentration at the surface of the carburised case.

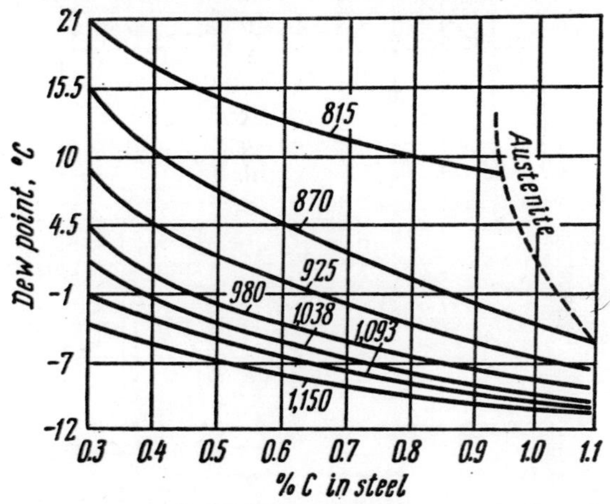

Fig. 185. Experimental curves showing equilibrium conditions between steel and endogas having various dew points

Due to diffusion of carbon into the metal and the interaction of the surface of the work with the diluent gas, the carbon concentration is reduced to the specified value in this period.

The dew point of the endogas is controlled by special instruments. Experimental equilibrium curves, for various steels in atmospheres having dew points from plus 20-21°C to minus 12-18°C (Fig. 185), have been plotted to facilitate the practical application of gas composition control by the dew point.

The required dew point is obtained by varying the proportions of gas and air entering the generator.

HIGH-SPEED GAS CARBURISING WITH INDUCTION HEATING

Carburising may be sharply accelerated by raising the temperature of the process. A wide application of high-temperature carburising of inherently fine-grained steels is impeded, however, by the fact that the service life of the furnace components is substantially reduced in this procedure. This disadvantage has been eliminated by using high frequency current for heating.

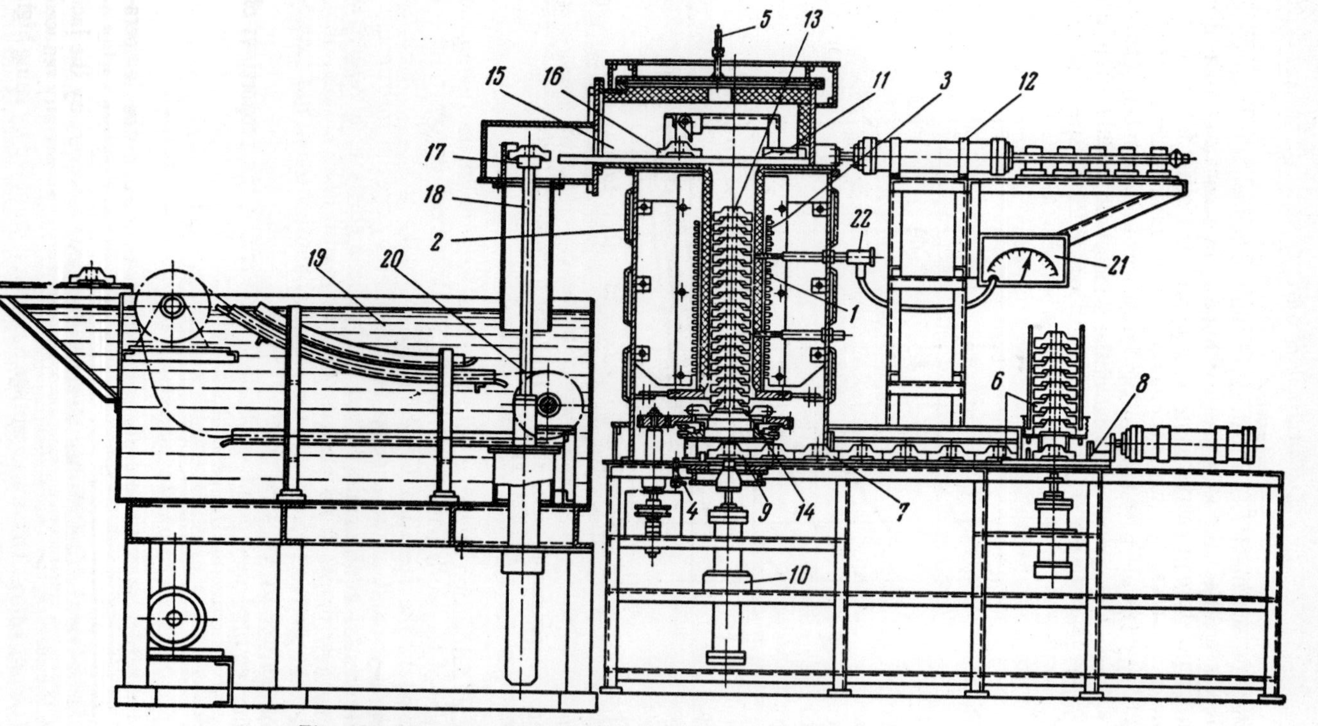

Fig. 186. Unit for high-speed gas carburising with induction heating

Fig. 186 schematically illustrates a unit developed by a group of engineers at the Likhachov Auto Plant in Moscow* for high-speed high-temperature gas carburising with induction heating. The heating inductor *1* is arranged inside the housing *2*. A ceramic sleeve *3* is provided to prevent overheating of the inductor and to align the workpieces with it. The carburising gas enters the furnace through pipe *4*; it passes through the working channel of the furnace and escapes through pipe *5* forming a flaming torch. The gear to be carburised *6* is fed into the furnace through the channel *7* by the pusher *8*. The last gear *9* is positioned centrally with the inductor and the hydraulic elevating cylinder *10*. At this time, the upper gear *13* on the stack is removed by the ejector *11*, operated from the cylinder *12*. After removing gear *13*, the stack of gears is lowered by cylinder *10* and the lowest gear rests on the hinged levers *14*. After the stack is lowered, ejector *11* continues to advance, pushing the removed gear into the precooling chamber *15*. When the ejector reaches its extreme forward position, gear *16* is pushed on to the plate *17*. Here, the gear is lowered into the quenching tank *19* by the action of the cylinder and piston rod *18*. Conveyer *20* removes the finished (hardened) gears from the tank. The temperature is checked by the thermocouple *21-22*.

The above-described process is fully automatic and the cycle is repeated every 1.5 to 3 minutes. The unit has an output of 20 to 40 hardened gears per hour.

The carburising temperature is taken in the range from 1,080° to 1,100° C. This provides for a case 0.8 to 1.0 mm deep in from 45 to 50 min of holding time. Provision is made for quenching the gears after precooling to 870° C. It must be noted that due to grain growth in carburising at high temperatures, better results are obtained if the gears are cooled to room temperature and then reheated to 850-870° C for hardening. Phase recrystallisation, occurring in heating, will refine the grain.

This process presents wide opportunities for arranging case hardening directly in the general production line of a workpiece as required by the sequence of operations.

A. Moshchinsky and G. Matiya proposed a carburising process in which the work is induction heated in a liquid medium. The work is to be immersed in a bath of liquid hydrocarbons or alcohols. At a carburising temperature of 1,000° C, a case from 0.5 to 0.7 mm deep may be obtained in from 10 to 20 minutes.

A number of investigators report that carburisation can be substantially accelerated if the work is subject to ultrasonic treatment.

---

* Inventors, K. Z. Shepelyakovsky, A. D. Assonov, and P. A. Lankin, see *Physical Metallurgy and Metal Treatment*, No. 3, 1955.

PRESSURE GAS CARBURISING *

Many experiments have shown that gas carburising at a high pressure (4 to 8 atm) will increase the carbon concentration at the surface without altering the depth of the diffusion layer.

The advantage of this procedure, in comparison with ordinary carburisation, is that well-developed hypereutectoid and eutectoid layers can be obtained in short holding times without appreciably increasing the total depth of the diffusion layer.

Several manufacturers in the U.S.A. offer shaft furnaces with provision for varying the furnace pressure. This enables the degree of saturation of the case with carbon to be regulated.

PASTE CARBURISING

It is sometimes expedient, in small-scale production, to carburise portions of a workpiece with special pastes applied at the surface areas which are to be saturated with carbon. After coating with the paste, the parts are placed in a box and heated to the carburising temperature (920-930° C). These pastes consist of low-sulphur coke or lamp black, carbonates, potassium ferrocyanide, and the salts of certain organic acids. Molasses, water glass, or glue is used as a binder. The compositions of various pastes of this type are given in Table 18.

Table 18

Compositions of Carburising Pastes (per cent)

Constituents	Type of paste			
	No. 1	No. 2	No. 3	No. 4
Dutch lamp black or coke	30-60	30-60	35	45
$BaCO_3$	—	—	15	20
Na_2CO_3	20-40	20-40	20	20
$K_4Fe(CN)_6$	5-10	5-10	15	15
Cyanplav ГИПХ*	5-10	5-10	—	—
Ferrochrome	—	—	15	—
Sodium oxalate	—	5-10	—	—
Cobalt oxalate	—	5-10	—	—

* Cyanplav ГИПХ consists of 43 per cent $Ca(CN)_2$, 3 per cent $CaCN_2$, 30-35 per cent NaCl, 14-16 per cent CaO, and 4-5 per cent C.

Paste carburising is much faster than pack carburising and, therefore, grain growth will be less. Quenching is accomplished, after carburising, directly from the box.

* This method was first proposed by a worker-inventor of the Moscow Ball Bearing Plant, P. A. Piterkin, in 1929.

LIQUID CARBURISING

This carburising process is performed in baths of molten salts containing 75 to 85 per cent sodium carbonate, 10 to 15 per cent sodium chloride, and 6 to 10 per cent silicon carbide (carborundum). The carburising effect in this bath is due to carbon monoxide evolved according to the formula

$$SiC + 2Na_2CO_3 = \underbrace{Na_2SiO_3 + Na_2O}_{\text{slag}} + 2CO + C$$

Sodium silicate and sodium oxide, produced by this reaction, rise to the surface of the bath and form a slag layer.

The relation between the depth of the diffusion layer and the holding time at 850° C in a liquid carburising bath is characterised by the following data:

Holding time in carburising, hrs	0.5	1.0	1.5	2.0	3.0
Depth of diffusion layer, mm	0.2	0.25	0.35	0.4	0.5

The principal advantages of salt bath carburising are the uniform heating, the possibility for direct quenching from the bath, and the small deformation of the work.*

The addition of 3 to 5 per cent ammonium chloride (NH_4Cl) enables the salt bath to be employed for saturating the work simultaneously with both carbon and nitrogen and accelerates the process to a certain extent.

HEAT TREATMENT AFTER CARBURISING AND PROPERTIES OF CARBURISED PARTS

Parts of carbon steel, requiring high mechanical properties, are usually subject to double-hardening, followed by tempering, after the carburising process. The following considerations make double-hardening a necessary procedure in these cases.

After carburising, the work may be considered to consist of two layers: an outer layer, which is a eutectoid or hypereutectoid steel and requires a hardening temperature of 750-780° C (above point Ac_1), and a core which is a hypoeutectoid steel of the initial composition containing from 0.1 to 0.25 per cent C and requiring grain refinement by heating to temperatures above point Ac_3 (for the core), i.e., 880-900° C. Thus, to improve the structure of the core and to impart optimum properties to the surface layers, a single heating to one temperature will evidently be insufficient. Therefore, the first hardening or normalising is conducted at a temperature of 880-900° C

* Due to the considerable loss in weight of the work immersed in a medium with a high specific weight (salt), deformation of the work as a result of its own weight is sharply reduced.

to improve the core structure of the work which has been overheated in carburisation. This heating will also dissolve the cementite network which will not reappear during quenching. The second hardening operation is conducted at 750-780° C to eliminate the effects of overheating and to impart a high hardness to the carburised case.

Heat treatment is completed by tempering at 150° to 180° C.

The chief disadvantages of this heat treatment are the complexity of the procedure, excessive warping of the work, and the possibility of oxidation and decarburisation. For these reasons, single hardening at 820° to 850° C is extensively used at present for inherently fine-grained carbon steels and alloy steels. Such a procedure will refine the grain in both the core and the case and, at the same time, warping and decarburisation will be considerably reduced.

In gas carburising, quenching direct from the furnace, after precooling to 840-860° C, is frequently practised.

The surface layer, after heat treatment, will have a martensite structure with excess carbides (Fig. 187).

The structure of the core depends upon the accepted heat treatment and the steel composition. If a complex heat treatment is applied, the low-carbon core will not harden completely after the second quenching from 750-780° C. It will consist of ferrite, which imparts toughness to the core and a small amount of martensite (for alloy steels) or a mixture of ferrite and pearlite-sorbite (for carbon steels).

In cases of single hardening or quenching direct from the carburising furnace, the heating temperature of the core will be in the vicinity of A_3. In low-carbon steels, austenite cannot be supercooled to the martensite point in quenching from this temperature and the core will have a ferrite and sorbite structure. After heating austenite in alloy steels to a temperature close to point Ac_3, it can easily be supercooled, without decomposition, to the martensite point. The core, in this case, will acquire a low-carbon martensite structure possessing high strength and ample toughness. Ferrite is undesirable in the core as it will considerably reduce the strength, ductility, and toughness.

The normal hardness at the surface of the carburised case should be within R_c 59-63.

When alloy steels are subject to single hardening, a large amount of retained austenite will be found in the carburised case. This sharply reduces the hardness. Retained austenite may be eliminated by a sub-zero treatment which transforms the greater part into martensite, thereby substantially increasing the hardness.

Retained austenite is frequently decomposed after carburising by high-temperature tempering at 630-640° C followed by a second hardening at a lower temperature, and low tempering.

Case-hardened steel has high wear resistance in general but the best results, in this respect, are obtained when the case consists of fine crystalline martensite with small included cementite grains. Retained austenite is objectionable in the surface of the case since pitting occurs at the austenite inclusions and the wear resistance of the case will be reduced.

Fig. 187. Structure of the carburised case of steel 12XH3A after hardening: *a*—normal structure, × 250, *b*—with an increased amount of retained austenite, × 500

In addition to increasing the hardness and wear resistance of steel, carburising, followed by hardening and tempering, substantially increases the resistance to alternating loads (by from 30 to 100 per cent). This is due to the appearance of compressive residual stresses in the carburised layer.

11-3. Nitriding of Steel

Nitriding is the process of saturating the surface of steel with nitrogen by holding it for prolonged period at a temperature from 480° to 650° C in an atmosphere at ammonia (NH_3). Nitriding increases the hardness of the surface to a very high degree (high hardness is retained even when the steel is subsequently heated to 600-650° C). It also increases the wear resistance, fatigue limit, and resistance to corrosion in such media as the atmosphere, water, steam, etc.

THE MECHANISM BY WHICH THE NITRIDED CASE IS FORMED

The iron-nitrogen equilibrium diagram is illustrated in Fig. 188. This diagram shows that the following phases are formed in the iron-

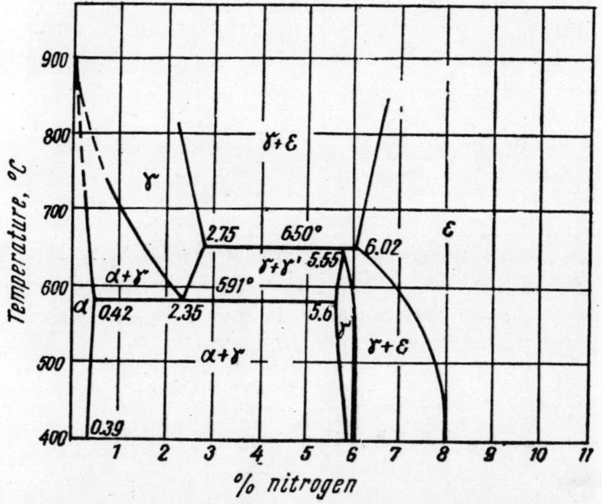

Fig. 188. The Fe-N equilibrium diagram

nitrogen system: 1) solid solution of nitrogen in α-iron (α-phase), at the eutectoid temperature (591° C), the nitrogen concentration in the α-phase will be 0.42 per cent and it will be reduced to 0.015 per cent at room temperature; 2) γ′-phase which is a solid solution on the basis of iron nitride Fe_4N (5.5 to 5.95 per cent N); 3) ε-phase, a solid solution on the basis of iron nitride Fe_2N (8 to 11.2 per cent N); the maximum nitrogen content in the ε-phase is the same as for iron nitride Fe_2N.

The γ-phase exists at temperatures above 591° C. It is a solid solution of nitrogen in γ-iron. At 591° C, the γ-phase undergoes eutectoid decomposition. The nitrogenous eutectoid is similar to pearlite. It contains 2.35 per cent N and consists of a mixture of the α- and γ-phases.

Nitriding is usually applied to alloy steels containing aluminium, chromium, molybdenum, and other elements. Nitrogen forms nitrides of the Cr_2N, Mo_2N and VN types with these alloying elements.

Nitrogen dissolves in both α- and γ-iron. An appreciable amount of nitrogen can be adsorbed by iron, however, only if the nitrogen is in the atomic state, obtained as the result of the decomposition of ammonia gas in which nitriding is conducted.

At the nitriding temperature, ammonia gas dissociates as follows:

$$2NH_3 \rightarrow 2N + 3H_2$$

The atomic nitrogen thus formed diffuses into the iron.

If nitriding is conducted at a temperature below the eutectoid point (591° C), the α-phase (solid solution of nitrogen in α-iron) will be formed on the surface at the initial moment of saturation. When maximum saturation of the α-phase for the given temperature is achieved, the next phase (γ'), stable at the given temperature, will begin to form. Upon further saturation with nitrogen, the ε-phase (nitride Fe_2N) will be formed.

As a result of diffusion at a temperature below 591° C, we will obtain a nitrided layer at the saturation temperature. The ε-phase, richest in nitrogen, will be at the surface. It will be succeeded by the γ'-phase and, finally, the α-phase which goes over to the core:

$$\varepsilon \rightarrow \gamma' \rightarrow \alpha \rightarrow \text{core}$$

When the temperature is lowered, the ε- and α-phases decompose and the excess γ'-phase is precipitated as shown in the equilibrium diagram (Fig. 188). Therefore, at room temperature we obtain a case in which the phases are arranged in the following order (from the surface to the core):

$$\varepsilon + \gamma'_{\text{excess}} \rightarrow \gamma' \rightarrow \alpha + \gamma'_{\text{excess}} \rightarrow \text{core}$$

If nitriding is done above the eutectoid temperature, for example, at 600° C, first the α-phase is formed (see Fig. 188). When the α-phase reaches maximum saturation, the γ-phase is formed. This is saturated next and the γ'-phase is formed at the surface and, finally, the ε-phase. Thus, at the diffusion temperature, we obtain a nitrided case which consists of the following phases (from the surface to the core):

$$\varepsilon \rightarrow \gamma' \rightarrow \gamma \rightarrow \alpha \rightarrow \text{core}$$

Subsequent slow cooling decomposes the ε- and α-phases to precipitate the excess γ'-phase (Fe_4N), while the γ-phase undergoes a eutectoid transformation, decomposing into a mixture of the α- and γ'-phases. Consequently, at room temperature, the nitrided case will consist of:

$$\varepsilon + \gamma' \rightarrow \gamma' \rightarrow \underbrace{\alpha + \gamma'}_{\text{eutectoid}} \rightarrow \alpha + \gamma'_{\text{excess}} \rightarrow \text{core}$$

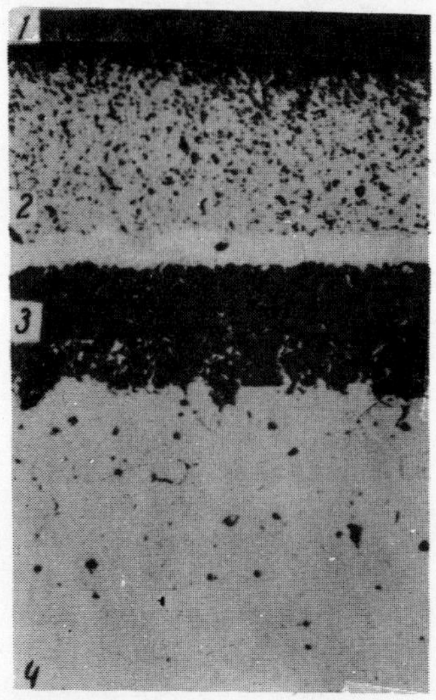

Fig. 189 illustrates the microstructure of the surface layer on iron, nitrided at 600° C (after slow cooling).

The transition from one phase to the next corresponds to a sharp change in nitrogen concentration in the diffused layer (Fig. 190). The diffusion layer is not very hard, notwithstanding the important structural changes that occur in nitriding iron.*

Nitriding is usually applied to medium-carbon alloy steels which acquire high wear resistance as a result of this treatment. This, naturally, raises the question of the effect of various alloying elements on

Fig. 189. Microstructure of the nitrided case, obtained by nitriding at 600° C, ×500: 1—$\varepsilon+\gamma'$, 2—γ', 3—$\alpha+\gamma'$ (eutectoid), 4—$\alpha+\gamma'_{excess}$ (Y. M. Lakhtin and F. I. Maslennikov)

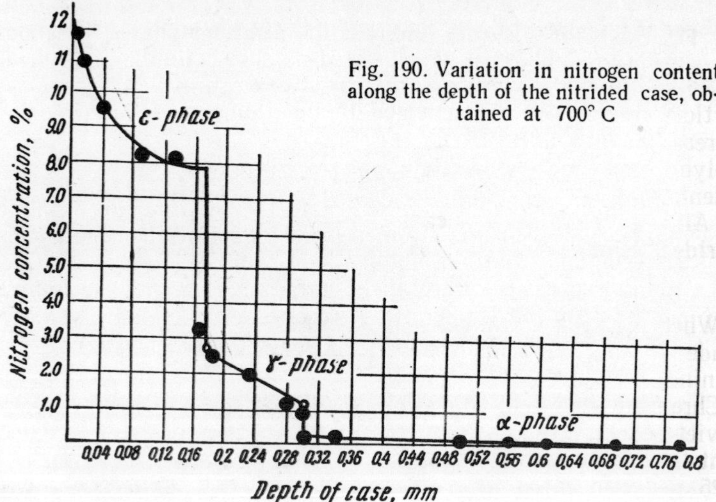

Fig. 190. Variation in nitrogen content along the depth of the nitrided case, obtained at 700° C

* Certain investigators, I. Y. Kontorovich, A. A. Savalova, Y. M. Lakhtin, and others, have shown that after heat treatment—quenching from the γ region (nitrogenous austenite), the nitride case acquires its high hardness due to the formation of nitrogenous martensite (see Fig. 58).

the structure of the nitrided case and on the nitriding procedure.

No essential changes are caused by carbon in the formation of the nitrided case. It must be pointed out, however, that in nitriding steel, the ε- and γ-phases are carbonitrides and not nitrides. They contain both carbon and nitrogen atoms at the same time. Nitrides of special elements are obtained in the nitriding of alloy steels. Highly dispersed

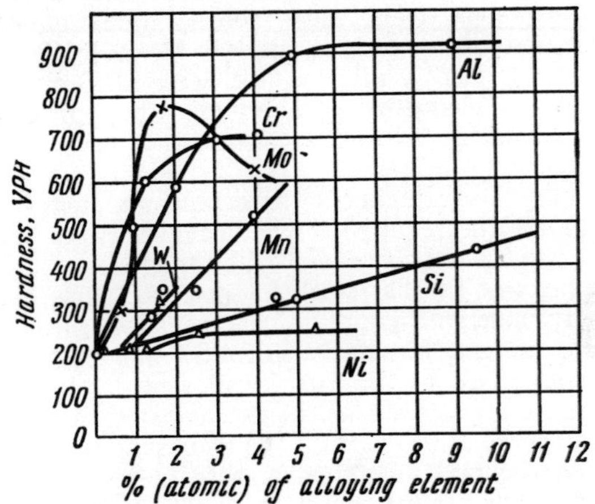

Fig. 191. Effect of alloying elements on the hardness of the nitrided case

particles of these nitrides interlock the slip planes and thus considerably increase the hardness of the nitrided layer. Aluminium, chromium, molybdenum, and vanadium increase the hardness to the greatest extent (see Fig. 191).

Alloying elements and carbon somewhat reduce the depth of the nitrided case.

NITRIDING PROCEDURES

When high hardness and wear resistance are the chief requirements made to a nitrided case, the work in question is made of a steel containing aluminium.

Chromium-molybdenum steel, Grade 38XMЮA, is employed in Soviet industry for this purpose. Its composition is: 0.35 to 0.42 per cent C, 1.35 to 1.65 per cent Cr, 0.7 to 1.10 per cent Al and 0.15 to 0.25 per cent Mo.

Aluminium, chromium and molybdenum in the steel impart an exceptionally high hardness and wear resistance to the nitrided case.

At the present time, Grade 38XMЮA is being replaced to some extent by Grade 38XBФЮA which is cheaper and does not contain the more critical molybdenum. This steel is alloyed with aluminium (of lower content, 0.6 per cent), tungsten, and vanadium (0.8 to 1.2 per cent W and 0.1 per cent V).

When nitriding is applied to increase the fatigue limit and the corrosion resistance, the parts may be manufactured of any grade of structural steel. Good results may be achieved in nitriding steels Grade 10, 20, 30, 40, etc., chromium steels (20X, 40X, etc.), chromium-nickel-tungsten steels, and more complex grades. Stainless and heat-resistant steels, such as Grade X13, 1X18H9, 4X14H14B2, and others, are frequently nitrided (see Secs. 13-10 and 13-11).

The following sequence of operations may be applied for nitriding steel Grade 38XMЮA, as well as structural steels, to increase the fatigue limit:

1. Preliminary heat treatment (hardening and tempering) is performed to impart the required mechanical properties to the core of the work, i.e., to increase its strength and toughness.

Rough blanks or the bar stock for machine parts of light cross-section usually undergo this preliminary heat treatment. The following procedure is used for steels 38XMЮA and 38XBФЮA: rough blanks are normalised at 930-970° C followed by high-temperature tempering; machine parts are heated to 900-950° C, quenched in oil or water, and tempered at 600-675° C. The tempering temperature should always exceed the maximum temperature in subsequent nitriding.

2. All required machining operations, including grinding, are done.

3. All areas, which are not to be nitrided, are protected by a thin layer of tin applied by an electrolytic method. The tin shields the treated areas by melting at the nitriding temperature when it is retained on the surface of the steel, due to its surface tension, as a thin film, impenetrable to nitrogen. The tin layer should be from 0.01 to 0.015 mm thick.

4. Nitriding.

5. Finish grinding or lapping is applied in accordance with the specified tolerances on the work.

Fig. 192 is a schematic diagram of a nitriding installation. Here ammonia is supplied from cylinder *1* through the reducer *2* to the dryer *3* and further, through gas meter *4* to the furnace *6*. From the furnace, dissociated ammonia is discharged to the absorber *11* filled with water. The water-type manometer *9* and the dissociation meter *10* are provided to control the process. The latter indicates the degree of dissociation of ammonia in the furnace. At nitriding temperatures from 500° to 520° C, the degree of ammonia dissociation is established at 15 to 25 per cent, at 600-650° C, 40 to 50 per cent ammonia is dissociated.

It is advisable to nitride thin-walled work of complex form, made of steel 38XMЮA, as well as other alloyed structural steels, at a temperature of 500° to 520° C. The time required depends on the case depth wanted. Curves showing the variation of hardness and case depth with the nitriding time are given in Fig. 193. These data indicate that the higher the nitriding temperature, the lower the hardness of the nitrided case formed in a given interval of time and the greater its depth. Nitriding of parts whose form is not particularly susceptible to warping, may be accelerated by raising the process temperature to 540-560° C. Another rapid method is a two-stage process in which nitriding is first conducted at 500-520° C and then

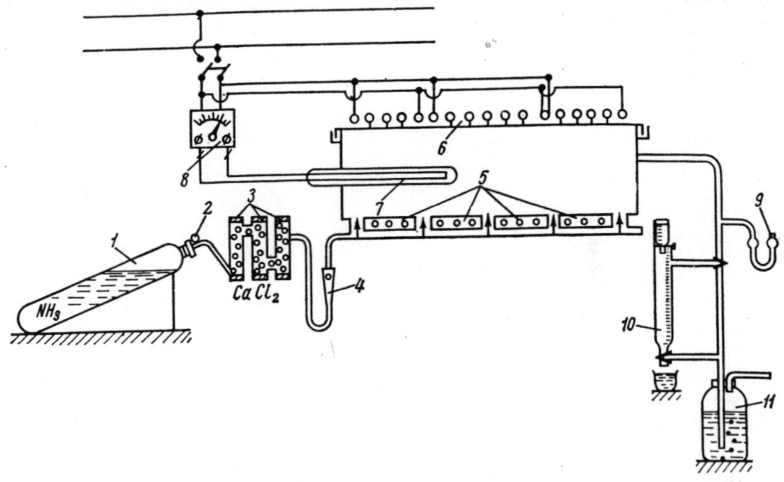

Fig. 192. Schematic representation of a nitriding installation (after S. F. Yuryev): 1—cylinder of ammonia, 2—reducer, 3—ammonia gas dryer, 4—gas meter, 5—furnace ports for gas entrance, 6—furnace, 7—thermocouple, 8—galvanometer, 9—manometer, 10—dissociation meter, 11—vessel with water

at 600-620° C. This latter method considerably reduces the time required for the process and retains the high hardness of the case.

After nitriding, the parts are cooled in a stream of ammonia gas to 200° C. Properly cooled work will be of a grey colour.

The application of induction heating enables the nitriding time to be sharply reduced. Nitriding for three hours at 500-550° C produces a case 0.2 to 0.25 mm thick having a hardness of 1,000 VPN (for steel 38XMЮA).

Steel may be nitrided in a liquid medium, using induction heating and also by passing the current directly through the work. Nitriding may be accomplished, for example, by induction heating of the work in a 25 per cent aqueous solution of ammonia. Here, a very hard nitride case, 0.1 to 0.15 mm deep, may be obtained on steel 38XMЮA at a temperature of 600° C in from 10 to 15 minutes. The possibility of saturating the surface of steel, heated in an electrolyte, with nitrogen (method proposed by I. Z. Yasnogorodsky) is also worthy of attention. A 10 per cent aqueous ammonia solution is used as the electrolyte. Electrolytic heating enables a hardened layer of practical depth to be obtained in a short time.

A process known as "ionitriding" is being employed to some extent. It is based on the application of a glow discharge. This process is somewhat faster than conventional nitriding.*

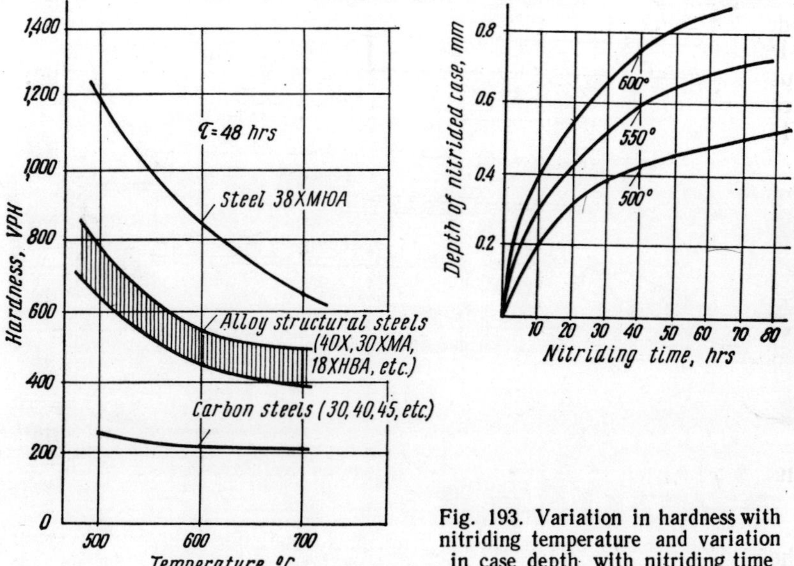

Fig. 193. Variation in hardness with nitriding temperature and variation in case depth with nitriding time

Certain changes in size are observed when steel is saturated with nitrogen. This is due to an increase in volume of the surface layer and also to deformation of the work. Deformation and size changes depend on the degree of saturation of the layer with nitrogen, depth of the nitrided case, nitriding temperature, and the size and shape of the work. The higher the nitriding temperature, the deeper the case, and the more complex the work, the more deformation may be expected. The change in size of parts having cases of equal depth will be more for the parts having thinner walls and a larger diameter.

Nitriding to increase the corrosion resistance is done at a temperature from 600° to 750° C. The nitriding time depends on the selected

* See E. Mitchell, "Developments in Surface Hardening", *Metal Treatment and Drop Forging*, Vol. 25, No. 158, 1958.

temperature. The higher the temperature, the less time will be required to obtain an effective corrosion-resistant layer. Anticorrosive nitriding ordinarily requires from 15 minutes for small workpieces to 6-10 hours for large ones.

STRUCTURE AND PROPERTIES OF THE NITRIDED CASE

We have seen that the saturation of steel with nitrogen results in considerable changes in the surface layer even before heat treatment (see Fig. 189). The microstructure of the nitrided case on steel 38XMЮA is illustrated in Fig. 194. A white, weakly-etched layer is seen at the surface. It is the ε-phase, rich in nitrogen. It has a nitrogen content from 9 to 11 per cent. Next is the α-phase, accompanied to a certain depth

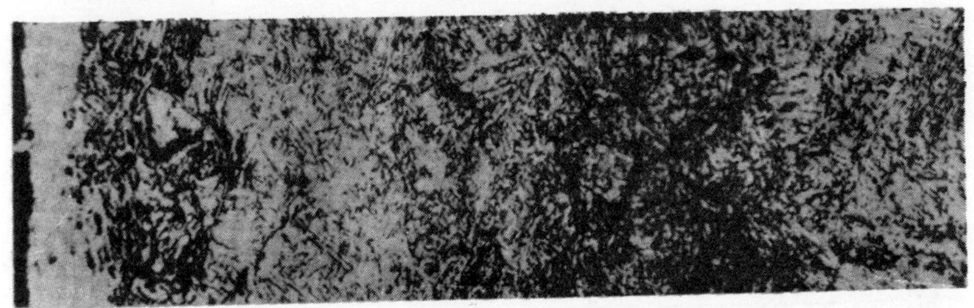

Fig. 194. Microstructure of the nitrided case obtained on steel 38XMЮA at 520° C, ×250

by the γ′-phase and dispersed nitrides of the alloying elements (aluminium, chromium, and molybdenum) which cannot be observed by micrography. In steel 38XMЮA, this part of the case has a characteristic sorbitic structure and differs from the core only in that it is more strongly etched due to the high nitrogen content.

Notwithstanding the considerable structural changes, the hardness of a nitride case on iron or carbon steel, after slow cooling, will not exceed 250-300 VPH. The introduction of alloying elements (aluminium, chromium, molybdenum, etc.) substantially increases the hardness of the case (Figs. 191 and 193).

After nitriding, the hardness of chromium (40X, etc.), chromium-nickel (30XH3, 37XH3A, etc.), and chromium-nickel-tungsten (18XHBA, etc.) steels, is increased to 600-800 VPH. The hardness of chromium-molybdenum-aluminium steels (of the 38XMЮA type), which are nitriding steels, is as high as 1,100-1,200 VPH.

A characteristic feature of a nitrided case is its retention of hardness after subsequent heating to 600-650° C. This provides excellent wear resistance of articles operating at high temperatures.

The higher the nitriding temperature and the longer the time, the more gradually the hardness will be reduced along the depth of the diffusion layer.

One of the most valuable functional properties of the nitrided case is its high wear resistance. It considerably exceeds that of carburised and hardened surfaces. Even after the short nitriding time, required for anticorrosive purposes, the resistance to mechanical wear increases by 3 or 4 times.

The formation of the high-nitrogen ε-phase (Figs. 189 and 194) in a continuous layer on the surface of a part sharply increases the corrosion resistance of steel in the following media: atmospheric air, fresh water, superheated steam, gasoline, etc. For example, no traces of corrosion whatsoever were found on nitrided machine parts immersed in fresh water for hundreds of hours.

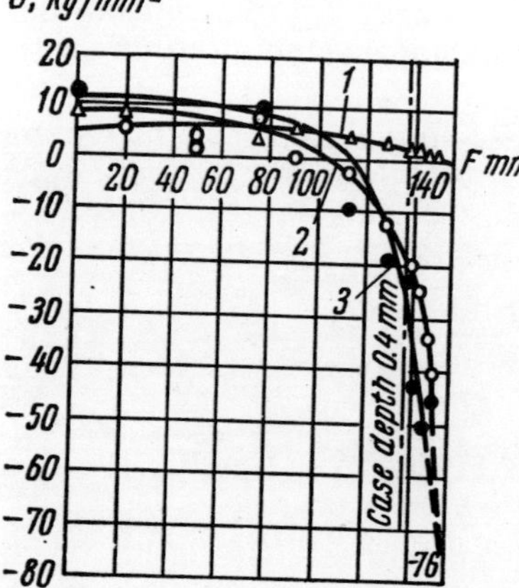

Fig. 195. Distribution of residual stresses along the thickness of cylindrical specimens (14 mm in diameter) after nitriding at 525° C for 40 hours (after B. F. Balashov):
1—radial, *2*—axial, *3*—annular

We may note that a nitrided surface polishes well and acquires a handsome metallic lustre which remains practically untarnished in the above-mentioned media.

The formation of the high-nitrogen phases at the surface of nitrided work is associated with large volume changes which lead to the appearance of compressive residual stresses in the hardened layer (Fig. 195). These stresses substantially increase the fatigue limit of the work as a whole. Nitriding unnotched laboratory specimens (7 to 10 mm in diameter), for example, increases their fatigue limit by 30-40 per cent. If the specimens have stress raisers (sharp notches), the fatigue limit will be 2 or 3 times more than for unnitrided specimens. Nitriding makes steel less sensitive to stress raisers that result from a poor surface finish, sharp changes in cross-section, holes, etc. Therefore, the fatigue limit of nitrided parts does not depend upon the surface finish to any practical extent. It must be noted, however, that the effect produced by nitriding is reduced if the cross-section of the specimen is increased. The fatigue limit of chromium-nickel-molybdenum crankshafts increases from 25 to 60 per cent after nitriding.

One great advantage of nitriding is that it increases the fatigue limit under corrosive conditions.

11-4. Cyaniding and Carbonitriding of Steel

Cyaniding and carbonitriding are processes in which both carbon and nitrogen are added to the surface layer of steel to increase its hardness and wear resistance. These methods are also effective for increasing the fatigue limit, particularly of medium- and small-size machine parts (gears, shafts, wrist pins, etc.). Cyaniding involves heating the steel in liquid or solid media; if the process is performed in a gas atmosphere it is called carbonitriding.

LIQUID CYANIDING

In this process, the work is heated in molten salt baths containing various prussiates. During heating, carbon and nitrogen are introduced into the surface layer of the steel.

As to techniques employed and results obtained, the liquid cyaniding of tool and structural steels may be classified into three chief groups:

a. *Low-temperature cyaniding* is employed to increase the life of high-speed steel tools (see page 336). This process is conducted at 550-560° C in a bath containing 30 to 50 per cent NaCN, 20 to 45 per cent Na_2CO_3, and 10 to 20 per cent NaCl. The time required depends upon the depth of the case and is usually between 5 and 30 minutes. Cutting tools are cyanided after having undergone complete heat treatment, grinding, and sharpening. No further heat treatment is required after cyaniding. A thin diffusion layer (0.02 to 0.04 mm) of high nitrogen content is produced by low-temperature cyaniding. This case is distinguished for its high hardness, wear resistance, and low coefficient of friction in metal cutting. These factors provide for high tool life in operation.

b. *Medium-temperature cyaniding* is used to obtain a thin diffusion layer (0.075 to 0.25 mm) which is wear resistant under small specific loads. In this procedure, the steel is heated to 750-860° C in molten salts containing 20 to 35 per cent NaCN. Soda ash (Na_2CO_3) and sodium chloride (NaCl) are employed as the neutral salts. Cyaniding time is determined by the depth of hardened case required and varies from 5 to 90 minutes. The resulting case contains 0.6-0.8 per cent C and 0.4-0.5 per cent N. Cyaniding is followed by quenching direct from the salt bath and low-temperature tempering. The case has a hardness of R_c 56-60. This process is applied to bolts, screws, nuts, small gears, and other machine parts of low- and medium-carbon steels.

c. *High-temperature (deep) cyaniding*. This process is accomplished in baths containing barium chloride, sodium chloride, potassium chloride, and 6 to 10 per cent sodium cyanide (6 to 10 per cent NaCN, 80 to 84 per cent $BaCl_2$, and not over 10 per cent NaCl).*

The process is carried out at 900-960° C from one to six hours in accordance with the required depth of the cyanided case (Table 19).

Table 19

Relationship Between Case Depth and Cyaniding Time

Cyaniding time, hours	1	2	3	4	5
Depth of case, mm	0.5-0.6	0.8-0.9	1.0-1.1	1.2-1.3	1.4-1.5

Baths containing $BaCl_2$ must be covered with a layer of ash-free graphite to reduce the degree of depletion of the bath in cyanide salt and heat loss. To maintain normal conditions, the bath must be periodically refreshed by alternately adding $BaCl_2$ and NaCN in the proportion 4 : 1. Cyaniding is followed by quenching direct from the bath and low-temperature tempering.

High-temperature cyaniding will produce a case from 0.5 to 1.5 mm deep (Table 19) containing 1.0 to 1.2 per cent C and 0.2 to 0.3 per cent N.

The properties of a deep cyanided case are near to those of a carburised case.

Deep cyaniding is a relatively new process and possesses many advantages over solid and gas carburising. Principal advantages include the high speed of the process, the simultaneous saturation with both carbon and nitrogen which provide a higher wear resistance, and the considerably less deformation of complex machine parts (gears, shafts, etc.).

Due to these features, deep cyaniding is being more and more extensively used in Soviet plants.

The following main chemical reactions occur in the process of low- and medium-temperature cyaniding.

The sodium cyanide in the bath reacts with the oxygen in the air and is oxidised according to the reaction:

$$1)\quad 2NaCN + O_2 \rightarrow 2NaCNO$$

$$2)\quad 2NaCNO + O_2 \rightarrow Na_2CO_3 + CO + 2N_{at}$$

$$\hookrightarrow 2CO \rightarrow CO_2 + C_{at}$$

* Recently, in a number of plants, expensive sodium cyanide (NaCN) is being successfully replaced in deep cyaniding by the cheaper cyanplav ГИПХ (43-49 per cent calcium cyanide $Ca(CN)_2$, 2-3 per cent calcium cyanamide $CaCN_2$, 30-35 per cent NaCl, 14-16 per cent CaO, and 4-5 per cent C). A mixture of salts $CaCl_2$+ +NaCl or $BaCl_2$+NaCl are recommended for use as neutral salts with cyanplav ГИПХ.

Carbon monoxide, formed by the decomposition of sodium cyanate (NaCNO), evolves atomic carbon, when it comes into contact with the iron. This atomic carbon is diffused in the iron. At the same time nitrogen, produced by the dissociation of sodium cyanate, is also introduced into the iron.

Somewhat different reactions occur in high-temperature liquid cyaniding. Here, the composition of the bath is selected so as to exclude the accumulation of a large amount of sodium carbonate which impedes the penetration of carbon into the steel. On the other hand, the bath should have a sufficiently high melting point since cyaniding is done at 900-950° C. Therefore, a large amount of barium chloride ($BaCl_2$) is included in the bath. This prevents the formation of Na_2CO_3 and activates the carburising of the steel.

In the presence of $BaCl_2$, the following reaction occurs:

$$2NaCN + BaCl_2 \rightarrow 2NaCl + Ba(CN)_2$$

The barium cyanide $Ba(CN)_2$ formed here is unstable and decomposes at high temperatures:

$$Ba(CN)_2 \rightarrow BaCN_2 + C_{at}$$

The carbon produced by this reaction diffuses into the iron. Thus, the gaseous phase CO is excluded in the production of atomic carbon. In the second stage of the reaction, barium cyanamide is oxidised to form CO and nitrogen:

$$BaCN_2 + O_2 \rightarrow BaO + CO + 2N_{at}$$

The different character of the reaction occurring in a high-temperature cyaniding bath leads to the penetration into the steel of mainly carbon and a limited amount of nitrogen.

CARBONITRIDING

Carbonitriding is the treatment of parts made of various tool and structural steels in a gaseous mixture consisting of a carburising gas and 20 to 30 per cent ammonia (NH_3).

In carbonitriding, as in liquid cyaniding, both carbon and nitrogen simultaneously saturate the surface of the steel but the process is slower.

Carbonitriding is classified as:

a. *Low-temperature carbonitriding* which is carried out at 540-560° C. This procedure is applied for high-speed steel tools. The process requires from 30 minutes to 3 hours. For example, twist drills, reamers, and core drills are carbonitrided in from one to two hours (depending on the diameter), taps in from 30 minutes to one hour, milling cutters (plain, form, and end mills) in from one to two hours, etc.

This time is sufficient to produce a case from 0.02 to 0.04 mm deep possessing high hardness (950-1,100 VHN) and wear resistance.

b. *Medium-temperature carbonitriding* at a temperature of 840-860° C in an atmosphere of the carburising gas (natural gas, gas produced in the pyrolysis and cracking of kerosene, town gas, etc.) and ammonia in a 3 : 1 proportion. The time required may be determined from the following data:

Depth of case, mm	0.2-0.3	0.4-0.5	0.6-0.7	0.8-1.0
Carbonitriding time, hours	1.5-2.0	3.0-4.0	5.0-6.0	8.0-10

Carbonitriding is followed by quenching either direct from the furnace or after precooling to 800-825° C or after reheating. The final operation is tempering at 160-180° C.

This type of carbonitriding is usually applied to complex machine parts that have a tendency to warp during heat treatment.

c. *High-temperature carbonitriding* at a temperature of 900-950° C. This procedure is used instead of gas carburising to save time and to prevent the deposition of soot and coke on the work. It also increases the wear resistance due to the introduction of nitrogen into the steel.

The rate of gas carburising at 930° C is from 0.15 to 0.18 mm per hour. An addition of 15 to 25 per cent ammonia to the carburising gas enables the rate to be increased to 0.20-0.25 mm per hour (to obtain a case 1.0 mm deep).

As in gas carburising, this carbonitriding procedure is followed by quenching and, finally, tempering at 160-180° C.

Carbonitriding is, without doubt, a highly progressive method for the heat treatment of structural steels. It has a number of essential advantages as compared to gas carburising. These include the considerable acceleration of the process, the exceptionally high wear resistance of the case obtained, and the lower costs due, chiefly, to the lower processing temperature (medium-temperature carbonitriding). This lower temperature of the process increases the service life of the furnace and its accessories and also reduces the fuel consumption.

PACK CYANIDING

In this procedure, the cyaniding is done in a mixture of 60 to 80 per cent charcoal and 40 to 20 per cent potassium ferrocyanide [$K_4Fe(CN_6)$]. Pack cyaniding is chiefly used at the present time to better the cutting properties of high-speed steel tools. The cyaniding temperature is from 540° to 560° C and the holding time is from 1.5 to 3 hours.

Though pack cyaniding produces a very hard, wear-resistant case, it has many disadvantages. It takes more time than liquid cyaniding and does not ensure stable properties of the work.

Various sections of the work may be cyanided selectively by coating them with certain pastes containing substances which easily evolve carbon and nitrogen.

Structures and properties of the cyanided case. The cyanided case on structural steel contains both carbon and nitrogen as mentioned above. This case has a very high hardness, reaching 800-900 VHN. Its structure is similar to that of a carburised case but, at the surface, the high nitrogen content forms a thin (0.01-0.05 mm) unetched layer (ε-phase). The nitrogenous γ'-phase may also be present, deeper in the case. After hardening, the case contains martensite in which carbon and nitrogen are dissolved.

In addition to increasing the hardness and wear resistance, cyaniding substantially increases the fatigue limit which is of great practical importance. The short time that the work is held in the cyaniding bath (15 to 30 minutes) enables the process to be combined with heating for hardening. Such heating in a cyaniding bath also excludes decarburising which sharply reduces the fatigue limit.

The principal disadvantage of cyaniding is the high cost and toxicity of the cyaniding salts and the necessity for corresponding measures to protect the personnel involved.

11-5. Diffusion Coatings

Saturation of the surface of steel by aluminium, chromium, beryllium and other elements is called *diffusion coating.*

Parts, whose surface is coated in this manner, acquire various valuable properties, including high heat resistance, corrosion resistance, increased wear resistance, and hardness.

Diffusion coating of steel by various metals, and also by silicon, is done by embedding the parts in the corresponding powdered mixtures (usually ferroalloys *), by immersing in the molten metal if the diffusing element has a comparatively low melting point (for example, zinc and aluminium), or by saturation from a gaseous medium.

Diffusion coating of iron (or steel) by metals by means of direct contact of the powdered medium with the work surface presents various difficulties due to the high temperatures and the long holding time required. These difficulties may be overcome to a great extent by saturation from a gaseous medium. Volatile chlorous compounds of the metals ($AlCl_3$, $CrCl_2$, $SiCl_4$, etc.) are used as the medium. They are obtained by the action of chlorine (or hydrogen chloride) on the metals or their alloys with iron (ferroalloys) at high temperatures. These chlorides dissociate on the surface of the work and the metal produced in the atomic state diffuses into the iron.

The rate of diffusion depends to a considerable extent upon the composition of the steel. Increased carbon content impedes diffusion and a redistribution of the carbon content is observed in the surface

* Ferrochromium, ferrosilicon, and ferroaluminium are alloys of iron and chromium, silicon, and aluminium, respectively.

layers. In certain cases, the carbon atoms are crowded out of the diffusion layer by aluminium or silicon; in other cases, on the opposite, carbon diffuses intensively toward the surface (for example, in saturation with chromium).

CALORISING

Calorising is a procedure in which aluminium penetrates into the surface of steel to impart to it high heat resistance (up to 850-900°C). This property of calorised parts is due to the formation of a dense surface film of aluminium oxide (Al_2O_3) which provides reliable protective action against oxidation and corrosion.

Several calorising techniques are employed.

Powder calorising. The parts are packed in a box where they are embedded in a mixture of aluminium or ferroaluminium powder and a small amount of ammonium chloride (0.5 to 2 per cent). Sometimes clay, alumina, or chamotte are added to prevent caking. The boxes are held for several hours at a temperature from 950° to 1,000°C. The time required for calorising low-carbon steel at a given temperature depends upon the specified depth of the diffusion layer. From 3 to 6 hours will be required to obtain a case 0.2-0.3 mm deep at 950°C; a case 0.4-0.5 mm deep will require from 12 to 16 hours.

Chemical processes that occur in powder calorising consist in dissociation of the ammonium chloride at a high temperature into ammonia and hydrogen chloride:

$$NH_4Cl \rightarrow NH_3 + HCl$$

Further, the hydrogen chloride reacts with the aluminium (or ferroaluminium) to form the volatile compound, aluminium chloride:

$$6HCl + 2Al \rightarrow 2AlCl_3 + 3H_2$$

Then the following reaction occurs upon contact with iron:

$$Fe + AlCl_3 \rightarrow FeCl_3 + Al_{at}$$

Atomic aluminium diffuses into the iron.

Calorising is sometimes followed by diffusion annealing at 900-1,000°C to reduce the brittleness of the calorised layer which is due to heavy supersaturation of the surface zones in the diffusion layer with aluminium (40 to 50 per cent). Diffusion annealing reduces the aluminium concentration at the surface (by diffusion deeper into the steel) thus reducing its brittleness. This is the most extensively used calorising procedure.

Dip calorising. This process entails the immersion of steel parts into a bath of molten aluminium saturated with iron (6 to 8 per cent) to prevent the aluminium from corroding (dissolving) the steel part. The bath temperature is 750-800°C and the process takes from 45 to

90 minutes. This will provide a calorised case from 0.2 to 0.35 mm deep. Calorising is followed by diffusion annealing at 900-1,000°C.

Disadvantages of this procedure are the embrittlement caused by the supersaturation of the layer with aluminium and the low service life of the crucibles in which the aluminium is melted.

In addition to the two calorising methods described above, there are several methods in which the steel is first simply mechanically coated with aluminium (by plating or metal spraying) which diffuses into the steel upon subsequent high-temperature annealing.

The calorised layer, obtained in powder mixtures or by the dip method, is a solid solution of aluminium in iron (Fig. 196). The aluminium content at the surface of the case is about 30 per cent.

An increase in the carbon content or that of alloying elements in the steel impedes the diffusion of aluminium (Fig. 197).

Calorising is applied to the fire grates of gas generators, tubes of thermocouples, components of pouring ladles, and to other parts operating at high temperatures.

New methods of rapid calorising in molten aluminium under a layer of flux are of great practical interest.

Work with a carefully cleaned surface is heated in a flux melt (for example, 40 per cent KCl, 40 per cent NaCl, 10 per cent Na_3AlF_6 and 10 per cent AlF_3) in a bath having a layer of molten aluminium at the bottom.

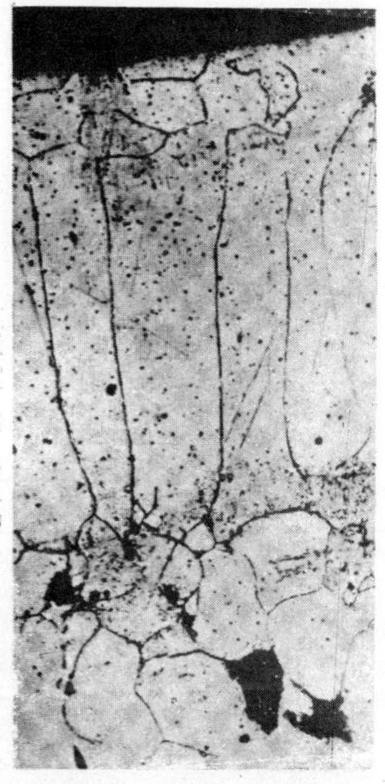

Fig. 196. Microstructure of the calorised case, × 500

Heating in the flux continues for 4 or 5 minutes and then the part is lowered into the molten aluminium which has a temperature of 720-780°C and is held there from several seconds to 10 or 15 minutes depending on the type of the part being treated.

Finally, the part is blasted with hot air to remove adhering aluminium.

The case obtained by this method comprises two parts: 1) a layer of pure aluminium 0.04 to 0.05 mm thick and 2) a diffusion layer 0.015 to 0.05 mm thick consisting mainly of α solution.

A rapid calorising procedure by means of metallisation is being used in the U.S.A. The work is heated to 200-250°C and aluminium is sprayed on its surface by an oxyacetylene spraying device. After this, the work is heated to 780-810°C and is held at this temperature for 10 to 15 seconds. Heating is accomplished either in a molten flux bath or by high-frequency induction. The layer obtained will be from 0.015 to 0.025 mm deep.

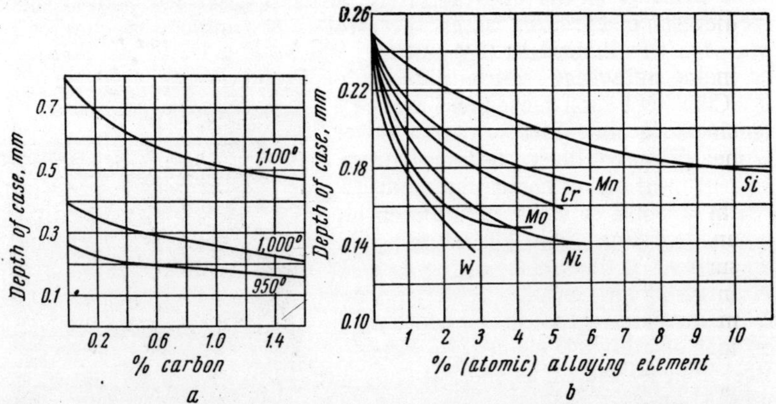

Fig. 197. Relationship between the depth of the calorised case and the content (*a*) of carbon, and (*b*) of alloying elements (Y. M. Lakhtin and P. I. Georgievsky)

Rapid calorising is applied to automobile valves and also to sheets and wire, instead of galvanising, as an anticorrosive measure.

CHROMISING OF STEEL

The saturation of the surface of steel machine parts by chromium provides a high resistance to corrosion by gases at temperatures up to 800°C, and high anticorrosive properties in such media as fresh water, sea water, and nitric acid. High-carbon steels are chromised to obtain hardness and wear resistance. There are three principal types of chromising techniques.

Pack chromising. The work is embedded in a mixture of finely ground ferrochromium and powdered alumina, kaolin, chamotte, etc. Ammonium chloride is added to this mixture (1 to 5 per cent) to form the active gaseous phase consisting of chromium chlorides. The process temperature is from 1,100° to 1,200°C. A case from 0.15 to 0.20 mm may be obtained in from 12 to 15 hours.

The following reactions characterise the chromising in a mixture of powdered ferrochromium and ammonium chloride:

$$NH_4Cl \rightarrow NH_3 + HCl$$
$$2HCl + Cr \rightarrow H_2 + CrCl_2$$

The interchange reaction, caused by the contact with the iron (work), releases atomic chromium:

$$Fe + CrCl_2 \rightarrow FeCl_2 + Cr_{at}$$

Atomic chromium diffuses into the work.

Volatile ferrous chloride produced by the interaction of chromium dichloride and iron is removed from the chamber.

Gas chromising. In this process, the work is heated to 950-1,050°C in an atmosphere of chromium dichloride and chromic chloride vapours ($CrCl_2$ and $CrCl_3$) obtained by passing hydrogen chloride vapours over heated chromium or ferrochromium.

At the chromising temperature, the vapours of chromium dichloride and chromic chloride contact the surface of the work. This causes an interchange reaction in which atomic chromium is released to diffuse into the iron:

$$CrCl_3 + Fe \rightarrow FeCl_3 + Cr_{at}$$
$$CrCl_2 + Fe \rightarrow FeCl_2 + Cr_{at}$$

Liquid chromising. Here, the work is held at 900-1,000°C in the molten salts $BaCl_2$, $MgCl_2$, $CaCl_2$, etc., to which is added either chromium dichloride ($CrCl_2$), in a proportion of 15 to 30 per cent by weight, or ferrochromium (20 to 25 per cent).

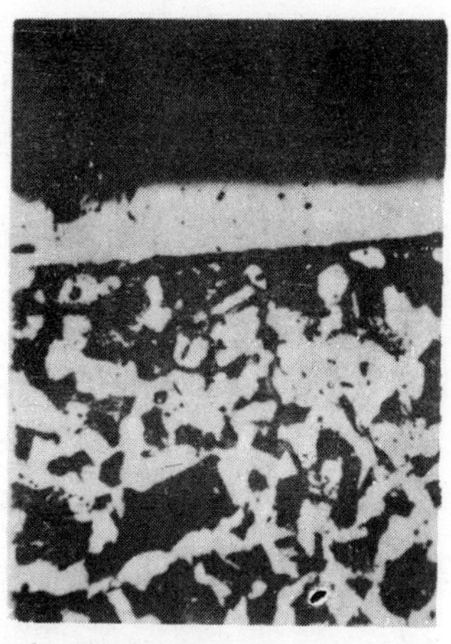

Fig. 198. Microstructure of a chromised case on steel containing 0.45 per cent C (after G. N. Dubinin)

Liquid chromising is based on the following reaction: $CrCl_2$ (in the molten salt) + Fe (work) → $FeCl_2 + Cr_{at}$. When commercial iron is chromised, the diffusion layer consists of a solid solution of chromium in α-iron. In high-carbon steels the layer will consist almost completely of the chromium carbide $(FeCr)_7C_3$ (Fig. 198). The formation of a continuous carbide layer is due to the diffusion of carbon from the inner layer to meet the chromium. Carbon has a higher rate of diffusion than chromium and, therefore, not all of it is used in forming the carbide layer; a portion forms a transition layer with a high carbon content. Due to the intensive carbon diffusion in low-carbon steels a transition layer consisting of chromium carbide may be formed to meet the chromium (Fig. 198). The carbide $(FeCr)_7C_3$ formed at the surface of the diffusion layer considerably increases its hardness. The chromised case obtained on iron has a hardness of 250-300 VHN, its hardness on high-carbon steel is 1,200-1,300 VHN. The depth of the case does not ordinarily exceed 0.15-0.20 mm. Carbon in steel impedes the diffusion of chromium into the iron. The addition of tungsten, molybdenum, and silicon to steel accelerates diffusion while chromium, nickel, and manganese reduce its rate (Fig. 200).

Fig. 199. Microstructure of the boron saturated case

Chromising is applied in the manufacture of components for steam power equipment, water and steam fixtures, valves, connection pipes, and machine parts subject to wear and exposed to corrosive media.

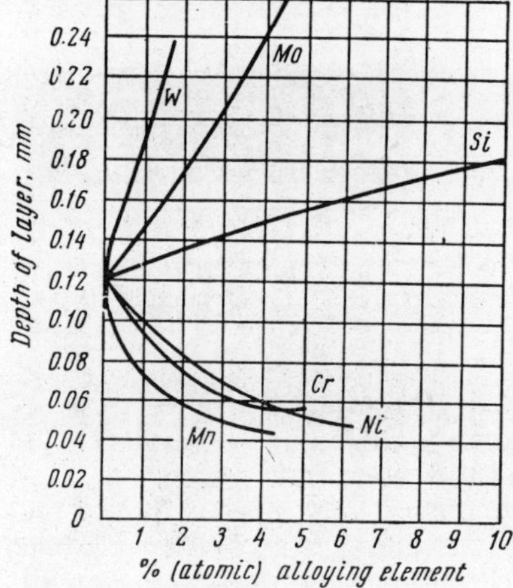

Fig. 200. Effect of alloying elements on the depth of the chromised case (Y. M. Lakhtin and P. I. Georgievsky)

SILICONISING

Siliconising is a process in which silicon diffuses from the surrounding medium to saturate the surface of steel (or malleable iron). Siliconised steel acquires a high resistance to the corrosive effects of sea water, as well as nitric, sulphuric, and hydrochloric acids, resistance to scaling at high temperatures (up to 750°C), and a somewhat higher wear resistance.

Two principal siliconising methods are employed.

Pack siliconising. The work is embedded in a layer of ferrosilicon (50 to 95 per cent Si), ammonium chloride (2 to 5 per cent), and neutral components (chamotte and quartz sand). The process is carried out at 1,100-1,200°C and takes from 2 to 24 hours. The case obtained is from 0.2 to 0.8 mm deep.

Siliconising in powdered mixtures is rarely applied.

Gas siliconising is accomplished in retort furnaces where the steel is heated in an atmosphere of silicon tetrachloride ($SiCl_4$). Upon interaction with the iron, this produces atomic silicon which diffuses into the metal:

$$3SiCl_4 + 4Fe \rightarrow 4FeCl_3 + 3Si_{at}$$

The retort is first charged with silicon carbide (carborundum) or ferrosilicon and then with the part. After heating the furnace to 950-1,050°C, chlorine or chlorine hydride is passed through the retort at a low rate. At the end of the process, the work is cooled with the furnace to 300-400°C. The time required is from 2 to 5 hours and a case from 0.5 to 1.25 mm deep will be produced. The siliconised case is a solid solution of silicon in α-iron. The surface zone of the case has a concentration of about 14 to 15 per cent Si.

The microstructure of the siliconised case is characterised by a bright unetched layer at the surface. This is a solid solution of silicon in α-iron. Pearlite segregations are frequently observed beneath the diffusion layer. Below this, the structure is that of the core. This type of structure is due to the expelling of carbon from the diffusion layer since it is little soluble in siliceous ferrite.

Notwithstanding its low hardness (200-300 VHN), the siliconised case is very brittle and may be machined by cutting tools only with great difficulty. It acquires a high wear resistance, however, after being saturated with oil at 170-200°C. A special feature of the case is its high porosity.

Components of machinery for the chemical, paper, and oil industries (pump shafts, pipelines, fittings, nuts, bolts, etc.) are most frequently siliconised.

BORON SATURATION

Steel, saturated with boron, acquires very high hardness, wear resistance, and corrosion resistance in various media.

Steel may be saturated with boron by embedding in ferroboron powder and heating to 950-1,050°C.

A more promising procedure involves the electrolysis of molten salts containing boron. The parts to be treated are immersed as cathodes in a bath of molten borax ($Na_2B_4O_7$). Baths of molten chlorous salts (NaCl, $BaCl_2$), to which powdered ferroboron or boron carbide has been added, are also suitable for this process.

Excellent results are obtained by a gaseous saturation method.* Here steel is saturated with boron in a medium of diborane (B_2H_6) mixed with hydrogen ($B_2H_6 : H_2 = 1 : 50$) or boron chloride (BCl_3), mixed with hydrogen, at a temperature of 850-900°C.

* This method was devised and developed by M. A. Pcholkina.

These boron treatments result in a case having a layer of iron mono-boride FeB at the surface, followed by iron sub-boride Fe_2B and then a solid solution of boron in iron (Fig. 199).

The borides in the diffusion layer form specific columnar crystallites (Fig. 199). Only a part of the layer is continuous. The hardest component is iron monoboride FeB (2,000 VHN); the hardness of iron sub-boride Fe_2B is somewhat lower (1,600-1,800 VHN). The more carbon and alloying elements in the steel, the less the depth of the case will be. High-alloy steels, including stainless and heat-resistant steels (Sections 13-10 and 13-11), give good results when they are saturated with boron.

Chapter 12

MINOR CONSTITUENTS AND ALLOYING ELEMENTS IN STEEL

12-1. Effects Produced by the Minor Constituents

A number of inevitable impurities (or minor constituents as they are called) are always present in steel. They include Si, Mn, S, and P.

Effect of Si *and* Mn *on the properties of steel.* Carbon steels do not normally contain more than 0.35-0.4 per cent Si and more than 0.5-0.8 per cent Mn.

The mechanical properties of steel are influenced only to a small extent by Si and Mn in these amounts. Both silicon and manganese are effective deoxidisers. Manganese reduces red-shortness in steel, i.e., brittleness at high temperatures, which is usually due to a high sulphur content.

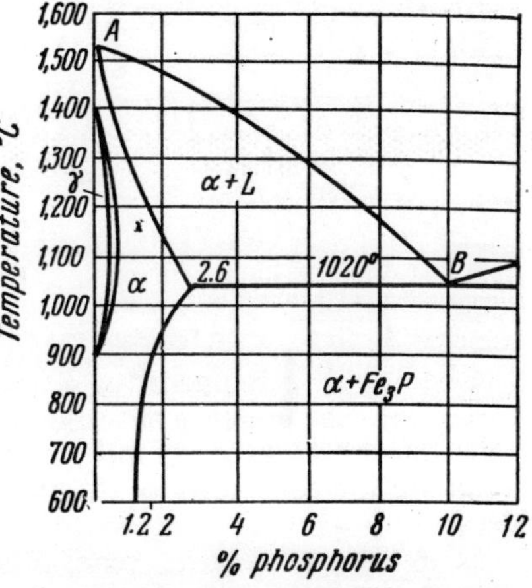

Fig. 201. The Fe-P equilibrium diagram

Phosphorus in steel. Phosphorus dissolves in both γ- and α-iron and forms the chemical compound iron phosphide Fe_3P (Fig. 201). The solubility of phosphorus in α-iron is 2.6 per cent at 1,020°C and 1.2 per cent at 600°C. Its solubility in γ-iron is considerably less (Fig. 201). Phosphorus dissolved in ferrite imparts cold-shortness (see Sec. 5-4). Fig. 202 illustrates the effect of phosphorus on the mechanical properties of low-carbon steel. An increase in the phosphorus content increases the tensile strength but sharply reduces the impact strength and ductility of steel. At a phosphorus content over 0.2 per cent, the impact strength is reduced practically to zero. The embrittlement effect of phosphorus is increased by an increase in the carbon content of steel. The detrimental effect of phosphorus is intensified by its tendency to segregate. This may result in zones very rich in phosphorus where embrittlement is especially pronounced.

The phosphorus content should not exceed 0.085-0.09 per cent in ordinary quality steels, 0.04 per cent in quality steels, and 0.035 per cent in high-quality steels. It must be noted, however, that the embrittling effect of phosphorus is sometimes utilised for various purposes and low-carbon steel with a high phosphorus content (up to 0.15-0.20 per cent) is produced. These steels have higher machinability since they form discontinuous chips in metal-cutting.

Sulphur in steel. Sulphur forms the chemical compound FeS with iron (36.4 per cent S), which is practically insoluble in solid iron. When the alloy contains 31.6 per cent S (85 per cent FeS), a eutectic

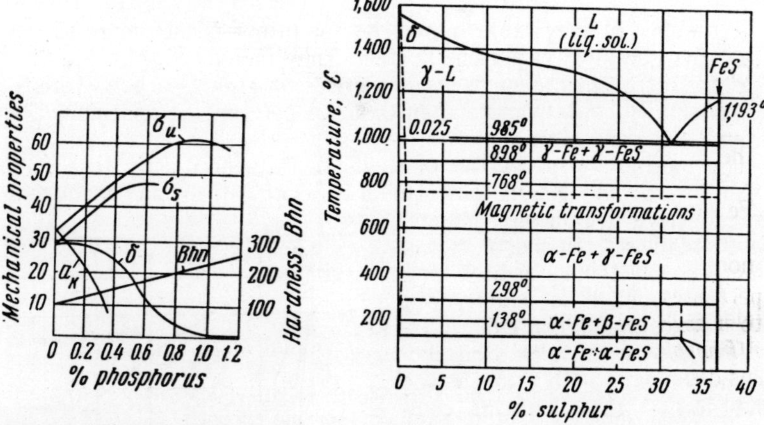

Fig. 202. Effect of phosphorus on the mechanical properties of steel

Fig. 203. The Fe-S equilibrium diagram

(Fe-FeS) with a comparatively low melting point is formed at 985°C (Fig. 203). The presence of this eutectic at the grain boundaries and the fact that it may actually melt at rolling and forging temperatures and break up the continuity of the steel grains is the reason for red-shortness*.

As mentioned above, manganese practically excludes red-shortness by forming the highly refractory compound MnS with the sulphur. In solid steel, MnS is found as separate greyish inclusions of more or less holohedral form. In worked steel, these inclusions are stretched out in the direction of metal flow.

Other properties are also unfavourably affected by considerable amounts of sulphur inclusions. For example, the mechanical properties and corrosion resistance may be appreciably reduced, as well as the fatigue limit. Sulphur has an unfavourable effect on the welda-

* This action of sulphur is intensified by the presence of oxygen (as FeO).

bility of steel, therefore its content must be at a minimum for steels used in highly stressed welded structures.

The sulphur content is strictly limited in practice. It should not exceed 0.04 per cent in quality steels or 0.03 per cent in high-quality grades. An exception is the so-called free-cutting steels in which the sulphur content is deliberately increased to improve the machinability. The sulphur content in such steels may reach 0.5 per cent (with a simultaneous increase in the manganese content). In this case the sulphur is present as MnS inclusions which are uniformly distributed in the steel.

12-2. Distribution of Alloying Elements in Steel

Upon being introduced into steel, alloying elements (Si, Mn, Ni, Cr, Mo, W, V, Ti, Nb, Al, etc.) may:

1) form solid solutions with iron;

2) dissolve in the cementite, by substituting for iron atoms in the cementite lattice or by forming independent (special) carbides;

3) form intermetallic compounds with iron (FeCr, Fe_3W_2, Fe_2W, Fe_3Mo_2, FeMo, FeV, Fe_3Ti, Fe_3Si_2, etc.) when their content is high.

Alloying elements may also be present in steel as oxides and other nonmetallic inclusions (MnO, SiO_2, Al_2O_3, TiO_2, V_2O_5, Fe_4N, etc.).

Table 20 shows the approximate distribution of certain alloying elements between the principal phases, in accordance with the heat treatment that structural steel has undergone.

Table 20

Relationship Between the Distribution of Alloying Elements in Steel and the Type of Heat Treatment Employed * (after V. A. Delle)

Heat treatment	Alloying elements	
	in ferrite	in the carbide phase
Annealing	Ni, Si, Al, Cu, [Mn], Cr, Mo, W	Mn, [Cr], [Mo], [W], Nb, V, Zr, Ti
Hardening	Ni, Si, Al, Cu, Mn, [Cr], [Mo], [W]	Nb, V, Zr, Ti
Hardening followed by tempering at 600° C	Ni, Si, Al, Cu, [Mn], Cr, Mo	Mn, [Cr], [Mo], [W], Nb, V, Zr, Ti

* The boxed-in elements indicate the phases in which they are chiefly found.

These data show that nickel, silicon, aluminium, and copper are always found in the solid solution.

Chromium, molybdenum, and tungsten are mainly in the carbide phase and are dissolved in ferrite to a lesser degree. Manganese is an element with only a weak tendency to form carbides. In the presence of elements having a stronger affinity to carbon, for example, chromium, the manganese will be found mainly in the ferrite. Vanadium, titanium, niobium, and zirconium are chiefly in the carbide phase.

Binary alloys of iron-alloying elements (effect of alloys on the polymorphism of iron). All alloying elements, with the exception of carbon, nitrogen, and boron, form substitutional solid solutions with iron. These three elements have a small atomic radius and form interstitial solid solutions.

Alloying elements dissolved in iron have a great influence on the position of the A_3 and A_4 points which determine the temperature range in which α- and γ-iron exist. As to their influence on the allotropic modifications of iron, alloying elements may be classified into two groups of two sub-groups each. Elements of the first group include nickel and manganese which lower the critical point A_3 and raise point A_4. As a result, the γ-phase range is extended on the equilibrium diagram of iron and the alloying element and the α-phase range is narrowed (Fig. 204). As shown in Fig. 204, *a*, point A_4 is raised by the alloying element to the solidus while point A_3, at a definite alloying element content (*a*), is lowered to room temperature. Therefore, alloys having an alloying element content exceeding "a" do not undergo the phase transformations ($\alpha \rightleftarrows \gamma$) and, at all temperatures, are a solid solution of the alloying element in γ-iron. These are called *austenitic alloys*.

Alloys which partly undergo the $\alpha \rightleftarrows \gamma$ transformation are called *semi-austenitic alloys*. The Fe-Ni and Fe-Mn equilibrium diagrams, given in Fig. 205, illustrate the above-described changes in the critical points.

The second sub-group of elements includes copper, carbon, nitrogen and others which extend the γ-phase range (Fig. 204) when their content is small but, due to their limited solubility in iron, the γ-phase range begins to narrow, at a certain concentration, and then completely disappears (Fig. 204, *b*).

A typical representative of this type of equilibrium diagram is found in the Fe-C system (see Fig. 108).

Elements of the first sub-group in the second group (Cr, W, Mo, V, Al, Si, etc.) lower the critical point A_4 and raise point A_3. As a result, at a definite concentration of the alloying element (see point *b* in Fig. 204, *c*), these critical points coincide and the γ-phase region is completely enclosed by a loop. At an alloying element content exceeding point *b*, the alloys will consist, at all temperatures, of a solid solution of the alloying element in α-iron. These are called *ferritic alloys.*

Alloys with only partial transformation of $\alpha \rightleftarrows \gamma$ are called *semi-ferritic.* Fig. 206 illustrates equilibrium diagrams for this type of alloy.

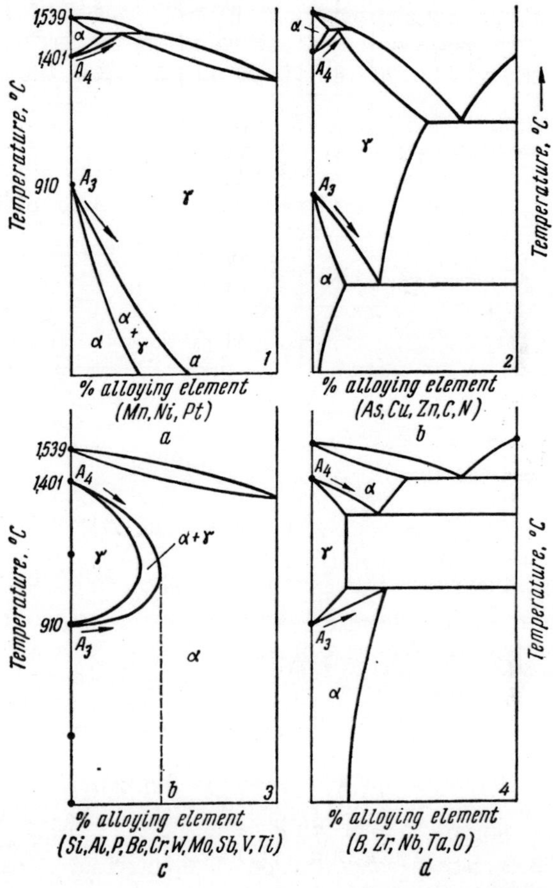

Fig. 204. Types of binary iron alloy equilibrium diagrams

The second sub-group (of the second group) consists of alloys of iron with boron, zirconium, niobium, and certain other elements which narrow the γ-phase even when present in very small amounts in steel. However, due to their low solubility in iron, two-phase regions appear in these alloys before the γ-phase region is completely enclosed (Fig. 204, *d*).

The reasons for these effects of alloying elements on the allotropic transformations have not as yet been made clear. It has, however, been noted that elements having a crystal lattice resembling that of

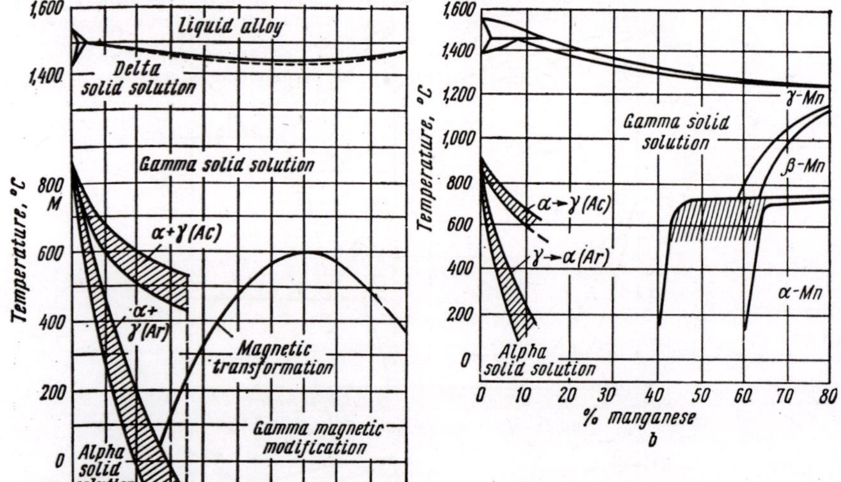

Fig. 205. Equilibrium diagrams:
a—Fe-Ni, b—Fe-Mn

γ-iron or having a small atomic radius, extend the γ-phase range. Elements with a lattice identical to that of α-iron narrow the γ-phase region and extend that of the α-phase.

Certain alloying elements (Cr, Mo, W, Si, and others) form chemical compounds with iron, when they are present in a sufficient concentration, and solid solutions based on these compounds.

The effect of an alloying element on the γ-phase range for a ternary system, Fe-C-alloying element, is shown in Fig. 207.

Fig. 207, *a* shows that nickel and manganese lower the line *GS* (A_3), while points *S* and *E* in the Fe-C system are shifted to the left, toward a lesser carbon concentration. The eutectoid transformation temperature A_1 is also lowered.

Elements which narrow the γ-phase range, such as chromium (Fig. 207, *b*), tungsten, molybdenum, silicon, etc., gradually enclose the γ-phase region and shift points *S* and *E* sharply toward a lesser carbon content. These same elements raise point A_1.

Influence of alloying elements on the properties of ferrite. Alloying elements, dissolved in ferrite, change its mechanical properties to a great extent. They, therefore, change the steel properties as well since ferrite is the basis for many structures observed in steel. Fig. 208

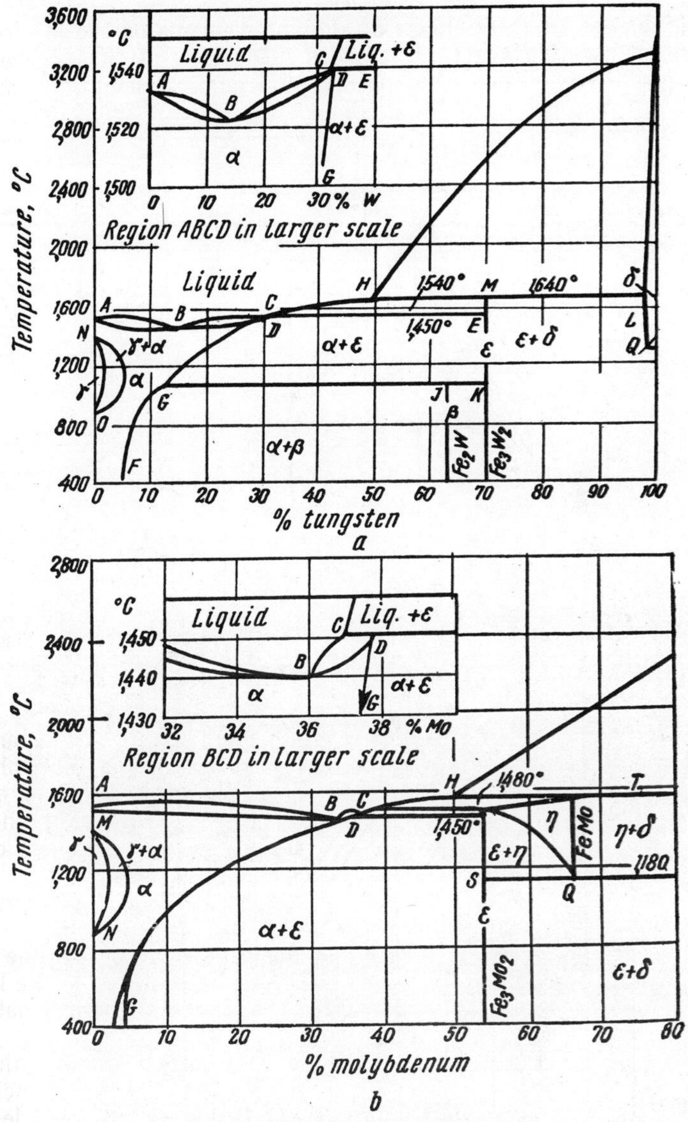

Fig. 206. Equilibrium diagrams:
a—Fe-W, *b*—Fe-Mo

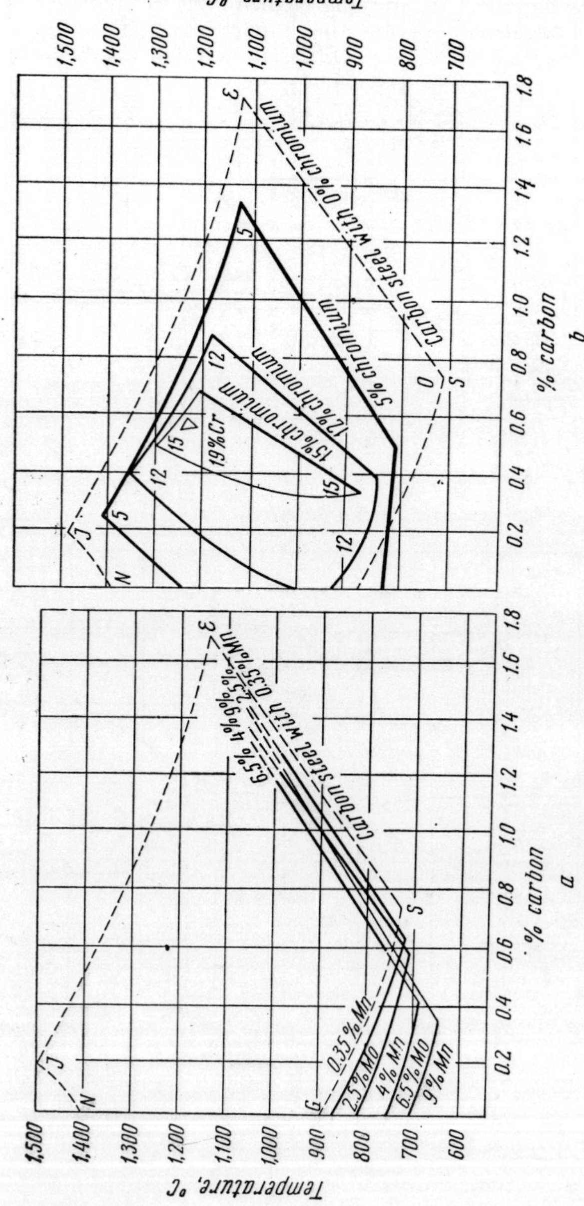

Fig. 207. Effect of alloying elements on the maximum solubility of carbon in austenite:
a —effect of manganese, *b*—effect of chromium

illustrates the effect of alloying elements on the tensile strength (*a*), relative elongation (*b*), and the impact strength (*c*) of ferrite. All elements increase the strength of ferrite without appreciably changing

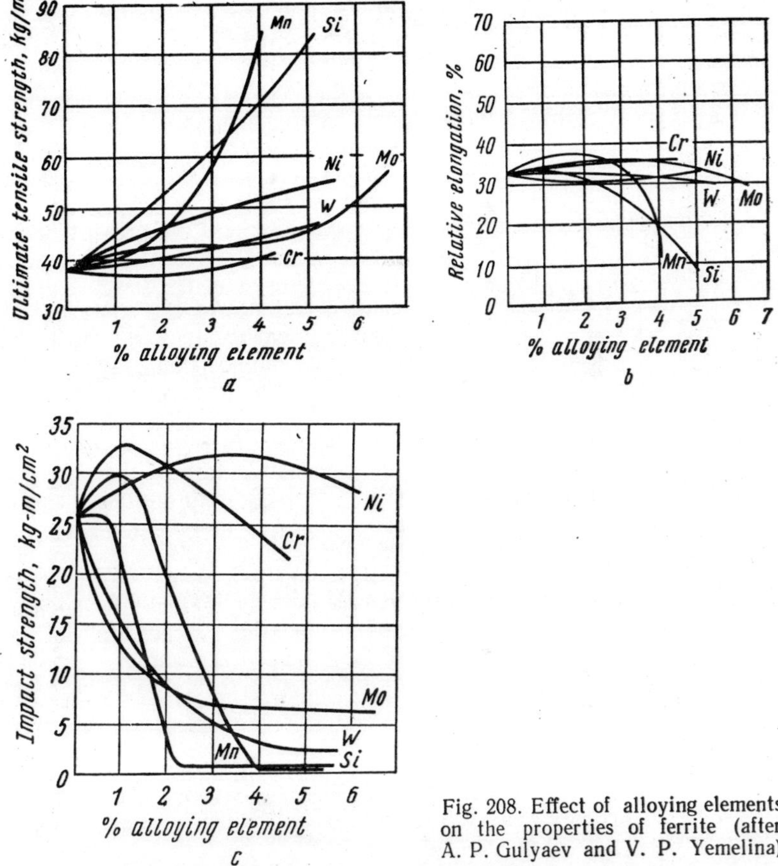

Fig. 208. Effect of alloying elements on the properties of ferrite (after A. P. Gulyaev and V. P. Yemelina)

the elongation but, when present in amounts exceeding 1.0 per cent, they considerably reduce the impact strength. An exception is nickel, which does not reduce the impact strength (up to 6 per cent Ni) and even retains it at low temperatures (to 80°C below zero).

Small additions of chromium (up to 1 per cent) also increase the impact strength.

Ferrite, containing nickel, manganese, or chromium, may be hardened by quenching from the temperature range in which the γ-phase exists. The acicular ferrite, formed after quenching, is of high hardness.

Carbide phase in alloy steels. All alloying elements may be divided into two groups by their relation to carbon: 1) elements which do not form carbides in steel (Ni, Cu, Si, and Co); these elements facilitate the formation of graphite and 2) the carbide-forming elements Mn, Cr, W, Mo, V, Ti, and Nb, which form stable carbides in steel.

As mentioned above, these elements of the second group are more liable to form carbides, if sufficient carbon is present, than to dissolve in the ferrite. Special carbides may be formed only when considerable proportions of carbon and the alloying elements are present in the steel. When their content is small, the alloying elements dissolve in the cementite to form the so-called alloyed cementite. In these cases, the cementite composition may be represented by the formula (Fe, $M)_3C$ in which M is the alloying element. For example, manganese dissolved in cementite forms the carbide (Fe, $Mn)_3C$, chromium forms (Fe, $Cr)_3C$, molybdenum forms (Fe, $Mo)_3C$, etc.

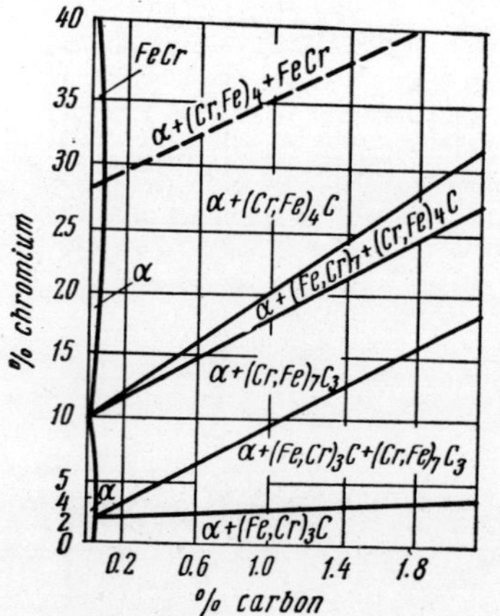

Fig. 209. The iron corner of the ternary diagram for Fe-Cr-C

Experiments have shown that manganese can replace all the iron atoms in the cementite lattice ($Fe_3C \rightarrow Mn_3C$), chromium can replace up to 25 atomic per cent, molybdenum up to 3 atomic per cent, tungsten up to 0.8-1.0 per cent, etc. Ti, Nb, and Zr are practically insoluble in cementite.

Fig. 209 is an isothermal cross-section of a portion of the ternary diagram for Fe-Cr-C at room temperature. This cross-section shows that alloyed cementite is formed if the chromium content of the steel does not exceed 2 per cent.

At a higher chromium content, a special carbide (Cr, $Fe)_7C_3$ is formed. At even larger chromium content (over 10-12 per cent) the carbide (Cr, $Fe)_4C$ [or (Cr, $Fe)_{23}C_6$] is formed.

When tungsten and molybdenum are added to steel in an amount exceeding the maximum saturation of these elements in cementite, the complex carbides (Fe, $Mo)_{23}C_6$ and (Fe, $W)_{23}C_6$ will be formed.

Double carbides, Fe_2W_2C and Fe_2Mo_2C, are obtained in steel with high molybdenum and tungsten content; for example, in high-speed steel.

At a vanadium content exceeding 0.5 per cent, the carbide VC is formed.

All carbides may be divided into two groups. The first group includes Fe_3C, Mn_3C, Cr_7C_3, $Cr_{23}C_6$, Fe_2W_2C, Fe_2Mo_2C, and others with a complex crystal lattice. For example, the carbide $Cr_{23}C_6$ has a complex cubic lattice (the unit cell contains 92 atoms of chromium and 24 carbon atoms) while Cr_7C_3 has a complex hexagonal lattice (the unit cell contains 56 chromium and 24 carbon atoms). A characteristic feature of the first group of carbides is their high solubility in austenite upon heating. This is of prime importance in the heat treatment of steel.

The second group includes W_2C, WC, Mo_2C, VC, NbC, TiC, ZrC, TaC, Ta_2C, and others. They are typical interstitial compounds (see page 122). These carbides have comparatively simple crystal lattices, built up of ions of metallic elements and with carbon ions occupying interstitial positions.

The interstitial compounds differ from the carbides of the first group in that they do not dissolve in the austenite when steel is heated under the actual conditions met with in practice.

All carbides are capable of dissolving various metallic elements to some extent. For example, the carbide Cr_7C_3 dissolves up to 55 per cent iron $[(Cr, Fe)_7C_3]$ at room temperature while the carbide $Cr_{23}C_6$ dissolves up to 35 per cent $[(Cr, Fe)_{23}C_6]$.

The double carbides Fe_2Mo_2C and Fe_2W_2C have a wide range of homogeneity. Besides this, each of these carbides dissolves molybdenum and tungsten.

Carbides of the interstitial compound type differ from those in the first group in that they are capable of dissolving large amounts of the basic metal. They form solid solutions with a carbon deficiency (so-called defect compounds or subtraction phases). Interstitial compounds with the same type of crystal lattice are mutually soluble. Therefore, their composition does not exactly correspond to the formulas assigned to them.

The type of carbide is frequently determined by the conditions of heat treatment. If steel, containing a large amount of tungsten and molybdenum, is very slowly cooled or is held for considerable time in tempering, interstitial compounds WC, W_2C, and Mo_2C will be formed. If the steel is cooled more rapidly (as is the case in actual practice), the double carbides Fe_2W_2C and Fe_2Mo_2C are formed.

Interstitial compounds are undesirable in steel since they deteriorate the properties of many steels (high-speed steels, steels for permanent magnets, etc.).

All carbides found in steel have a high hardness and melting point. Interstitial compounds are the hardest types. The more dispersed the carbides are in a steel, the higher its strength and hardness will be because the particles of these phases offer considerable obstacles to slip in the ferrite thus increasing the resistance to plastic deformation.

12-3. Effects of the Alloying Elements on Phase Transformations in Steel

Effects on processes occurring in the heating of steel. When steel is heated, the eutectoid is transformed into austenite and the carbides dissolve in this austenite. The formation of austenite described earlier for carbon steels is true for most alloy steels. It must be noted, however, that the introduction of alloying elements shifts the temperature of the equilibrium points A_1 and A_3. This must be taken into consideration in assigning a heating procedure for the heat treatment of an alloy steel.

Nickel and manganese lower the critical points A_1 and A_3, while they are raised by Cr, Mo, W, V, Ti, and others.

The introduction of alloying elements has a large influence on the rate at which austenite is formed in steel.

Nickel increases the rate of carbon diffusion in austenite and refines the initial structure of the eutectoid. Thereby, it increases the rate of austenite formation. The transformation of pearlite into austenite in an alloy steel, containing carbide-forming elements, proceeds slower than in a carbon steel. Upon continuous heating, this process covers a large temperature interval and takes a great deal of time for its completion. This may be explained by the fact that most carbide-forming elements impede carbon diffusion in the austenite and by the difficulty in dissolving the carbides in the austenite. It must be pointed out that after the formation of austenite in alloy steel and its homogenisation in respect to its carbon content, it (austenite) will remain inhomogeneous as far as alloying element distribution is concerned because these elements have a low rate of diffusion in austenite. Extra time is required to equalise the concentration of alloying elements in the austenite (Fig. 210).

In actual practice, when complex alloy steels containing carbide-forming elements are heated under ordinary conditions, the process of dissolving the carbides and the homogenisation of the solid solution are not completed.

All alloying elements, except manganese, impede austenite grain growth in heating. Carbide-forming elements are the most effective in this respect.

This susceptibility to austenite grain growth is reduced strongly by V, Ti, and Zr, moderately by W, Mo, and Cr, and weakly by Si and Ni. Mn, P, and C increase this tendency.

Effect of alloying elements on the isothermal decomposition of supercooled austenite. Almost all alloying elements, with the exception of cobalt, titanium, and aluminium, when dissolved in the austenite, increase its stability in the pearlite transformation range. This can be seen in Fig. 211 where diagrams of isothermal austenite decomposition are given for steels with various Ni, Cr, and Mo content. An increase in the amount of nickel, chromium, and molybdenum shifts the isothermal decomposition curve to the right, i.e., these elements increase austenite stability. The presence of carbide-forming elements (Cr, Mo, etc.) in steel, alters the form of the isothermal curve. The stability of austenite in steels, alloyed with these elements, is reduced with higher degrees of supercooling (below point A_1). It reaches a minimum at about 600-650°C and then increases again. However, this

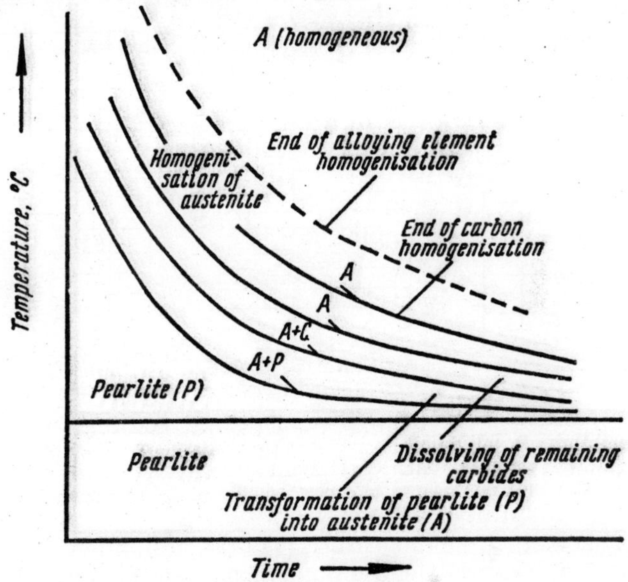

Fig. 210. Transformation of pearlite into austenite at constant temperature

increase in austenite stability does not grow continually to the martensite point as in carbon or nickel steels (Fig. 211). Below 450-500°C, the stability is reduced again, reaching a minimum at about 300-400°C. The existence of an intermediate stage of accelerated austenite decomposition (or unstable zones) is the most essential feature of isothermal austenite decomposition in most alloy steels. The product of austenite decomposition in this zone is acicular troostite (bainite).

A distinguishing feature of the intermediate transformation in alloy steels is the fact that it usually is not completed. The higher the isothermal holding temperature, the less acicular troostite (bainite) will be formed.

Upon subsequent cooling, the undecomposed austenite undergoes martensitic transformation while a certain portion of it is supercooled to room temperature.

The most typical diagram of austenite transformation in the temperature range of intermediate transformation and subsequent cooling is given in Fig. 212. Here, temperature t_1 corresponds to the upper limit of the intermediate range (400-450°C) while t_2 is the lower limit (slightly above point *M*).

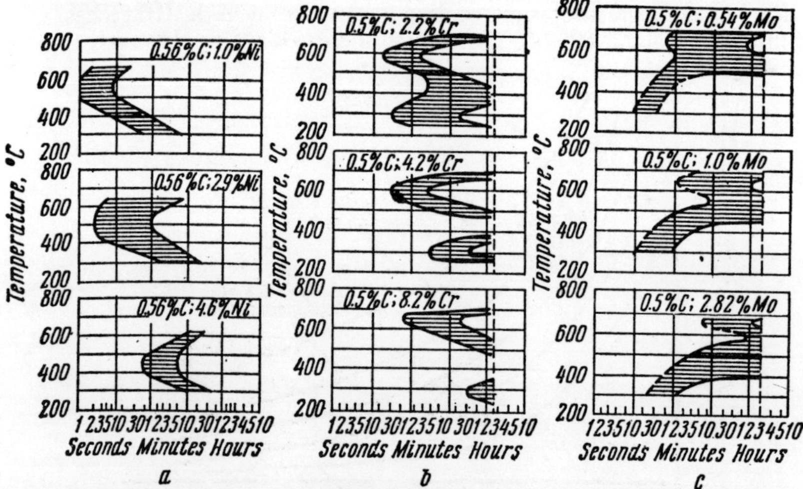

Fig. 211. Effect of alloying elements on the isothermal decomposition of austenite (after V. I. Zyuzin):
a—effect of nickel, *b*—effect of chromium, *c*—effect of molybdenum

Fig. 212 shows that the final structure of the steel, after isothermal holding in the intermediate transformation range of temperatures, consists of upper or lower acicular troostite (bainite), martensite, and undecomposed (retained) austenite.

The amount of martensite formed depends upon the temperature of the isothermal reaction and the composition of the steel. The higher the temperature of isothermal decomposition in the acicular troostite (bainite) temperature range, the more retained austenite will undergo a martensitic transformation in cooling (Fig. 212). In low-carbon alloy steels, the austenite which was not decomposed in the isothermal reaction is almost completely transformed into martensite upon subsequent cooling (Fig. 212, *b*). It is quite the opposite in high-carbon alloy steels. Here the undecomposed austenite is very stable and undergoes almost no martensitic transformation upon subsequent cooling (Fig. 212, *c*). A large amount of retained austenite is found in the structure of these steels.

If the steel is held isothermally for a long period at the upper limit (400-500°C) of the temperature range, after completing the intermediate transformation the retained austenite frequently undergoes a pearlite transformation (Fig. 215). In this case, the transformation is completed and results in the formation of acicular troostite (bainite) and pearlite (troostite).

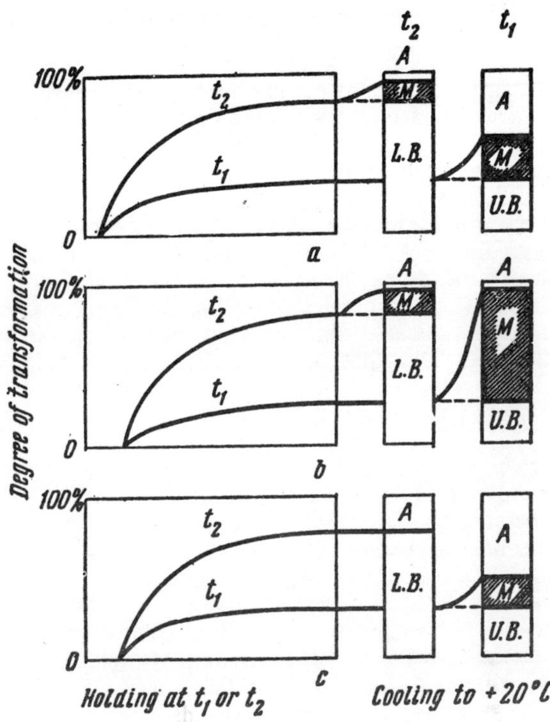

Fig. 212. Kinetic curves of supercooled austenite transformation in the intermediate (bainite) range, and final structure of steel (after A. P. Gulyaev)

It must be noted that the introduction of several alloying elements into steel increases the stability of a supercooled austenite in the pearlite region to a much greater degree than could be expected by simply summarising the action of the separate elements.

The reasons for this higher stability of supercooled alloyed austenite, particularly in the pearlite transformation range, have not yet been established.

Many investigators (R. F. Mehl, A. Hultgren, R. I. Entin and others) have found that the decomposition of alloyed austenite in the pearlite region produces ferrite and carbide (either alloyed cementite, or a special carbide whose composition and structure are in equilibrium for the given steel)*. Therefore, a diffusion redistribution, not only of carbon but of the alloying elements as well, must occur between the ferrite and carbide before a stable ferrite-carbide mixture can be formed. In the general case, it may be assumed that the carbide-forming elements are concentrated in the carbide phase while those which do not form carbides are concentrated in the ferrite.

In all probability, the retardation of austenite decomposition in the pearlite zone may be explained by the low diffusion rate of the alloying elements in austenite and the reduced rate of carbon diffusion under the influence of the alloying elements.

Data, obtained by R. I. Entin for certain alloy steels, show that the retardation of supercooled austenite decomposition may be associated with the reduced rate of the $\gamma \rightleftarrows \alpha$ transformation.

An acceleration of alloyed austenite decomposition may be observed in the intermediate (bainite) transformation range. This is due to the fact that the diffusion of alloying elements is excluded at low temperatures and austenite decomposes into ferrite and carbide of the cementite type which have the same concentrations of alloying elements as the original austenite. Therefore, to obtain the ferrite-carbide mixture in the intermediate range only carbon diffusion is required and this transformation is not accompanied by redistribution of the alloying elements.

By increasing the stability of supercooled austenite, alloying elements dissolved in the austenite reduce the critical cooling rate and thus increase the hardenability of steel. This enables oil to be used as a quenching medium (instead of water) and high mechanical properties to be obtained in large components.

Alloyed austenite transformation in continuous cooling. The decomposition of alloyed austenite at various cooling rates differs substantially from austenite decomposition in carbon steel, due to the existence of a second zone of accelerated austenite decomposition.

---

* Professor M. Y. Blanter is of the opinion that the decomposition of an alloyed austenite always yields a carbide of the cementite type first, without regard for the type of stable carbide which the steel should contain under equilibrium conditions. In cases when a special carbide is stable, after complete saturation of the cementite by the carbide-forming element, a carbide transformation, producing the stable carbide will occur. In the resulting mixture of ferrite and cementite, the concentration of alloying elements is approximately the same as in the original austenite. The redistribution of the concentrations of alloying elements proceeds during austenite decomposition and after it is completed. Elements, which do not form carbides, diffuse from the carbides to the ferrite while carbide-forming elements diffuse from the ferrite to the carbide.

M. Y. Blanter explains the high stability of supercooled alloyed austenite by the influence of the alloying elements on the diffusion of carbon in austenite, by the fact that they retard the relief of residual stresses caused by phase changes in the austenite (due to the difference in specific volumes of the decomposition products and of the initial austenite), and by the position of the critical point A_1 (see *Physical Metallurgy and Metal Treatment*, No. 4, 1955).

Figs. 213 and 214 give diagrams showing the isothermal decomposition of alloyed austenite and the effect of the cooling rate on the transformation temperature and on the percentages of structural components obtained in alloy steels. Fig. 213 shows that only the pearlite decomposition of austenite occurs at low cooling rates if the rate of transformation in the pearlite zone exceeds the rate of decomposition in the zone of intermediate transformation. In this case, the usual ferrite-carbide mixture (pearlite or sorbite) will be produced. At higher cooling rates (Fig. 213), in addition to the pearlite transformation, there will also be a transformation into acicular troostite (bainite), and the portion of the austenite, which was not decomposed in the zones of pearlite and intermediate transformation, will be transformed into martensite. In this case, the structure of the cooled steel will consist of pearlite (sorbite or troostite), acicular troostite (bainite), martensite, and a certain amount of retained austenite (Fig. 213, *c*). Partial transformation of the austenite, in the zones of pearlite and intermediate transformation, alters its composition by enriching it with carbon. This lowers the martensite point and increases the amount of retained austenite (Fig. 213, *b* and *c*). Upon a further increase in the cooling rate, intermediate transformation is suppressed and only partial decomposition occurs in the pearlite zone. The main part of the austenite is supercooled to point *M* and undergoes martensitic transformation. After being cooled, the structure of the steel will consist of troostite+martensite+retained austenite (Fig. 213, *c*). At cooling rates exceeding the critical value, austenite will be subject only to martensitic transformation and the steel will acquire a structure of martensite+retained austenite.

If the rate of the intermediate (bainite) transformation exceeds that of decomposition in the pearlite zone, it will require a high cooling rate to suppress the intermediate reaction. Fig. 214 shows that pearlite (sorbite) is formed in austenite transformation at low cooling rates. At higher cooling rates, in addition to the pearlite transformation, acicular troostite (bainite) will also be formed and a part of the austenite, supercooled to point *M*, will form martensite. The resulting structure of the steel will contain pearlite (sorbite), acicular troostite, martensite, and retained austenite. At still higher cooling rates, transformation in the pearlite temperature range will be suppressed and only intermediate transformation will occur to form acicular troostite (bainite) and martensite (Fig. 214, *b*). Partial austenite decomposition with the formation of acicular troostite may lead to the retaining of a large amount of austenite, much more than would be found in the steel after cooling at a rate exceeding the critical value (Fig. 214, *c*).

Fig. 215 illustrates isothermal and thermokinetic decomposition of supercooled austenite for a steel containing 0.45 per cent C and 1.0 per cent Cr. It presents a more proper conception of super-

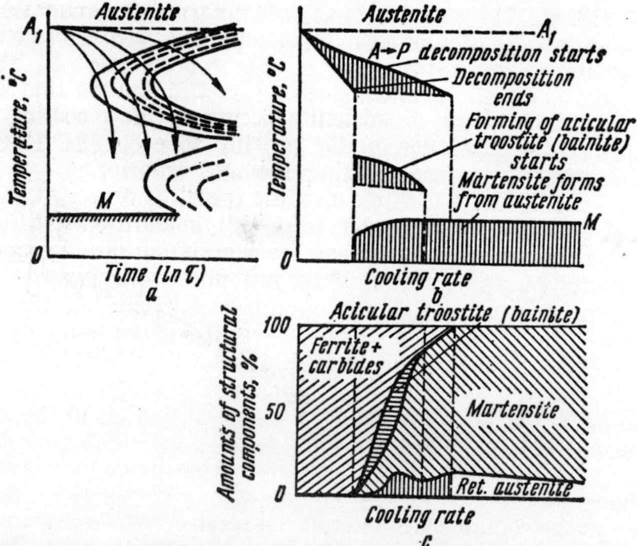

Fig. 213. Transformation of alloyed austenite of moderate stability at intermediate (bainite) transformation temperatures (after A. A. Popov):

a—kinetics of isothermal austenite transformation, *b*—effect of cooling rate on transformation temperature, *c*—effect of cooling rate on the number of the structural components

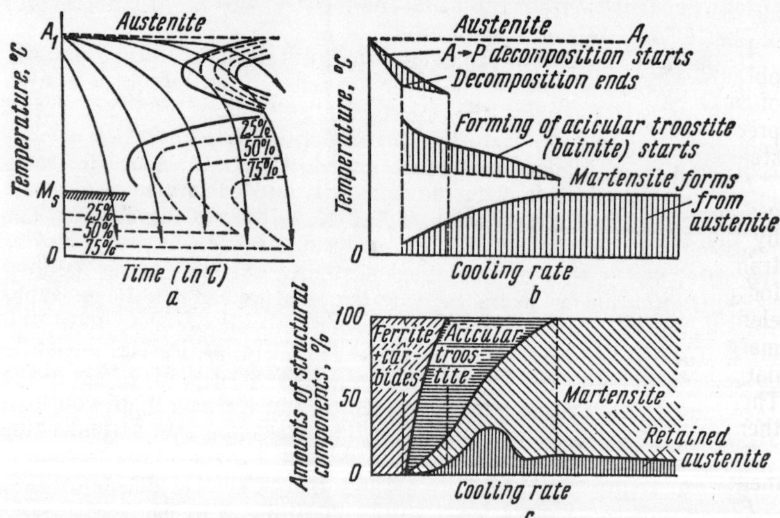

Fig. 214. Transformation of alloyed austenite of low stability at intermediate (bainite) transformation temperatures:

a—kinetics of isothermal austenite transformation, *b*—effect of cooling rate on transformation temperature, *c*—effect of cooling rate on the number of the structural components

cooled austenite transformation in alloy steels upon continuous cooling.

Thermokinetic diagrams of most alloy steels differ from those of carbon steels (Fig. 136) in that the former have a zone of intermediate transformation corresponding to the formation of acicular troostite (bainite). This reaction is not completed and at a lower temperature is resumed as a transformation of the martensitic type.

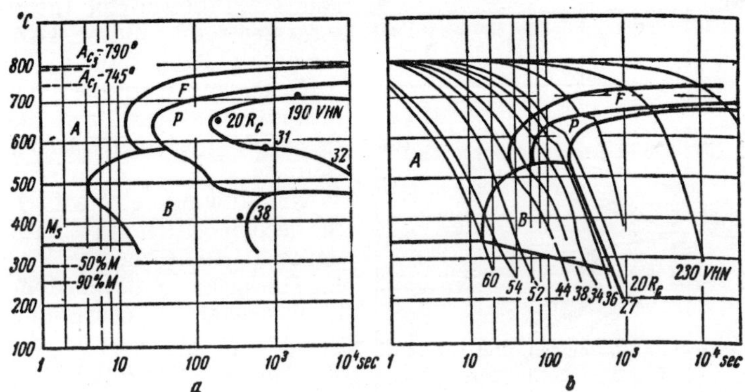

Fig. 215. Isothermal and thermokinetic diagrams of supercooled austenite decomposition in steel Grade 45X

It is clear from the aforegoing, that the structure of alloy steels obtained by various cooling rates is much more complex than that of carbon steels. The structure becomes more complex if excess ferrite precipitates from a hypoeutectoid steel or carbide from a hypereutectoid steel (Fig. 215).

Effect of alloying elements on the martensitic transformation. No changes are contributed to the kinetics of the martensitic transformation by the addition of alloying elements to a steel. All features of this transformation as observed in carbon steel (see Sec. 8-5) are the same for an alloy steel. It must be noted, however, that all alloying elements, with the exception of cobalt and aluminium, lower the martensite points M and M_f (Fig. 216). In doing this, alloying elements, naturally, increase the amount of retained austenite in hardened steel. The exceptions, cobalt and aluminium, raise the martensite points and, therefore, reduce the amount of retained austenite.

The martensite hardness remains almost unaffected by alloying elements.

Effect of alloying elements on transformations in tempering. Alloying elements retard the processes occurring in the tempering of hardened steel. Even a large amount of alloying elements has no appreciable effect on the first stage of martensite decomposition (100° to 150°C).

This stage is characterised by the appearance of a large amount of carbide nuclei whose growth ceases very soon.

The second stage of martensite decomposition (further slow precipitation of carbon from the martensite and a certain growth of the carbide particles due to carbon diffusion) is greatly influenced by alloying elements.

The carbide-forming elements, chromium, vanadium, titanium, molybdenum, and tungsten, as well as silicon, retard the second stage and retain the state of the supersaturated solid solution at higher tempering temperatures.

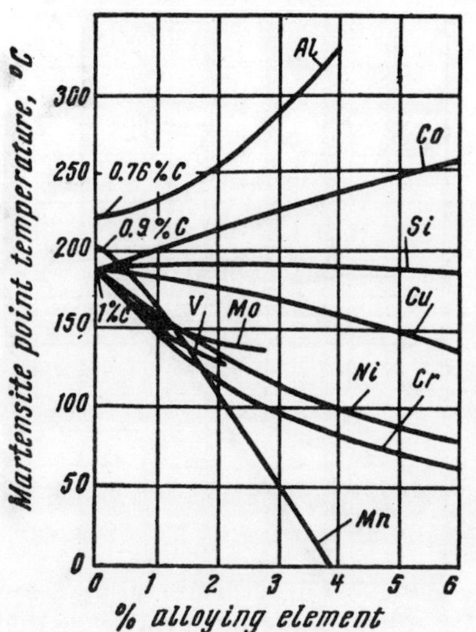

Fig. 216. Effect of alloying elements on the martensite point (after V. D. Sadovsky, V. I. Zyuzin, and S. I. Baranchuk)

Nickel and manganese have an insignificant effect on the second stage of martensite decomposition.

All alloying elements retard the decomposition of retained austenite to some degree by raising the transformation temperature. Manganese, chromium, and silicon are the most effective in this respect. The influence of nickel, molybdenum, tungsten, and vanadium is comparatively weak.

Alloying elements also have a considerable effect on the coagulation of carbides in tempering.

The growth of carbide particles may be substantially changed as a result of the influence of alloying elements on the carbon diffusion rate in the α solution. It has been established that the carbide-forming elements, vanadium, molybdenum, titanium, and chromium, as well as silicon, retard coagulation. Therefore, at the same tempering temperatures, alloy steel containing these elements will have more dispersed particles of the carbide phase than plain carbon steel. Nickel and cobalt have no appreciable effect on the rate of coagulation. It is necessary to point out that in tempering an alloy steel at a low temperature, cementite will be formed without regard for the type of carbide that is stable under the given conditions. This cementite will have the same concentration of the alloying element as the martensite since the diffusion of the alloying elements is very limited at these low temperatures. Noticeable diffusion of the alloying elements is possible at temperatures above 400-450°C. This redistributes these elements between the ferrite and carbide and refers chiefly to carbide-for-

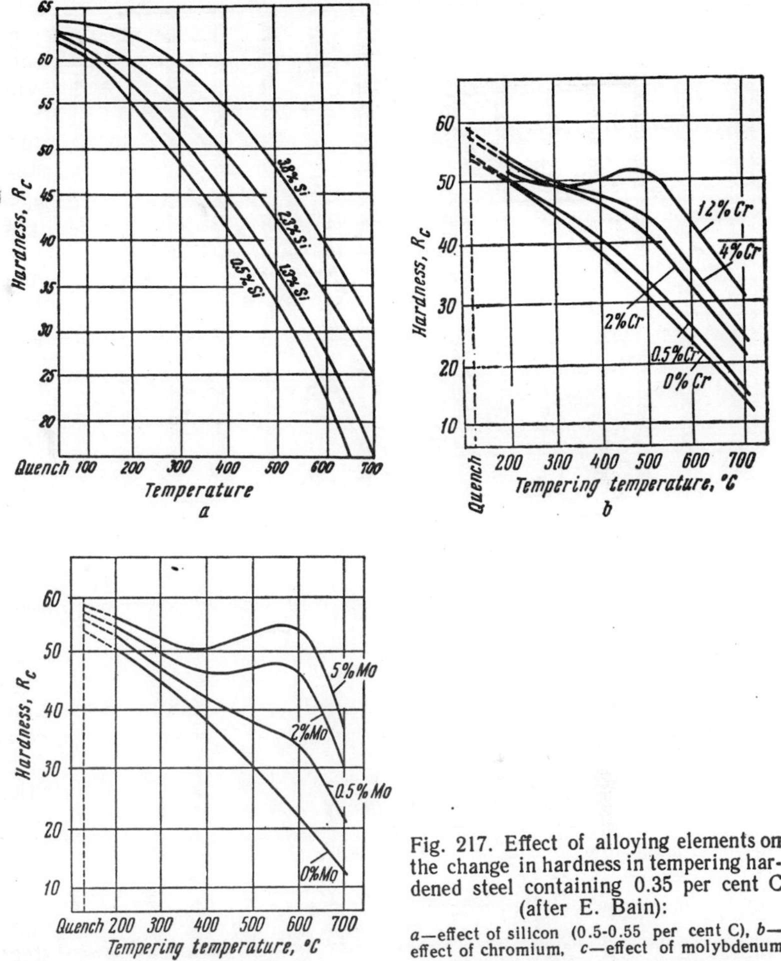

Fig. 217. Effect of alloying elements on the change in hardness in tempering hardened steel containing 0.35 per cent C (after E. Bain):

a—effect of silicon (0.5-0.55 per cent C), *b*—effect of chromium, *c*—effect of molybdenum

ming elements which pass over to the carbide from the ferrite. When a special carbide is the equilibrium compound for a given steel, then a carbide transformation will occur, after the cementite is saturated by the alloying element, to yield the special carbide $(Fe, M)_3C \rightarrow \rightarrow (M, Fe)_xC_y$.

In retarding transformation processes in tempering, alloying elements also have a large influence on the mechanical properties of the tempered steel. The effects of chromium, molybdenum, and silicon

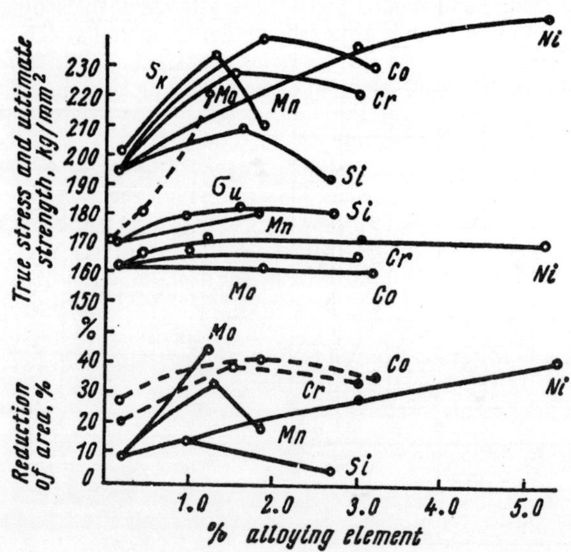

Fig. 218. Effect of alloying elements on the mechanical properties of steel with 0.35 per cent C, hardened and then tempered at 200° C (after S. T. Kishkin and S. Z. Bokstein)

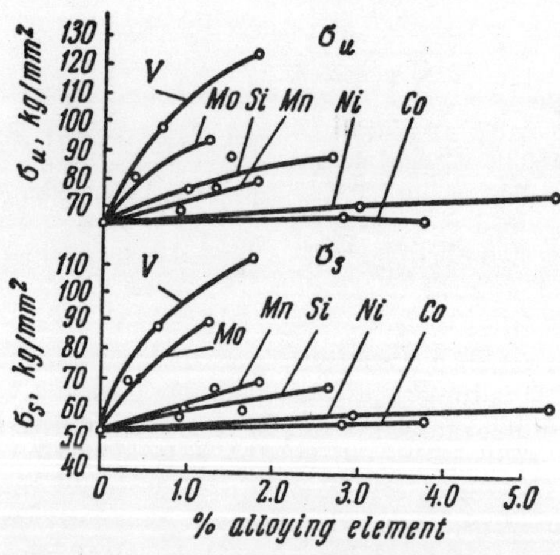

Fig. 219. Effect of alloying elements on the mechanical properties of steel after high tempering (after S. Z. Bokstein)

on the variation in the hardness of hardened steel with the tempering temperature are shown in Fig. 217.

All alloy steels, and particularly those alloyed with carbide-forming elements, are harder than plain carbon steel tempered at the same temperatures.

The effect of alloying elements on the reduction in hardness in tempering is explained by the fact that they retard the precipitation of carbon from the martensite; at higher tempering temperatures they retard the coagulation of the carbides. The increase in hardness at high tempering temperatures in steels, containing large amounts of such elements as tungsten, molybdenum, or chromium, is probably associated either with the transformation of retained austenite into martensite or with the precipitation of a large amount of carbide particles in the martensite. These particles lock the slip planes. Slower hardness reduction at high tempering temperatures in a silicon steel is due to retardation of carbide coagulation and to hardening of the ferrite by the silicon it dissolves.

Nickel hardens the ferrite in the same manner and with the same results. The hardening effect of such elements as silicon, nickel, and manganese is much lower than that of carbide-forming elements.

The effect of alloying elements on the mechanical properties of tempered martensite is illustrated in Fig. 218. Their effect on the mechanical properties after high tempering is shown in Fig. 219.

EMBRITTLEMENT IN TEMPERING ALLOY STEELS

Two types of temper brittleness may occur in the tempering of alloy steels (Fig. 220). As has been indicated (see page 214), the *impact strength is appreciably reduced* when a hardened steel is tempered in the range from 250° to 400°C irrespective of the rate of subsequent cooling. A distinctive feature of this type of embrittlement is its nonrecurring character, i.e., if brittleness is eliminated by heating above 400°C, then a subsequent heating in the range 250-400°C will not reduce the impact strength.

The most probable reason for this type of embrittlement is the precipitation of carbides from martensite during tempering.

A second type of embrittlement is found in certain steels if they are slowly cooled after tempering in the range 500-600°C (or after too long a holding time at 500-550°C). It does not occur when cooling is carried out rapidly from the tempering temperature. Fig. 220 shows that impact strength is not reduced in rapid cooling but increases normally with the tempering temperature. An essential feature of this type of embrittlement is its recurrence.

Brittleness due to slow cooling after tempering in the range from 500° to 600°C may be eliminated by a second tempering followed by rapid cooling. It can be restored by additional tempering followed by slow cooling. This embrittlement is most frequently observed in steels containing large amounts of phosphorus, manganese, silicon, chromium, vanadium, or nickel (in conjunction with chromium or manganese). The addition of molybdenum or tungsten to steel considerably reduces its tendency to be embrittled by tempering.

In all probability, all steels are susceptible to temper brittleness to some extent. This tendency, however, cannot be revealed by impact tests at a single temperature (+20°C). It will be necessary to determine

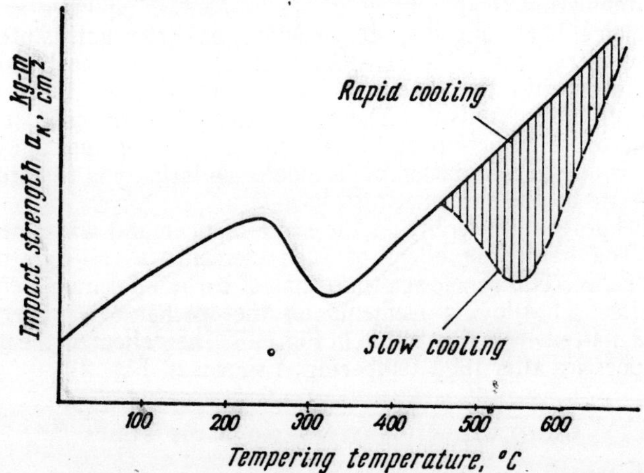

Fig. 220. Variation in impact strength with the tempering temperature for steels susceptible to temper brittleness (schematic)

the impact strength at various temperatures so as to find the cold-shortness temperature threshold (critical temperature of brittleness) and its variations with the steel composition and tempering conditions.

The true reason for the second type of temper brittleness has not yet been established but it is probably due to the precipitation of some kind of particles—carbides (for example, Fe_3C), nitrides, phosphides, or others, at the grain boundaries in a form which has an unfavourable effect on the ductility. If the steel is cooled rapidly (quenched) from the high tempering temperatures, these components remain in solution and no embrittlement occurs.

P. B. Mikhailov-Mikheyev is of a different opinion. On the basis of numerous experiments, he suggests that this embrittlement is due to the diffusion of dissolved atoms to the grain boundaries. This leads to supersaturation of the grain surface layers by these elements without the precipitation of finely dispersed phases (carbides, phosphides, etc.)*.

* See *Physical Metallurgy and Metal Treatment*, No. 2, 1956.

12-4. Structural Classes of Alloy Steels

In their equilibrium state (annealed), in accordance with the positions of the eutectoid point S and point E (maximum saturation of austenite by carbon) in the Fe-C diagram, alloy steels, as well as carbon steels, may be classified as: 1) hypoeutectoid, 2) hypereutectoid, and 3) ledeburitic or eutectic. It must be noted that alloying elements shift the points S and E toward a lesser carbon content. This is quite clear from Fig. 221. In the structural diagram for alloying elements that contract the zone in which the γ-phase exists (Fig. 221, *b*), we find

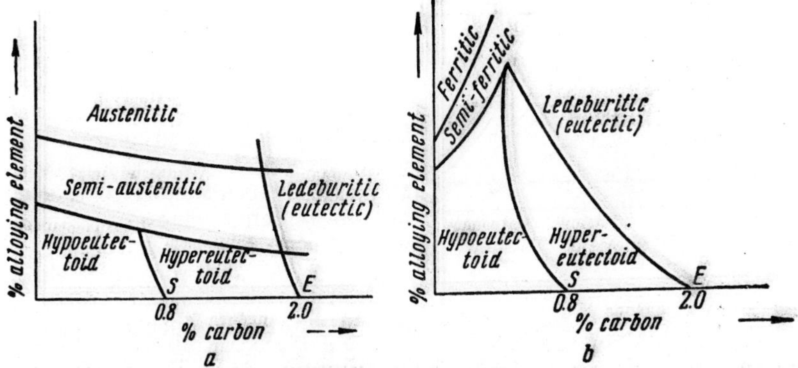

Fig. 221. Structural class distribution in the ternary system iron-carbon-alloying element (after A. P. Gulyaev):
a—alloying element which extends the γ-phase zone, *b*—alloying element which contracts the γ-phase zone

a ferritic steel zone; elements which extend the γ-phase range produce an austenitic steel zone (Fig. 221, *a*). The $\alpha \rightleftarrows \gamma$ transformation does not occur in these steels. It should be noted, however, that these three classes are actually observed only in alloy steels which contain carbide-forming elements. Increasing the silicon or nickel content, in conjunction with an increase in carbon content, will not lead to the formation of the eutectic class of steels (because of graphitisation). Semi-austenitic, semi-ferritic, austenitic, and ferritic classes of steel may also be obtained if more of the alloying element is added.

On the basis of the structure obtained when specimens of a small cross-section are cooled in air, the three chief classes of steel are: 1) pearlitic, 2) martensitic, and 3) austenitic. Which class is obtained is determined by the stability of the supercooled austenite. The higher the alloying element content, the more stable the supercooled austenite will be and the lower the martensitic transformation temperature. Therefore, at the same cooling rate (in air), different structures will be observed in steels of various composition (Fig. 222).

These two criteria, the effect of alloying elements on the points *S* and *E* of the Fe-C equilibrium diagram and the structure obtained in cooling small specimens in air, have been accepted as the basis for the following general structural classification of alloy steels:

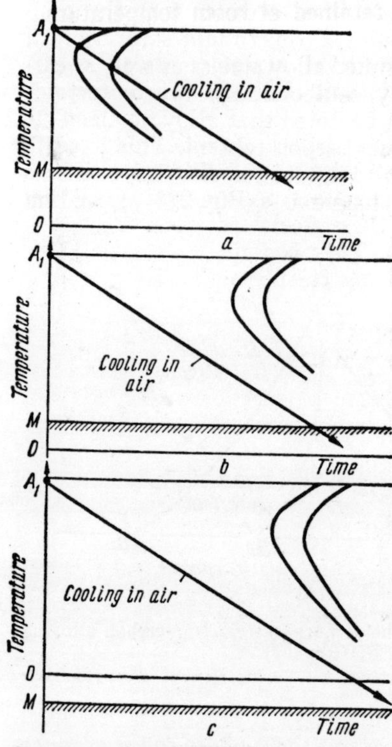

Fig. 222. Diagrams of isothermal austenite decomposition for steels of the *a*—pearlitic, *b*—martensitic, *c*—austenitic classes

1. *Ferritic steels.* Steels of this class contain a negligible amount of carbon and a great deal of alloying elements that extend the zone in which the α-phase exists. Such elements are Cr, Mo, W, V, Si, etc.

Ferritic steels belong to the non-hardening type since the $\alpha \rightleftarrows \gamma$ transformation does not occur when they are heated and only grain growth is observed. The structure of such a steel consists of alloyed ferrite and a certain amount of carbides (Fig. 223, *a*).

2. *Pearlitic steels.* These steels contain a comparatively small amount of alloying elements (not over 5-6 per cent) and, after cooling in air, the austenite decomposes in a high temperature range into the ferrite-cementite mixture—pearlite, sorbite, or troostite (Fig. 223, *b*). Excess ferrite (in hypoeutectoid steels) or carbides may also be present in the structure in accordance with the carbon content.

Most structural (hypoeutectoid) and tool (hypoeutectoid and hypereutectoid) steels belong to this class.

3. *Martensitic steels* contain a large amount of alloying elements which extend the γ-phase zone (such as Mn, Ni, and others). The presence of a considerable amount of alloying elements increases the austenite stability (the isothermal curve is shifted to the right) and cooling in air is sufficient to supercool the austenite to the martensite point (Fig. 222, *b*). After cooling in air, these steels will have a structure containing alloyed martensite (Fig. 223, *c*) in which there may be inclusions of excess carbide.

4. *Austenitic steels* contain up to 20 or 30 per cent alloying elements (chiefly Ni, Cr, and Mn). Due to the exceptionally high austenite sta-

bility and the lowering of the martensite point to sub-zero temperatures (Fig. 222, *c*) the austenite will be retained at room temperature (Fig. 223, *d*).

Austenitic steels include chromium-nickel stainless steels, certain heat-resistant steels, non-magnetic wear-resistant (high-manganese) steels, and others.

Steels whose structure contains primary carbides (carbides precipitating from the liquid alloy), without regard for the matrix metal (i.e., either of the pearlitic or martensitic classes) are frequently called carbidic or ledeburitic steels. Usually, these steels cannot be considered as belonging to any definite class. They contain a considerable amount of carbon and carbide-forming elements, such as Cr, W, V, Mo, and others.

The primary carbides in cast ledeburitic steel form the eutectic called ledeburite (Fig. 223, *e*). As the result of forging, the excess carbides acquire the form of separate globules (Fig. 223, *f*) more or less uniformly distributed in the matrix. High-speed steel is a typical representative of this class.

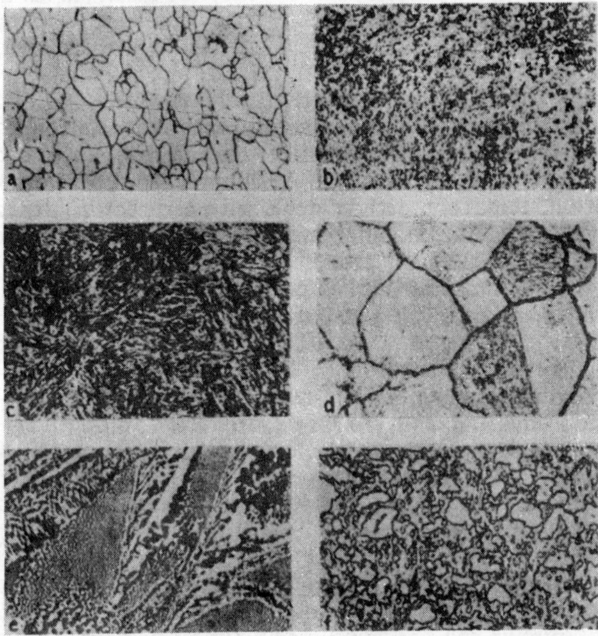

Fig. 223. Microstructures of the various classes of alloy steels:
a—ferritic (0.10 per cent C and 27 per cent Cr), × 250, *b*—pearlitic (0.12 per cent C, 1.0 per cent Cr, 1.8 per cent Ni), × 250; *c*—martensitic (0.18 per cent C, 1.6 per cent Cr, 4.28 per cent Ni, 1.05 per cent W), × 250; *d*—austenitic (0.1 per cent C, 18 per cent Cr, and 8 per cent Ni), × 250; *e*—ledeburite in cast high-chromium steel (1.48 per cent C and 12.5 per cent Cr), × 800; *f*—ledeburite in steel of the same composition, × 400

Chapter 13

STEEL

13-1. General Classification

All steels which find application in the U.S.S.R. at the present time may be classified on the basis of the following principal characteristics:

I. As to their chemical composition, they are classified as: a) plain carbon steels, and b) alloy (or special) steels. Besides carbon and the inevitable minor constituents (Mn, Si, S, P, N, and others), alloy steels contain certain other specially-added elements, such as Mn (> 1 per cent), Si (>0.5 per cent), Ni, Cr, Mo, W, V, Ti, and others which impart the required properties to the steel.

II. As to the method of manufacture, they are classified as:

a) Ordinary and improved quality steels. These are smelted either in converters (acid or basic) or in large open-hearth furnaces. In composition, these are carbon steels with a carbon content from 0.6 to 0.65 per cent. The manufacturers of these steels guarantee either their specified mechanical properties or chemical composition. They contain slightly more harmful impurities than the higher quality grades (sulphur up to 0.055 per cent and phosphorus up to 0.05-0.085 per cent).

b) Quality steels, smelted in basic open-hearth furnaces. Here, stricter conditions are observed in regard to the composition of the charge and smelting and pouring procedures. Much higher requirements are specified for these steels than the ordinary quality grades in respect to their chemical composition, presence of nonmetallic inclusions and other defects, as well as to their mechanical properties.

c) High-quality steels, smelted in acid or basic open-hearth furnaces or in electric furnaces. The composition of these steels is maintained within narrow limits; they contain only very small amounts of harmful impurities (sulphur and phosphorus) and almost no nonmetallic inclusions, their mechanical properties are strictly specified. These steels are poured into ingot moulds of the smallest permissible size.

III. As to their purpose, all steels may be classified into the following groups:

a) Structural steels with the sub-groups:

1) Constructional steels* designed for structural members and steel structures. This sub-group includes ordinary quality carbon steels possessing good weldability and satisfactory mechanical properties in the as-received condition, i.e., without additional heat treatment.

2) Engineering steels designed for the manufacture of components of industrial equipment. The main requirements made to these steels are high strength under static and dynamic loads and good fabrication properties (pronounced hardenability, minimum strain, satisfactory machinability, etc.). It is also important that their mechanical and fabrication properties vary within narrow limits in different melts or batches of the same steel. This is necessary to provide machine parts of uniform quality in cases when an established manufacturing process is strictly conformed to. This is a vital factor for mass production plants.

Carbon, alloy, quality, and high-quality grades find applications as engineering steels. Ordinary quality carbon steels are also used for low-stress applications. Most engineering steels undergo heat treatment.

b) Tool steels intended for the manufacture of cutting and measuring tools, as well as dies. Tool steels must have a high hardness and wear resistance. They must have the high strength required by the cutting edge of a tool to prevent its destruction by the high stresses due to metal cutting and to prevent breakage in those parts of the tool that are subject to large torque and bending stresses, as well as ample toughness to withstand impacts and vibration. Steel designed for the manufacture of hot-forging dies must have satisfactory mechanical properties at high temperatures.

This sub-group includes carbon steels (over 0.65-0.75 per cent C) and quality and high-quality alloy steels with a high carbon content.

c) Steels with special physical properties which suit their application. These include magnetic and nonmagnetic steels, steel with a definite coefficient of expansion, stainless and heat-resistant steels, etc. In their chemical composition, these special-purpose grades are quality and high-quality alloy steels.

DESIGNATION OF STEELS

The designation system for steels, standardised in the Soviet Union, utilises numbers and Russian letters.

Ordinary quality carbon steels are identified by the letters Ст. followed by the numbers 1, 2, 3, etc. (such as Ст. 1, Ст. 2, Ст. 3, Ст. 5, etc.). Increasing numbers indicate an increase in tensile strength and carbon content and a decrease in ductility.

* It is difficult to set a sharp boundary between constructional and engineering steels.

Quality structural steels are designated by the numbers 05, 08, 10, 20, 30, 45, etc., which indicate the average carbon content in hundredths of one per cent.

Carbon tool steel is designated by the Russian letter У and a number indicating the average carbon content in tenths of one per cent. Examples are У7 (containing 0.7 per cent C), У8, У9, У10, У11, У12, etc.

According to the U.S.S.R. Standard, GOST, the various alloy steels are designated by a combination of Russian letters and numbers. The letters indicate the presence of a certain alloying element. Accordingly, Г indicates manganese, С—silicon, Х—chromium, Н—nickel, М—molybdenum, В—tungsten, Ф—vanadium, Ю—aluminium, Т—titanium, К—cobalt, and Д—copper. The number preceding the letter indicates the carbon content in hundredths of one per cent while the number after the letter indicates the approximate percentage of the alloying element, if its content exceeds 1.5 per cent. In all cases, the letter A at the end of the designation indicates that the steel is of high quality (free of harmful impurities) and that high requirements are specified in respect to metallurgical quality control.

For example, steel 15Х is a chromium steel containing ~ 0.15 per cent C and ~ 1 per cent Cr and steel 12Х2Н4А is a high-quality chromium-nickel steel containing ~ 0.12 per cent C, ~ 2 per cent Cr, and ~ 4 per cent Ni.

13-2. Ordinary and Improved Carbon Structural Steels (see also Appendix I)

Table 21 lists the composition and mechanical properties of certain ordinary and improved steels.

Ordinary and improved hot rolled carbon steels are divided into the following three groups:

1) Steels of Group I are supplied with their mechanical properties guaranteed (БСт. 0, БСт. 3, БСт. 4, БСт. 5, БСт. 6, Ст. 0, Ст. 1, Ст. 2, Ст. 3, Ст. 4, Ст. 5, Ст. 6 and Ст. 7).

Normally these steels do not undergo heat treatment and they are utilised in the as-received condition. The mechanical properties of these steels are determined by the treatment they are subject to by the user.

2) Steels of Group II are supplied with their chemical composition guaranteed (Б09, Б16, Б23, Б33, М12кп, М18, М21, М26, etc.). Ordinary steels are made either by the open-hearth (М) or Bessemer (Б) processes. In comparison with open-hearth steel, Bessemer steel has a higher gas saturation, contains more sulphur and phosphorus, and, at the same carbon content, has lower ductility and higher strength and hardness. Bessemer steels have a greater susceptibility to ageing and cold-shortness. They may be recommended only for statically loaded structures.

In respect to the degree of deoxidation, ordinary quality steels may be classified as killed, semi-killed, and rimmed.

Killed steels are deoxidised in their manufacture and have a low gas content. They are used in making machine parts subject to dynamic loads. Rimmed steels are among the cheaper varieties since the losses in their manufacture are insignificant (a rimmed steel ingot does not have a pipe). At the same time, they possess essential disadvantages which include: 1) high gas saturation, 2) lower impact strength, and 3) high tendency to ageing.

3) Steels of Group III are of improved quality (Table 21). They are produced, as a rule, in open-hearth furnaces and are degasified and deoxidised. They are supplied with guaranteed composition and mechanical properties.

Table 21

Examples of Ordinary and Improved Quality Carbon Steels

Group	Grade	Chemical composition limits, per cent					Mechanical properties	
		C	Si	Mn	P_{max}	S_{max}	σ_u, kg/mm²	δ_{10}, per cent
		Bessemer steels						
I	БСт.3 (БСт. 3кп)	—	—	—	—	—	38-46	22-21
	БСт. 4 (БСт. 4кп)	—	—	—	—	—	42-52	22-20
		Open-hearth steels						
I	Ст. 3 (Ст. 3кп)	—	—	—	—	—	40-47	23-21
	Ст. 4	—	—	—	—	—	42-52	21-19
	Ст. 6	—	—	—	—	—	60-72	13-11
		Bessemer steels						
II	Б09 (Б09кп)	0.12	0.05	0.25-0.55	0.08	0.06	—	—
	Б16	0.12-0.2	0.12-0.32	0.5-0.8	0.08	0.06	—	—
	Б33	0.26-0.4	0.12-0.32	0.6-0.9	0.08	0.06	—	—
		Open-hearth steels						
II	М09кп	0.06-0.11	0.05	0.3-0.5	0.045	0.05	—	—
	М16	0.12-0.19	0.12-0.3	0.4-0.65	0.045	0.05	—	—
	М18	0.14-0.22	0.12-0.3	0.4-0.65	0.045	0.05	—	—
	М31	0.27-0.35	0.15-0.32	0.5-0.8	0.045	0.05	—	—
		Open-hearth improved steels						
III	М09	0.06-0.11	0.11-0.3	0.3-0.5	0.04	0.04	32-40	28
	М18а	0.14-0.21	0.11-0.3	0.4-0.65	0.04	0.04	46-50	24-22
	М31а	0.27-0.35	0.15-0.32	0.5-0.8	0.04	0.04	50-62	18-16

Ordinary and improved steels are extensively employed in construction and general engineering.

For example, rolled bars, structural shapes, sheets, and plates of steels Ст. 3, M18кп, and БСт. 4 are widely used for construction components and trusses. These same steels are expediently used in engineering for manufacturing various less important machine parts (bushings, liners, levers, rods, yokes, etc.), not subject to heat treatment, as well as parts to be carburised, such as wrist pins of low-power engines, gears, worms, etc. Steels Ст. 5, M26, M31, and БСт. 5 are used for reinforcement and various machine parts (shafts, pivots, nuts, washers, etc.), not subject to heavy loads in operation.

Improved steels (Table 22) are widely employed for welded (M16C) and rivetted (MСт. 3 мост) bridge structures, in ship-building (Ст. 1С, Ст. 2С, Ст. 3С, Ст. 4С), for prestressed concrete constructions, etc.

Table 22

Special-Purpose Carbon Steels

Grade	Chemical composition limits, per cent					Mechanical properties			Applications
	C	Mn	Si	S_{max}	P_{max}	σ_u, kg/mm²	σ_s, kg/mm²	δ_{10}, per cent	
M16C	0.12-0.20	0.40-0.70	0.12-0.25	0.045	0.040	38	23	24-22	Bridge-building
MСт. 3 мост	0.14-0.22	0.40-0.70	0.15-0.30	0.050	0.045	38	24	24-22	
Ст. 1С	0.07-0.12	0.35-0.50	—	0.050	0.050	32-40	—	28	Ship-building
Ст. 2С	0.09-0.15	0.35-0.50	—	0.050	0.050	32-42	21	26	
Ст. 3С	0.14-0.22	0.35-0.60	0.12-0.35*	0.050	0.050	38-47	22	23, 22, 21	
Ст. 4С	0.18-0.27	0.40-0.70	0.12-0.35	0.050	0.050	42-52	24	21, 20, 19	
M71	0.64-0.77	0.60-0.90	0.13-0.28	0.050	0.040	min 80	—	—	Rails
M76	0.69-0.82	0.70-1.0	0.13-0.28	0.050	0.040	min 80	—	—	
M67	0.60-0.75	0.60-0.90	0.13-0.28	0.050	0.040	min 70	—	—	

* Rimmed steels 1C, 2C, 3C, and 4C show only traces of Si.

13-3. Quality Carbon Structural Steels

Table 23 lists the composition of quality carbon hypoeutectoid steels while Table 24 indicates their mechanical properties. These steels are used as bar stock, forgings, sheet, plate and other semi-finished products.

Table 23

Compositoin of Quality Carbon Steels*

Grade	Carbon content, per cent	Grade	Carbon content, per cent
05кп	⩽0.06	45	0.42-0.49
08кп	0.05-0.10	50	0.47-0.55
10	0.07-0.13	55	0.52-0.60
15	0.12-0.18	60	0.57-0.65
20кп	0.17-0.24	65	0.62-0.70
20	0.17-0.24	70	0.67-0.75
25	0.22-0.29	75	0.72-0.80
30	0.27-0.34	80	0.77-0.85
35	0.32-0.39	85	0.82-0.90
40	0.37-0.44		

* Rimmed steels (кп) contain: Si ⩽ 0.03 per cent (all others contain 0.17-0.37 per cent Si), 0.35-0.65 per cent Mn (steels Grade 05кп, 08кп, 08, 10, 15, 20кп, and 20) or 0.50-0.80 pcr cent Mn (steels Grade 25, 30, 35, 40, 45, 50, 55, 60, 65, 70, 75, 80, and 85), P ⩽ 0.040 per cent, and S ⩽ 0.045 per cent.

Quality carbon steels are smelted in basic open-hearth furnaces.

Low-carbon steel Grade 08кп (or 05кп) has a ferrite structure and high ductility. To acquire high ductility, the steel, in addition to minimum carbon content, must contain the least possible amount of silicon which intensively hardens the ferrite. Therefore, these steels are of the rimming type and contain no more than 0.03 per cent Si. To obtain a smooth surface, the ferrite grain of steel 08кп (or 05кп) should be as fine as possible. Steel 08кп (or 05кп) is designed for a great variety of cold formed parts of sheet steel (headlights, fenders and bodies of cars, etc.).

Steels 10 and 15, having comparatively low strength, high ductility, and good weldability, are applied for machine parts carrying relatively small loads and in whose manufacture welding, forming, and other similar processes are required. They are also used as carburising steels. Steels 10 and 15, for example, are used for machine parts that are to undergo carburising or cyaniding (guide bushings, screws, gears for light loads, spline shafts, stops, pins, etc.), as well as for welded structures, in the form of plates or pipe. Steel Grades 20, 25, 30, 40, 45, and 50 have a higher tensile strength but lower fabrication properties, weldability, and ductility (in comparison to low-carbon steels). Their machinability, however, is higher.

Steels 20 and 25 have found application for small machine parts which are not subject to high stresses (bolts, nuts, screws, washers, etc.) while steels 30 and 35 are for machine parts subject to moderate stresses (cross-members, tie-rods, bolts, nuts, etc.). Steels 20, 25, 30, and 35 are ordinarily used in the as-normalised condition.

Table 24

Mechanical Properties of Quality Carbon Steels

Grade	σ_s, kg/mm²	σ_u, kg/mm²	δ_s, per cent minimum	ψ, per cent minimum	a_k, kg-m/cm² minimum
08кп	20	33-41	35	60	—
08	21	34-42	34	60	—
10	22	36-45	32	55	—
15	24	40-49	29	55	—
20кп	25	43-53	27	55	—
20	26	44-54	26	55	—
25	28	48-58	24	50	10
30	30	52-62	22	50	10
35	32	56-66	21	45	9
40	34	60-72	19	45	9
45	36	64-76	17	40	8
50	38	68-80	15	40	7
55	40	71-83	13	35	—
60	42	73-85	12	35	—
65	43	76-88	11	35	—
70	44	78-90	10	30	—
75	90	110	8	30	—
80	95	110	8	30	—
85	100	115	7	30	—

Steels 40 and 45 are used in making machine parts requiring high strength (connecting rods, gears, pivots, shafts, bushings, tractor crawler pins, etc.). They are used either as-normalised or after hardening followed by high-temperature tempering (structural improvement). Steels Grade 50, 55, and 60 are used, after hardening and tempering or after normalising, for manufacturing such machine parts as gears, tyres, flat springs, snap rings, various other springs, leaf springs, mill rolls, etc. Coil and leaf springs of steel 65 and 70 are hardened by heating to 830-840°C, quenching in oil, and tempering at 350-450°C.

Due to the low stability of the supercooled austenite (Fig. 224, *a*), carbon steels have a high critical cooling rate and a low hardenability (see Fig. 158). Carbon steels are usually quenched in water. However, even in this case, through or full hardening can be achieved only for parts up to 15 or 20 mm in diameter (or thickness). Therefore carbon steels possess high mechanical properties after structural improvement only in small cross-sections (Fig. 225) and can be recommended only for small parts. Variations in the mechanical properties of steel 45 with the tempering temperature are indicated in Fig. 165, *a*.

Carbon steels with a higher manganese content have higher hardenability. They include steels Grade 15Г, 20Г, 30Г, 40Г, 50Г, 60Г, 65Г, and 70Г, which contain from 0.8 to 1.0 per cent Mn, and steels 10Г2, 30Г2, 35Г2, 40Г2, 45Г2, and 50Г2 which contain from 1.4 to 1.8 per

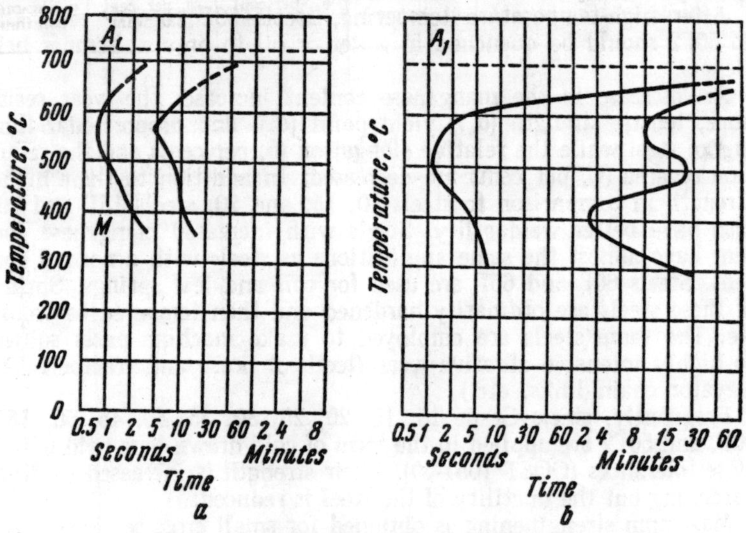

Fig. 224. Isothermal decomposition (TTT) diagrams:
a—for steel Grade 30, *b*—for steel Grade 30Г2

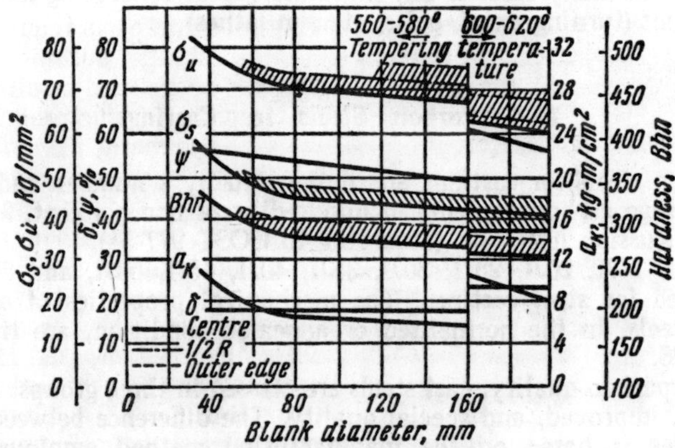

Fig. 225. Mechanical properties of steel Grade 40 of various cross-section

cent Mn (GOST 1050-52). The addition of manganese to the steel increases the stability of the supercooled austenite (Fig. 224, *b*), lowers the critical cooling rate, and increases hardenability.

Oil is the quenching medium used in many cases for hardening steels alloyed with manganese. Such steels must be carefully heated as they are very sensitive to overheating.

After high-temperature tempering, steels 50Г to 70Г and 30Г2 to 50Г2 should be quenched in water or oil to prevent temper brittleness.

An increase in the manganese content increases the wear resistance, tensile strength (σ_u), yield point (σ_s), and proportional limit (σ_p) of steel while the relative elongation (δ, per cent) and the reduction of area (ψ, per cent) are decreased. In addition to their higher strength in comparison to steels 10, 15, and 20, steels 15Г and 20Г also have better weldability. Steels with increased manganese content have almost the same applications as steels with a normal content. Steels 60Г and 65Г are used for coil and leaf springs. Springs of these steels are ordinarily hardened and then tempered to troostite. The same steels are employed to make machine parts subject to highly intensive abrasive wear (teeth of horse and tractor rakes, elevator chain links, etc.).

Frequently, steels Grade 10, 15, 20, 25, 30, 35, 40, 45, 50, 15Г, 50Г, and 50Г2 are applied in the form of cold drawn bars held within close tolerances (GOST 1051-50). Their strength is increased by work hardening but the ductility of the steel is reduced.

Maximum strengthening is obtained for small cross-sections (wire, sheets, etc.). Cold working is the most highly productive and most economical method for obtaining machine parts with smooth accurate surfaces. In many cases it can eliminate labour-consuming machining operations (turning axles, bolts, etc., in lathes).

13-4. Carbon Steels for Castings

Carbon steels for castings are designated by a number indicating the average carbon content, in hundredths of one per cent, followed by the Russian letter Л* (according to GOST 977-58).

Grades 15Л, 20Л, 25Л, 30Л, 35Л, 40Л, 45Л, 50Л, and 55Л are employed for steel castings. The mechanical properties of some of these steels, in the normalised or annealed condition, are listed in Table 25.

In respect to quality, cast steels are classed in three groups: normal quality, improved, and special quality. The difference between these categories is based on the manufacturing method employed, the amount of harmful impurities they contain, and requirements as to micro- and macrostructure.

* Abbreviation of the word "litaya", i.e., "cast".

Table 25

Mechanical Properties of Steels for Castings

Grade *	σ_s, kg/mm²	σ_u, kg/mm²	δ, per cent	ψ, per cent	α_k, kg-m/cm²
15Л	20	40	24	35	5
25Л	24	45	19	30	4
35Л	28	50	15	25	3.5
45Л	32	55	12	20	3.0
55Л	35	60	10	18	2.5

* The number indicates the average carbon content as for the steels listed in Table 23.

Steel castings are heat treated to obtain the required properties, to correct the cast structure, to eliminate chemical inhomogeneity, and to relieve foundry stresses.

Experience shows that the best combination of mechanical properties is obtained by hardening and tempering. Good results are achieved by homogenising with subsequent ordinary annealing and by normalising followed by tempering at 600-700°C. Somewhat lower but quite satisfactory mechanical properties are obtained by an ordinary annealing operation with phase recrystallisation.

13-5. Free Cutting Steels

Free cutting (also free machining) steels are extensively applied for machine parts, subject to comparatively light loads (bolts, nuts, screws, etc.) and produced on automatic screw machines or multiple-spindle automatics. These steels are intended for uses where easy machining is the primary requirement (Table 26).

Table 26

Chemical Compositions and Mechanical Properties of Free Cutting Steels *

Grade	Chemical composition, per cent				Mechanical properties of hot rolled steel			
	C	Mn	S	P	σ_u, kg/mm²	δ, per cent	ψ, per cent	Bhn
A12	0.08-0.16	0.6-0.9	0.08-0.2	0.08-0.15	42-57	22	36	160
A20	0.15-0.25	0.6-0.9	0.08-0.15	0.06	46-61	20	30	168
A30	0.25-0.35	0.7-1.0	0.08-0.15	0.06	52-67	15	25	185
A40Г	0.35-0.45	1.2-1.55	0.18-0.3	0.05	60-75	14	20	207

* The silicon content is within 0.15-0.35 per cent in all cases.

The distinguishing features of free cutting steels (high machinability and high quality surface finish after machining) are due to the higher sulphur and phosphorus content.

Sulphur exists in free cutting steel in the form of MnS. The manganese sulphide forms inclusions stretched out in the direction of rolling. These inclusions promote the formation of short brittle chips, reduce the friction on the surface being machined, and enable a satisfactory surface finish to be obtained at high cutting speeds. Phosphorus dissolves in the ferrite and increases its brittleness. This also makes the chip more brittle and enables a smooth bright surface to be obtained in cutting.

The hardness and brittleness of this steel are substantially increased when the phosphorus content exceeds 0.15 per cent but machinability remains unchanged. The addition of 0.15-0.35 per cent lead to steel considerably improves the machinability.

The tool life achieved in machining free cutting steels is from 2 to 2.5 times higher than when carbon steels of the same carbon content are machined. It must be noted, however, that free cutting steels have lower dynamic strength characteristics and are more susceptible to corrosion.

Free cutting steels are frequently supplied in the cold drawn (work hardened) form. These cold drawn steels have a higher tensile strength and hardness but less ductility when compared to the steels listed in Table 26.

13-6. Low-Alloy Constructional Steels

Low-alloy steels find the most extensive applications for all types of steel structures. They possess comparatively high mechanical properties without the need of additional heat treatment, good weldability, and are quite resistant to corrosion.

Table 27

Compositions and Mechanical Properties of Certain Low-Alloy Constructional Steels

Grade	Chemical composition, per cent						Mechanical properties		
	C	Si	Mn	Cr	Ni	Cu	σ_u, kg/mm²	σ_s, kg/mm²	δ, per cent
15ХГСНД	0.12-0.18	0.4-0.7	0.4-0.7	0.6-0.9	0.3-0.6	0.2-0.4	52	35	18
18Г2С	0.14-0.23	0.6-0.9	1.2-1.6	$\leqslant$0.3	$\leqslant$0.3	$\leqslant$0.3	60	40	14
12ХГ	$\leqslant$0.14	0.25-0.50	0.4-0.8	0.4-0.7	$\leqslant$0.3	$\leqslant$0.3	46	33	15
10ХНДП*	$\leqslant$0.12	0.2-0.4	0.3-0.6	0.5-0.8	0.3-0.6	0.3-0.5	—	—	—

* Contains 0.03-0.15 per cent P as well.

The addition of alloying elements which strengthen the ferrite increases the mechanical properties of a steel (Fig. 226). Nickel and copper, in addition, lower the temperature of plastic-to-brittle transition. This property is very essential for construction steels as the loss in impact strength must not exceed 40 per cent at 50°C below zero.

Besides strengthening construction steel, which is ordinarily used in the as-rolled condition, the addition of copper, nickel, chromium, and also phosphorus increases the corrosion resistance.

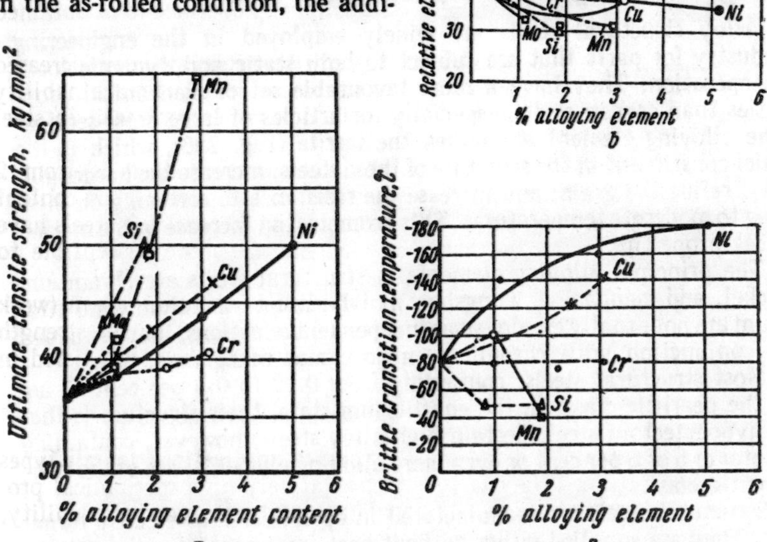

Fig. 226. Effect of alloying elements on the mechanical properties and brittle transition temperature of low-carbon steel in the equilibrium state (after V. A. Delle and L. A. Frumer)

Table 27 lists the compositions and properties of low-alloy steels, in the bar, structural shape, plate and sheet form, designed for rivetted and welded structural members, bridges, industrial and civil buildings, railroad cars, etc. (according to GOST 5058-49).

Steel 15ХГСНД is intended for cranes, all-steel gondola cars, oil tanks and containers, bridges, etc. Steels 18Г2С and 25Г2С are used for reinforcing bars in reinforced-concrete construction while steel 12ХГ, for grooved piles.

Low-alloy constructional steels are very sensitive to stress concentration and, therefore, bar stock and structural shapes must have as clean a surface as possible.

These steels are supplied in the hot-rolled condition. Specified mechanical and fabricating properties are provided for by the chemical composition, rolling procedure, and cooling after rolling.

Steel 14XΓC is used for manufacturing large-diameter welded gas mains. It contains 0.11-0.16 per cent C, 1.0-1.3 (or 0.75-1.05) per cent Mn, 0.4-0.7 per cent Si, 0.5-0.8 per cent Cr, $\leqslant$0.3 per cent Ni, and $\leqslant$0.3 per cent Cu.

13-7. Alloy Structural Steels

Alloy structural steels are widely employed in the engineering industry for parts that are subject to both static and dynamic loads in operation. They have a more favourable set of mechanical properties than carbon steels, especially for articles of large cross-section. The alloying element strengthen the ferrite (Fig. 226), which is the chief constituent in the structure of these steels; increase the hardenability; refine the grain; and increase the resistance to softening on heating to moderate temperatures. This promotes an increase in the mechanical properties.

The principal alloying elements in structural steels are chromium, nickel, and manganese. Tungsten, molybdenum, vanadium, and titanium are not usually employed as independent additions, they are added in conjunction with chromium, nickel, and manganese.

Most structural steels, containing from 0.12 to 0.5 per cent C, are of the pearlitic class. In the equilibrium state, their structure is that of hypoeutectoid steel. Certain high-alloy steels, however, containing a total of 5 or 6 per cent or even more alloying elements, are of the martensitic class.

Structural steels are manufactured in open-hearth or electric furnaces. They are supplied either without heat treatment or with one consisting of annealing or normalising followed by high tempering (at 640-660°C). In accordance with their chemical composition and guaranteed mechanical properties, they are classified as quality and high-quality steels (according to GOST 4543-57).

The characteristics of the more widely employed structural steels are described below.

Chromium and chromium-vanadium steels. Chromium is one of the cheapest of alloying elements. Its addition increases austenite stability and reduces the critical cooling rate, thus improving the hardenability. Chromium impedes grain growth in heating to some extent and increases the resistance to softening at elevated temperatures. The addition of chromium increases the tensile strength and yield point of steel after hardening and high tempering. The toughness (a_k) is lowered slightly while the ductility (δ, ψ) is practically unaffected.

Chromium steels are extensively applied in many fields of industry. Those used in the Soviet Union (GOST 4543-57) contain from 0.15 to 0.5 per cent carbon and from 0.7 to 1.0 per cent chromium (15X,

15XA, 20X, 20XA, 30X, 30XA, 35X, 38XA, 40X, 45X, 45XA, 50X, and 50XA).

The most widely used of this series are the carburising steel 15X (15XA) and steel 40X (38XA) which usually undergoes structural improvement treatment.

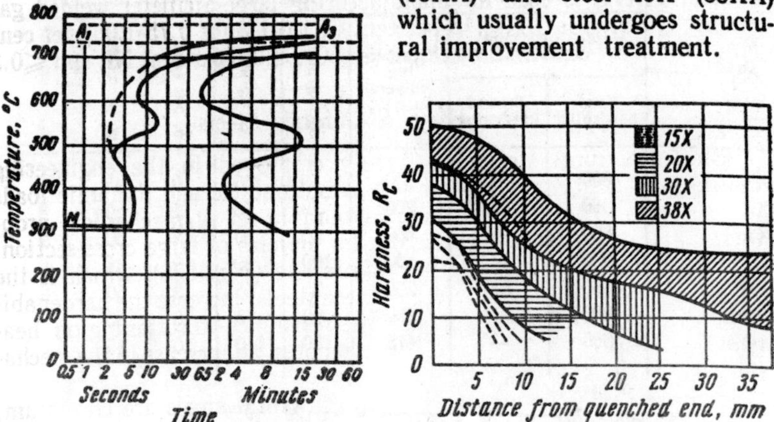

Fig. 227. Isothermal austenite decomposition (TTT) diagram for steel 40X

Fig. 228. Hardenability bands for chromium steels

Steels 15X and 15XA find application in the automobile and tractor industry for valve tappets, wrist pins, idler studs, etc. In the machine tool industry, they are used for gears operating at high speeds and medium loads, claw clutches, bushings, worms, and other parts subject to carburising, cyaniding, etc.

Steels 40X and 38XA are usually subject to structural improvement (high tempering) and are used for shafts and worms operating under medium loads, bushings, clutches and other components of machine tools and fixtures. In the automobile and tractor industries, they are applied for steering knuckles, axle shafts, transmission main shafts and many others.

Heat treating procedures and mechanical properties of steels 15X and 40X are listed in Table 28. Fig. 227 gives the isothermal austenite decomposition diagram for steel 40X while the hardenability bands for various chromium steels are shown in Fig. 228.

Fig. 228 shows that the more chromium in a steel, the higher its hardenability will be. Chromium-carbon steels (30X, 35X, 40X, and 45X) are hardened by quenching in oil.* This reduces warping to a great extent. Carburised steels (15X and 20X) are usually quenched in water. After high tempering, steels 40X and 38XA are quenched in oil or water to prevent temper brittleness. The effects of the component

*Parts of steel 38XA and 40X with cross-sections exceeding 50 or 60 mm are quenched in water.

Table 28

Mechanical Properties of Chromium and Chromium-Vanadium Structural Steels

Grade	Heat treatment			Mechanical properties (minimum)				
	Hardening temperature, °C		Tempering temperature, °C	σ_u, kg/mm²	σ_s, kg/mm²	δ, per cent	ψ, per cent	a_k, kg-m/cm²
	First hardening	Second hardening						
15X	860	780	200	70	50	10	45	7
15XA	860	780	200	70	50	11	50	8
38XA	860	—	550	95	80	12	50	9
40X	850	—	500	100	80	9	45	6
15XΦ	860	780	200	75	55	12	50	8
40XΦA	880	—	650	90	75	10	50	9
50XΦA	860	—	475	130	110	10	45	—

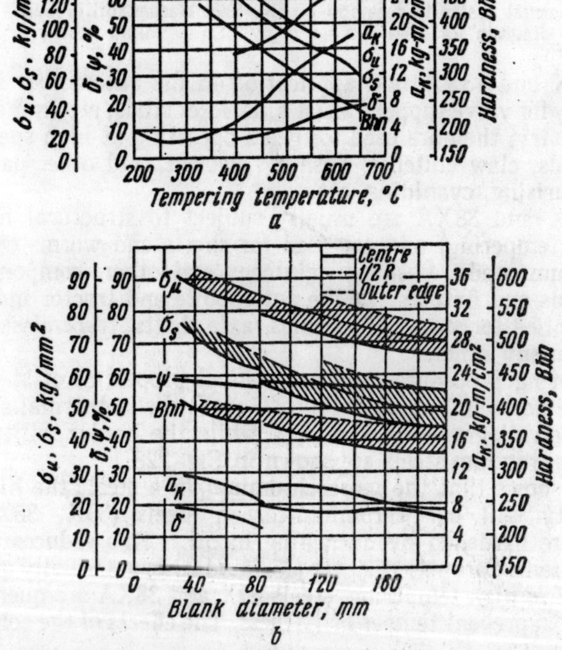

Fig. 229. Variation in the mechanical properties of steel 40X:
a—with the tempering temperature, *b*—with the blank diameter

size and tempering temperature on the mechanical properties of steel 38XA are shown in Fig. 229.

The addition of 0.1 to 0.2 per cent V to chromium steels (steels 15XΦ, 40XΦ and 50XΦA) reduces the susceptibility to overheating, improves the mechanical properties, especially the impact strength, and increases the resistance to softening on heating. Carburised steel 15XΦ is used for gears, camshafts, wrist pins, etc. After carburising, the steel retains a sufficiently fine grain, both in the core and the case. Subsequent heat treatment usually consists of single quenching, directly after carburising, in water (after precooling).

The heat treatment of steel 40XΦ consists in quenching in oil followed by high tempering (Table 28). It is used for crankshafts, bushings, bolts, cross-members, spindles, shafts, etc. Steel 50XΦA is used for critical springs. Hardening is followed by a medium tempering.

Antifriction bearing steels (GOST 801-58) are a special group of chromium structural steels.* The compositions of these steels are listed in Table 29.

Table 29

Chemical Compositions and Applications of Chromium Steels for Antifriction Bearings

Grade	Chemical composition, per cent [1]		Applications
	C	Cr	
ШХ6	1.05-1.15	0.4-0.7	Balls up to 13.5 mm and rollers up to 10 mm in diameter
ШХ9	1.0-1.1	0.9-1.2	Balls from 13.5 to 22.5 mm and rollers from 10 to 15 mm in diameter
ШХ15	0.95-1.05	1.3-1.65	Balls over 22.5 mm and rollers from 15 to 30 mm in diameter

[1] 0.17-0.35 per cent Si, 0.2-0.4 per cent Mn, $<$ 0.02 per cent S, and $<$ 0.027 per cent P.

In the operation of the antifriction bearings, the material of the rings, balls and rollers is subject to abrasive and chemical wear**, as well as to high specific loads of an alternating type which may lead to fatigue failure.

The high carbon content in these steels (Table 29) enables an adequately high strength to be obtained after hardening and tempering, as well as a high fatigue limit (for bending loads, $\sigma_w \approx 80$ kg/mm^2), high contact fatigue limit, and ample abrasion resistance.

* In composition, antifriction, bearing steels are tool steels (see Sec. 13-8) but they are structural steels in their application.

** Chemical wear here means corrosion.

Steel alloyed by chromium has excellent hardenability, and a high and uniform hardness can be attained. The initial structure of these steels consists of fine-grained pearlite. There should be no nonmetallic inclusions, porosity, gas holes and other metallurgical defects. The steel must be free of carbide segregations which may cause breaking out and surface cracks that will lead to failure of the bearing.

The hardening procedures for the bearing components depend upon their form and size. Small balls are oil-quenched; large ones are water-quenched (3.5-5 per cent aqueous soda solution). Hardening is followed by low tempering at 140-160°C. After heat treatment, the steel will have a fine crystalline martensite structure with dispersed carbides. The hardness will be R_c 61-65.

Good results are obtained from martempering, comprising heating to 830-850°C and quenching in a bath at the temperature 140-190°C. The hardness after this procedure should be R_c 61-63.

The dimensions of precision bearings are stabilised by a subzero treatment (+10° to —30°C) with subsequent tempering or ageing at 125-130°C for 5 to 25 hours.

Silicon and chromium-silicon steels. Silicon steels are applied to a great extent for making leaf and coil springs which require high elastic and fatigue limits. Steels 55C2 and 60C2 are most often specified for leaf springs (Table 30).

Table 30

Chemical Compositions of Steels for Leaf and Coil Springs

Grade	Chemical composition, per cent						
	C	Mn	Si	Cr_{max}	Ni_{max}	S_{max}	P_{max}
55C2	0.50-0.60	0.6-0.9	1.5-2.0	0.3	0.5	0.05	0.05
60C2A	0.56-0.64	0.6-0.9	1.6-2.0	0.3	0.5	0.04	0.04

Spring leaves of steels 55C2 and 60C2 are quenched in oil. This is followed by tempering at 460°C (420-520°C). After this treatment, the steel will acquire the mechanical properties listed in Table 31.

Table 31

Steels for Leaf and Coil Springs. Mechanical Properties after Heat Treatment

Grade	σ_u, kg/mm²	σ_s, kg/mm²	δ, per cent	ψ, per cent
	minimum			
55C2	130	120	25	30
60C2A	160	140	20	20

An excellent heat-treating procedure for coil and leaf springs of steel 55C2 is austempering which consists in heating to 850-880°C and then holding for 5 to 7 minutes in the quenching medium at 340-380°C. The hardness will be R_c 38-41. Austempering increases the elastic properties, ductility, and fatigue limit of steel 55C2.

Provision must be made for effective measures to prevent decarburisation in hardening springs since this substantially reduces the fatigue limit. Leaf springs are frequently shot-peened (surface cold working) to increase their endurance limit.

The chromium-silicon steels 33XC and 40XC (1.2 to 1.6 per cent Si and 1.3 to 1.6 per cent Cr) find numerous applications in industry. After hardening (heating temperature 900°C and oil quenching) and a high tempering (33XC at 630°C and 40XC at 540°C), these steels acquire the mechanical properties listed in Table 32.

Table 32

Mechanical Properties of Chromium-Silicon Structural Steels

Grade	σ_u, kg/mm²	σ_s, kg/mm²	δ, per cent	ψ, per cent	a_k, kg-m/cm²
	minimum				
33XC	95	80	13	50	8
40XC	125	110	12	40	5

Tempered at 250-270°C, steel 33XC will acquire a tensile strength of 150 kg/mm² as compared to 170 kg/mm² for steel 40XC after the same treatment. Chromium-silicon steels possess high strength, moderate toughness, and low hardenability. They are susceptible to overheating and temper brittleness and are used for small machine parts subject to high stresses and wear. Steel 33XC is also used to make thin-walled tubes.

Molybdenum and chromium-molybdenum steels. The properties of steel are improved by the addition of molybdenum. It increases the hardenability, reduces the tendency for overheating, excludes temper brittleness, and eliminates the danger of graphitisation after long service at elevated temperatures of such articles as boilers, fire-box components, etc. Molybdenum steel is used for heat-resistant applications. Steel containing 0.15 per cent C and 0.5 per cent Mo, having a tensile strength at room temperature near that of carbon steel ($\sigma_u \approx$40-45 kg per sq mm), has a strength of 35-38 kg/mm² at 550°C. Upon prolonged load application (1,000 hrs), it can carry up to 19 kg/mm². Under the same conditions the tensile strength of carbon steel will be about 20 kg/mm² (at 550°C) and 8 or 9 kg/mm² (for 1,000 hrs). The low-carbon steels 16M, 12MX, and 15XM (0.8 to 1.1 per cent Cr and 0.4 to 0.55 per cent Mo), as well as 12X1MΦ (0.2-0.35 per cent Mo) have good welding properties and find applications in the manufacture of boiler drums, surface-heated boiler tubes, superheaters and steam piping. Medium-carbon steel 30XM (0.8-1.1 per cent Cr and 0.15-0.25 per cent Mo) is used for fastening parts, high-pressure fittings, and for critical components of turbines, such as rotors, shafts, and discs. Steel 25X2MA (1.5-1.8 per cent Cr and 0.25-0.35 per cent Mo) finds use as fastening parts for high-pressure boilers and steam lines.

Chromium-molybdenum steel, type 30XMA, has exceptionally high mechanical properties and weldability. It is applied for high-stressed weldments. Steel 30XM, a modification, is also used for axles, rotors, gears, perforators, etc. This steel will through-harden in cross-sections up to 50 mm. The high cost of molybdenum restricts the application of these steels at the present time.

Chromium-manganese-titanium steels. Manganese, added to a chromium steel, increases the austenite stability, reduces the critical cooling rate and improves the hardenability. At the same time, however, it increases the sensitivity to overheating. If the hot working of chromium-manganese steel (inherently fine-grained type No. 6-8) is carefully performed and if titanium is added, which forms stable carbides and reduces the susceptibility to overheating, together with the chromium, a set of very high mechanical properties may be obtained.

The most successful of this series are the carburised grades 18XГT and 30XГT, developed by the Likhachov Automobile Plant in Moscow.

Steel 18XГT has the following composition: 0.16-0.24 per cent C, 0.8-1.1 per cent Mn, 1.0-1.3 per cent Cr, and 0.08-0.15 per cent Ti.

Data on the variation in the mechanical properties of steel 18XГT with the tempering temperature are given in Fig. 230.

The carburised steel 18XГT is a high-quality substitute for expensive chromium-nickel steels and finds wide application in the automobile and tractor industries for such components as shafts, gears, large-size antifriction bearings, and others.

Steel 18XГT is carburised at 920±10° and is hardened at 870°C by quenching in oil which is followed by tempering at 200°C. If carburising is done in through-type retort furnaces, the work is quenched direct from the carburising furnace after precooling to 840±10°.

Machine parts made of steel 18XГT are deformed to a minimum extent by heat treatment. Due to the small amount of austenite retained after carburising, quenching, and tempering, the case will have a hardness of R_c 58-61 while the hardness of the core will be R_c 28-36. This steel machines easily. A disadvantage is its relatively low hardenability (50-40 mm). At the present time, steel 30XГT, having a higher carbon content, is replacing other steels. This newer grade can be recommended both for carburising and for structural improvement (high tempering). Steel 30XГT has better hardenability than steel 18XГT and a stronger core (hardness R_c 32-46). This enables the case depth to be reduced without danger of its failure due to plastic deformation of the core. This reduction in case depth substantially decreases the time required for carburisation. Case-hardened steel 30XΓT has a higher fatigue limit than steel 18XΓT.

In addition to steel 18XΓT, steel 18XΓM is also used for gear manufacture. The latter contains from 0.2 to 0.3 per cent Mo.

Chromium-silicon-manganese steel (cromansil type). Steels of this group have found very extensive application in Soviet engineering industries and have replaced expensive chromium-molybdenum steels to a great extent. This steel contains from 0.17 to 0.39 per cent C (ave-

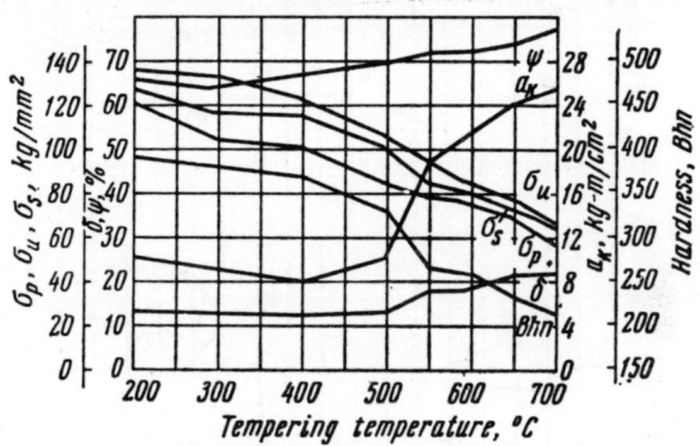

Fig. 230. Variation in the mechanical properties of steel 18XΓT with the tempering temperature

rage content), 0.9 to 1.2 per cent Si, 0.8 to 1.1 per cent Mn, and 0.8 to 1.0 per cent Cr. Table 33 lists the heat treatment procedure and mechanical properties of cromansil-type steel.

Table 33

Mechanical Properties of Chromium-Silicon-Manganese Structural Steels

Grade	Heat treatment		Mechanical properties (minimum)				
	Hardening temperature, °C	Tempering temperature, °C	σ_u, kg/mm²	σ_s, kg/mm²	δ, per cent	ψ, per cent	a_k, kg-m/cm²
20XΓC	880	500	80	64	12	45	7
30XΓC	880	550	110	95	10	45	5
30XΓCA	Austempering: heat to 880° C, quench in mixture of potassium (KNO_3) and sodium ($NaNO_3$) nitrates or molten sodium (NaOH) and potassium (KOH) hydroxides at 280-310° C		165	140	10	40	5

Alloying with a combination of chromium and manganese provides an austenite stability sufficient for oil-quenching (Fig. 231). The hardenability of cromansil-type steel is not very high but this is not an essential factor since the steel is applied chiefly for relatively thin weldments. It must be noted that this steel should be quenched in oil or water after high tempering (at 500-550°C) to prevent the development of recurring temper brittleness.

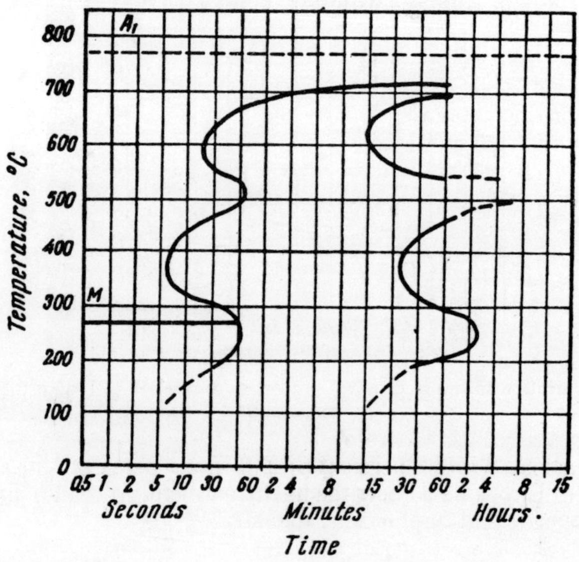

Fig. 231. Isothermal supercooled austenite decomposition (TTT) diagram for steel 30ХГСА

The application of austempering enables high mechanical properties to be obtained (Table 33). It also reduces the notch sensitivity and increases the resistance to combined stresses. It is for this reason that bright austempering is most frequently applied. The work is heated in molten chlorides and quenched in molten alkalis (NaOH-KOH).

Steel of this type with a low carbon content (20ХГСА) has good weldability and high ductility; it can be easily formed and bent. This steel is used in the form of plates and pipe for heat-treated weldments.

Steel 30ХГСА with a higher carbon content is available in the Soviet Union as plates, pipe, bars, shapes, and forgings. This steel has satisfactory weldability and formability.

An advantageous property of cromansil-type steels is their fair machinability notwithstanding the high hardness.

Steel 30XΓCHA is finding extensive application at the present time. It contains 0.28-0.33 per cent C, 0.9-1.2 per cent Cr, 1.0-1.3 per cent Mn, 1.5-1.8 per cent Ni, and 0.9-1.2 per cent Si.

This steel acquires the following mechanical properties after hardening and low tempering: $\sigma_u=170$ kg/mm², $\sigma_s=145$ kg/mm², $\delta=11$ per cent, $\psi=45$ per cent, and $a_k=7$ kg-m/cm². These properties considerably exceed those of steel 30XΓCA. The addition of nickel increases the toughness and resistance to brittle failure because the nickel dissolves in the solid solution and increases the dispersity of the obtained structure.

Steels with boron. The amount of boron which may be added to structural steels ranges from 0.001 to 0.005 per cent.

Boron increases the density of the ingot and improves the ability to undergo hot plastic deformation.

Boron increases the hardenability of structural steel.* The strength (σ_u, σ_s) is also increased while ample ductility and toughness (δ, ψ, a_k) are retained.

The favourable effect of small boron additions on the properties enables steels with high mechanical properties to be developed on the basis of low-alloy chromium-manganese and chromium grades.

The boron steels 20XΓP (0.17-0.24 per cent C, 0.8-1.1 per cent Cr, 0.7-1.0 per cent Mn, and 0.002-0.005 per cent B) and 35XPA (0.3-0.38 per cent C, 0.8-1.1 per cent Cr and 0.002-0.005 per cent B) have been recommended in the U.S.S.R. to replace the molybdenum steels 20XHM and 35XMA which are employed in the automobile industry for various types of gears.

After carburising, quenching from 860-880°C in oil, and tempering at 200°C, steel 20XΓP will acquire the following mechanical properties (according to specification 4MTУ 4804-54):

σ_u, kg/mm²	σ_s, kg/mm²	δ, per cent	ψ, per cent	a_k, kg-m/cm²
100	80	9	50	8

Steel 40 XP is used after hardening at 860°C followed by tempering at 540°C. After structural improvement, this steel develops the following properties:

σ_u, kg/mm²	σ_s, kg/mm²	δ, per cent	ψ, per cent	a_k, kg-m/cm²
100	80	12	50	9

* The addition of boron increases the hardenability of steels containing 0.1-0.4 per cent C, hardenability is unaltered in hypereutectoid steels.

The carburised steels 20XГHP and 15X2H2TPA have found applications in the manufacture of large heavily-loaded machine parts.

It is necessary to note that boron increases the susceptibility of a steel to overheating, as well as the deformation (warping) in hardening.

As a rule, carburised boron steels must be inherently fine-grained. The use of coarse-grained steels often leads to the forming of hardening cracks.

Another feature to be noted is that boron is effective only when added to fully killed steels.

Nickel steels. Nickel, when added to a steel, substantially increases the austenite stability (see Fig. 211) and thereby reduces the critical cooling rate and increases the hardenability. Nickel increases the strength and hardness characteristics without any appreciable reduction in ductility and toughness. The improvement in mechanical properties is based on the strengthening of the ferrite and the formation of more dispersed structures after hardening. Due to the high cost of nickel and the more effective application of chromium-nickel steels, straight nickel steels find limited use.

Chromium-nickel steels have a high hardenability (through-hardening can be achieved in parts over 75-100 mm in diameter or thickness) and very high mechanical properties as well. They are extensively employed for large machine parts of complex form operating under vibrating and dynamic loads.

The most widely used in the Soviet Union are the carburised steels 12XH3A and 12X2H4A (Table 34) which are found in such heavily loaded components as gears, shafts, ball-joint pins, spindles of precision machine tools, axles, rollers, etc.

Table 34

Heat Treatment and Mechanical Properties of Chromium Nickel Structural Steels

Grade	Heat treatment					Mechanical properties (minimum)				
	Hardening			Tempering						
	Temperature, °C		Quenching medium	Temperature, °C	Cooling medium	σ_u, kg/mm²	σ_s, kg/mm²	δ, per cent	ψ, per cent	a_k, kg-m/cm²
	First	Second								
12XH3A	860	780	Oil	180	Water or oil	100	85	12	55	10
12X2H4A	860	780	Oil	180	Air	120	100	10	50	9
37XH3A	820	—	Oil	530	Water or oil	110	90	10	50	6
18X2H4BA	950	850	Air	160	Air	120	105	12	55	11

The TTT diagram for steel 12X2H4A is shown in Fig. 232, *a* while Fig. 233 illustrates the hardenability bands for steels 12XH3A and 12XH4A.

Due to the high stability of the austenite after carburising followed by cooling the carburising boxes in air, steels 12XH3A and 12X2H4A acquire a partially hardened structure that makes subsequent machining quite difficult. To improve machinability after carburising, it is advisable to apply a high tempering at 650°C with several hours of holding time.

The carburised case of chromium-nickel steels retains a large amount of austenite which lowers the hardness even below R_c 55-56. The retained austenite may be decomposed by a high tempering or by sub-zero treatment*.

Instead of double hardening (Table 34), a single hardening operation, at 780-850°C, is frequently applied after carburising. This also provides quite satisfactory mechanical properties.

A typical chromium-nickel steel, usually subject to structural improvement, is Grade 37XH3A. It contains 0.37 per cent C, 1.4 per cent Cr, and 3.25 per cent Ni. The high austenite stability of this steel (see Fig. 232, *b*) imparts high hardenability so that this steel may be recommended for making large components (heavy forgings, shafts, piston rods, high-strength crankshafts, etc.). Fig. 234, *a* illustrates the variation in the mechanical properties of steel 37XH3A with the tempering temperature. Chromium-nickel steels are most susceptible to temper brittleness after high tempering (500-550°C) followed by slow cooling. Therefore, in such cases, the steel is cooled (quenched) in oil or water (for large articles). Steel 37XH3A is used, not only in the structurally-improved condition, but after hardening followed by low tempering at 220°C. As a result of the latter treatment, the steel acquires a high tensile strength in conjunction with ample ductility and toughness.

When tungsten and molybdenum are added to chromium-nickel structural steel, they increase the tensile strength and yield point without affecting the toughness. They also increase the resistance to softening on heating and reduce the susceptibility to temper brittleness. Steels of this group possess practically full hardenability in the largest cross-sections.

The most extensively used is the chromium-nickel-tungsten steel, Grade 18X2H4BA (18X2H4MA). This steel is used either after carburising or after hardening and tempering. Steel 18X2H4BA contains, on an average, 0.18 per cent C, 1.5 per cent Cr, 4.25 per cent Ni, and 1 per cent W. It is a typical steel of the martensitic class.

---

* High tempering is applied as an intermediate operation between the first and second hardening or after carburising and before hardening. Sub-zero treatment is applied after hardening.

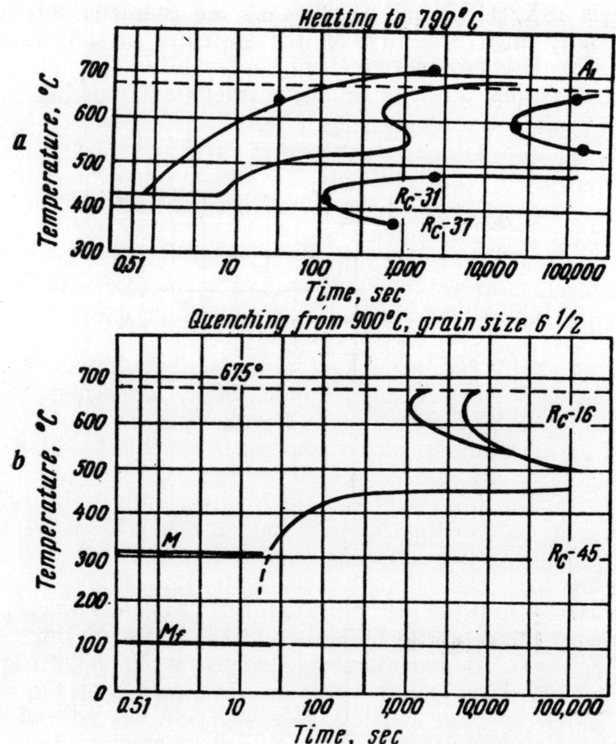

Fig. 232. Isothermal decomposition (TTT) diagrams for chromium-nickel steels: *a*—for steel 12X2H4A, *b*—for steel 37XH3A

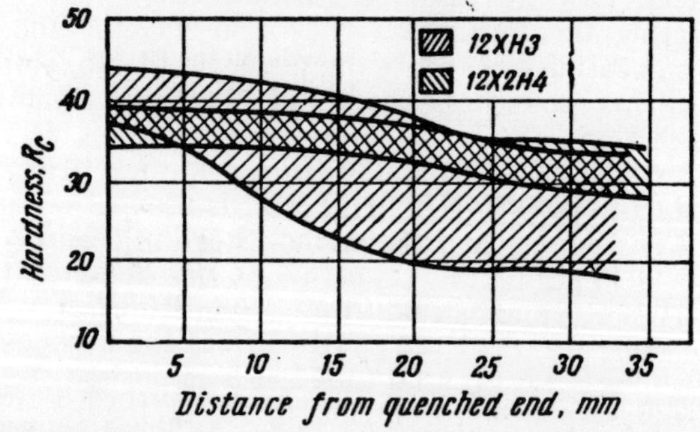

Fig. 233. Hardenability bands for steels 12XH3A and 12X2H4A

Steels 18X2H4BA and 18X2H4MA are characterised by the exceptionally high stability of the austenite in the pearlite range (Fig. 235). This provides excellent hardenability, excludes annealing, and enables these steels to be air-hardened (see Table 34).

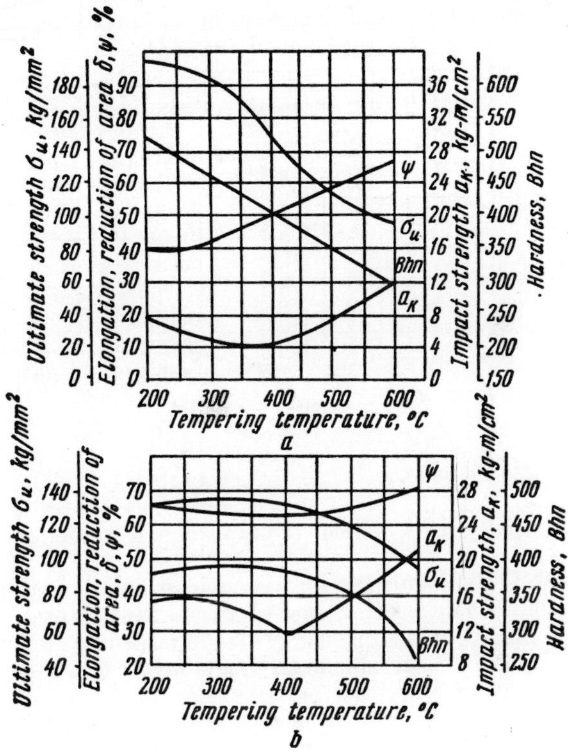

Fig. 234. Effect of the tempering temperature on the mechanical properties of chromium-nickel steels:
a—steel 37XH3A, *b*—steel 18X2H4BA

Instead of annealing, these steels are tempered at a high temperature (640°C) to reduce their hardness (to *Bhn* 269-217). Large forgings undergo normalising (cooling in air from 950°C) and tempering at 640°C to refine the structure and to reduce the hardness.

Grade 18X2H4BA is chiefly used as a carburised steel. It is necessary to point out that in carburising this steel, the use of an excessively active carburiser or raising the hardening temperature will cause a large amount (up to 60-80 per cent) of austenite to be retained in the case, thereby sharply reducing its hardness. Retained austenite in the case may be reduced by hardening after high tempering (640-650°C). A sub-zero treatment after quenching will achieve the same results.

Steel 18X2H4BA is also applied extensively after quenching followed by low tempering and after structural improvement. Data indicating the variation in the mechanical properties of steel 18X2H4BA with the tempering temperature are given in Fig. 234, *b*.

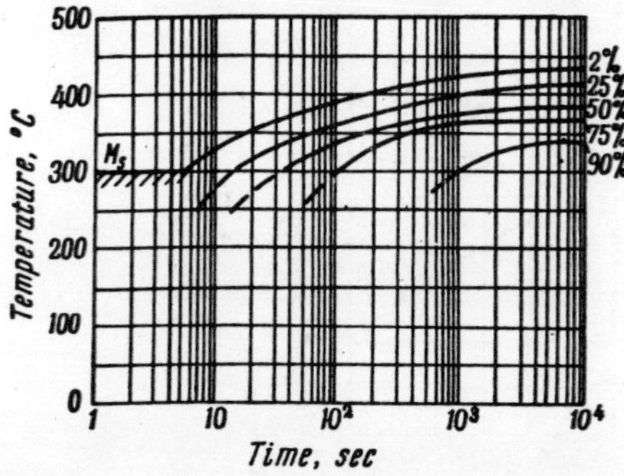

Fig. 235. Isothermal supercooled austenite decomposition (TTT) diagram for steel 18X2H4BA

This steel acquires a most expedient combination of properties after hardening in air followed by low tempering (see Table 34).

The fatigue limit of steel 18X2H4BA is frequently elevated by nitriding (for example, crankshafts of powerful engines).

Steel 18X2H4BA is applied for the most critical parts of machines, such as various gears, pins, axles, crankshafts, and others which require high dynamic and fatigue strengths.

This steel is distinguished for its high strength, and its ample ductility and toughness. It is one of the best structural steels. Its capabilities for air-hardening and through-hardening in large cross-sections render it indispensable in the manufacture of massive heavily-loaded complex machine parts.

Soviet industries also employ steels 25XHBA, 45XHMΦA (0.45 per cent C, 1 per cent Cr, 1.5 per cent Ni, 0.15 per cent V and 0.25 per cent Mo), and 40XHMA (0.4 per cent C, 0.75 per cent Cr, 1.5 per cent Ni, and 0.20 per cent Mo).

Steel 45XHMΦA provides good service under high torsional stresses. It is used for making large shafts and heavily loaded springs.

The addition of 0.2 per cent Mo to steel 40XHMA decreases its susceptibility to temper brittleness. This steel has deep hardness penetration after quenching and is used as a substitute for steel 37XH3A in the manufacture of large machine parts for which increased strength ($\sigma_u \geqslant 100$ kg/mm²), in conjunction with ample ductility and toughness ($\psi \geqslant 55$ per cent and $a_k = 10$ kg-m/cm²), is specified. The final heat treatment for this grade of steel consists in hardening (heating temperature 850°C, quenching in oil) and tempering at 570-640°C, followed by oil-quenching.

13-8. Tool Steels

Carbon and alloy steels designed for the manufacture of cutting and measuring tools and dies are known as tool steels.

High-speed steels comprise a special group in the general class of tool steels.

Tool steels are smelted in open-hearth and electric furnaces and belong to high-quality classes. In their as-received condition, carbon and alloy tool steels are usually annealed to a granular pearlite (cementite) and they thereby possess low hardness, good machinability and ductility, and decreased susceptibility to overheating.

STEELS FOR CUTTING TOOLS

This group includes high-carbon eutectoid and hypereutectoid steels.

After hardening and low tempering, the cutting edges of tools must have a high hardness (R_c 60-65), considerably exceeding that of the material to be machined. They must have a wear resistance sufficient to retain the size and form of the cutting edges in metal cutting and ample strength combined with a certain degree of toughness to prevent tool breakage in operation. Table 35 indicates the

Table 35

Chemical Compositions of the Most Extensively Used Steels for Cutting Tools

Grade	Chemical composition, per cent				
	C	Mn	Si	Cr	W
	Carbon steels *				
У8А	0.75-0.84	0.15-0.30	0.15-0.30	0.2	—
У10(У10А)	0.95-1.04	0.15-0.30	0.15-0.30	0.2	—
У12(У12А)	1.15-1.24	0.15-0.30	0.15-0.30	0.2	—
	Alloy steels				
X	0.95-1.1	⩽0.4	⩽0.35	1.3-1.6	—
9XC	0.85-0.95	0.3-0.6	1.2-1.6	0.95-1.25	—
ХВГ	0.9-1.05	0.8-1.1	0.15-0.35	0.9-1.2	1.2-1.6

* In addition to the listed carbon steels, GOST 1435-54 specifies the following grades: У7А, У8ГА, У9А, У11А, and У13А.

composition of steels for cutting tools while Table 36 indicates heat-treating procedures and applications of these steels.

Table 36

Heat Treating Procedures for Cutting-Tool Steels

Grade	Annealing temperature, °C	Hardening		Hardness after quenching, R_c	Tempering temperature, °C	Hardness after tempering, R_c	Applications
		Temperature, °C	Quenching medium				
У10 У12 (У10А, У12А)	750-770	770-790	Aqueous solution of alkalis (salts) or water followed by oil	63-65	150-160	61-62	Lathe and planer single-point tools, drills, files, round threading dies, taps, etc.
		780-800	Molten salt (150-160° C) *	61-63			
		790-810	Oil **	61-62			
Х	770-790	835-850	Oil	62-65	160-180	61-63	Drills, reamers, taps, etc.
		840-850	Molten salt (150-160° C)	61-64			
9ХС	780-800	850-870	Oil	62-64	170-190	61-62	Milling cutters, reamers, core drills, broaches, threading dies, etc.
		860-875	Molten salt (150-200° C)	61-63			
ХВГ	770-790	820-840	Oil	63-65	170-185	62-63	Milling cutters, reamers, taps, broaches, etc.
		830-850	Molten salt (150-160° C)	62-64			

* Holding time at 150-160° C for tools up to 5-12 mm in diameter—1-2 minutes.
** For tools up to 5 mm in diameter (or thickness).

Carbon tool steels are characterised by the low stability of the supercooled austenite. Therefore, they have a high critical cooling rate and a low hardenability. Through hardening can be achieved only in parts up to 12 or 15 mm in thickness or diameter. Consequently, these steels (У7А, У8А, У10А, У11А, and У12А) may be recommended for small-size tools, which are quenched in oil or molten salts, and for comparatively large tools (15 to 30 mm in diameter) in which the cutting section is only on the surface layer (files, core drills, short reamers, etc.).

When larger tools are hardened (of a diameter over 30 mm) the layer with a high hardness is so thin, even upon quenching in water, that the tools are not fit for cutting purposes.

Great disadvantages of carbon steels are their narrow range of hardening temperatures and the necessity for rapid quenching in water or aqueous alkali solutions (salts). This leads to considerable deformation and warping and even to the formation of cracks.

Therefore, tools of complex form with sharp changes in section and with a large length-to-diameter ratio should not be made of carbon steels. Warping and crack forming may be reduced somewhat by quenching in water only to 200-250°C with subsequent retarded cooling in oil. Stepped quenching (martempering) is advisable for small-size tools (Table 36). Good results may be obtained in applying induction hardening to certain types of tools.

Advantages of carbon steels are their cheapness, low hardness (*Bhn* 170-180), good machinability and formability in the annealed state, and the fact that they retain a tough unhardened core due to their low hardenability. This last factor improves their resistance to breakage under vibration and impacts.

Carbon steels are only applicable for tools operating at low cutting speeds since their hardness is substantially reduced at temperatures above 190-200°C.

Alloy steels for cutting tools are of the eutectoid and hypereutectoid type. After hardening, they acquire high hardness and wear resistance but not particularly high resistance to self-tempering. They are applicable only in cases when the cutting edges are not heated above 200-225°C.*

Compared to carbon steels, the alloyed grades have somewhat more wear resistance and less susceptibility to overheating. They harden deeper and their tendency to deform and warp is less since they are quenched in oil and retain a larger amount of austenite.

Alloy steels may be recommended for tools having a tendency to warp in hardening (long thin taps, reamers, etc.) and for large-size tools (for example, taps and reamers) over 35 or 40 mm in diam-

---

* Tools of steel 9XC can operate satisfactorily when heated to 225-250° C.

eter. Such large tools made of carbon steels would have a hardened layer too thin for practical purposes.

The most extensively used of these steels for cutting tools are Grades 9XC and XBΓ. Steel 9XC has an increased hardenability. Either stepped or isothermal quenching (martempering or austempering) may be applied, and this steel retains its high hardness at temperatures up to 225-250°C. Thus its cutting properties are 10 to 15 per cent higher than those of carbon steel. No appreciable deformation is observed after martempering or austempering tools made of steel 9XC.

Disadvantages of this grade are the tendency to decarburise in heating and the high hardness (*Bhn* 217-247) in the annealed state which impairs machinability.

Due to the presence of tungsten carbides in steel XBΓ, it has higher hardness and wear resistance than steels X and 9XC (Table 36) but is more expensive.

High-speed steels. High-speed steels are distinguished for their high red-hardness, i.e., their capability to retain their structure (martensite), hardness, and wear resistance at the high temperatures generated on the cutting edges when machining at high cutting speeds. High-speed steels are designed for the manufacture of high-production tools with high wear resistance which must retain their cutting properties at temperatures up to 600-620°C.

The composition of the best known Soviet high-speed steels is given in Table 37.

Table 37

Compositions of High-Speed Steels (GOST 5952-51)

Grade	Chemical composition, per cent			
	C	Cr	W	V
P18	0.7-0.8	3.8-4.4	17.5-19	1.0-1.4
P9	0.85-0.95	3.8-4.4	8.5-10	2.0-2.6

Steel P18 excels steel P9 in red-hardness. It will retain a hardness over R_c 60 after being repeatedly heated to 610-620°C. Steel P9 can withstand heating up to 605-615°C.*

* Besides the listed grades, high-performance high-speed steels are applied to a certain extent. They are P18Φ (0.8 per cent C, 4 per cent Cr, 18 per cent W and 2 per cent V) and P10Φ5K5 (1.50 per cent C, 4.5 per cent Cr, 10 per cent W, 4.2 per cent V and 6 per cent Co). Best cutting properties are found in steels P10Φ5K5 and PK10, which differ from steel P18 in that they contain up to 10 per cent cobalt.

In regard to its structure in the annealed condition, high-speed steel is of the cementitic type. After normalising, it acquires a martensitic structure and may be classed as a martensitic steel.

The phase composition of the P18 steel consists of α solution, complex carbides (Fe_2W_2C, $Cr_{23}C_6$, and VC), and carbides of the cemen-

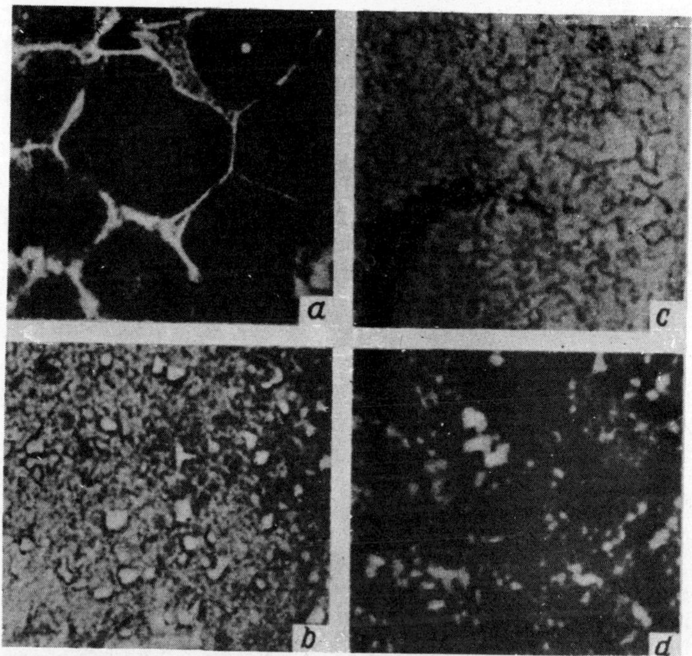

Fig. 236. Microstructure of high-speed steel P18, × 250:
a—eutectic (ledeburite) in cast high-speed steel, *b*—annealed steel, *c*—hardened steel, *d*—hardened and tempered steel

tite type. About 50 to 70 per cent of the chromium in the steel is in the α solution. The rest of the chromium and almost all of the tungsten and vanadium are in the carbides. The structure of cast high-speed steel includes a complex eutectic, resembling ledeburite (Fig. 236, *a*), located along the grain boundaries. Hot working breaks up the eutectic and refines the carbides.

In worked and annealed high-speed steel, both the eutectic and secondary carbides are uniformly distributed in a matrix of sorbitic pearlite (Fig. 236, *b*).

If the steel has not been hot worked to a sufficient degree, the primary (eutectic) and secondary carbides will not be uniformly distributed. Neither will they be uniform in size and in the form in which they precipitate from the matrix (network, bands or stringers along the direction of working, etc.).

The mechanical properties of the steel are lowered to a great extent by this lack of carbide uniformity. It also affects the red-hardness and makes it difficult to obtain uniform high hardness after quenching. Large blanks usually undergo additional working to eliminate carbide nonuniformity.

When high-speed steel is heated above point Ac_1 (820-860°C), pearlite is transformed into austenite. Further heating above the Ac_1 range will dissolve the secondary carbides in the austenite to increase its degree of alloyage. Due to the presence of a considerable amount of excess (primary) carbides, not dissolved in the austenite, high-speed steel retains its fine grained structure even when heated to high temperatures (to 1,200-1,240°C for steel P18 and to 1,190-1,220°C for steel P9). Heating to higher temperatures will cause appreciable grain growth. The transformation of the austenite in cooling depends upon the amount of alloying elements dissolved in the austenite in heating and upon the cooling rate. Diagrams of isothermal decomposition of austenite, formed at 900° and at 1,290°C, are given in Fig. 237. The first diagram shows the character of the austenite decomposition in annealing while the second indicates the transformations which must be taken into consideration in hardening procedures. The diagrams show that the higher the heating temperature, i.e., the larger the amount of alloying elements in the austenite, the higher its stability will be in the pearlite range, as well as in the range of intermediate (bainite) transformation.

High-speed steel usually undergoes isothermal annealing, after forging, to improve its machinability, reduce the hardness, and to prepare the structure for hardening. For annealing, the steel is heated to 830-850°C, then it is cooled to a temperature of 720-750°C, which corresponds to the range of pearlite decomposition (Fig. 237), and held at this temperature from 4 to 6 hours to complete the austenite decomposition. After this, the steel is cooled with the furnace to 600°C at a rate of 40° to 50°C per hour and then, finally, it is cooled in air.

The steel must be hardened and tempered to impart red-hardness to the tool (Fig. 238). Recommended heating temperatures for hardening high-speed steel are given in Table 38. Such high temperatures are required to dissolve the carbides to the greatest possible extent and to obtain an austenite highly alloyed with tungsten and vanadium. This enables a martensite to be obtained, after hardening, that has a high resistance to self-tempering, i.e., ample red-hardness. These indicated temperatures should not be exceeded as this will considerably lower the mechanical properties.

In heating for hardening, the tools are preheated to 800-875°C, as a rule, to prevent cracks (Fig. 238).

Holding time at the hardening temperature should be sufficient for dissolving that part of the carbides which can pass into solution at this temperature. Experience shows that the holding time in a molt-

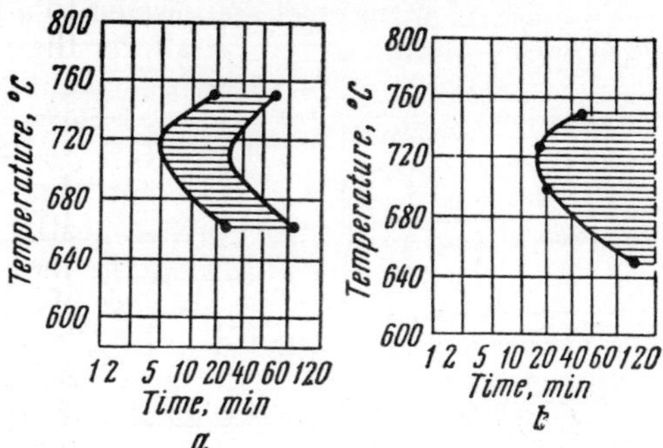

Fig. 237. Isothermal austenite decomposition (TTT) diagram for steel P18: *a*—heating to 900° C, *b*—heating to 1,290° C

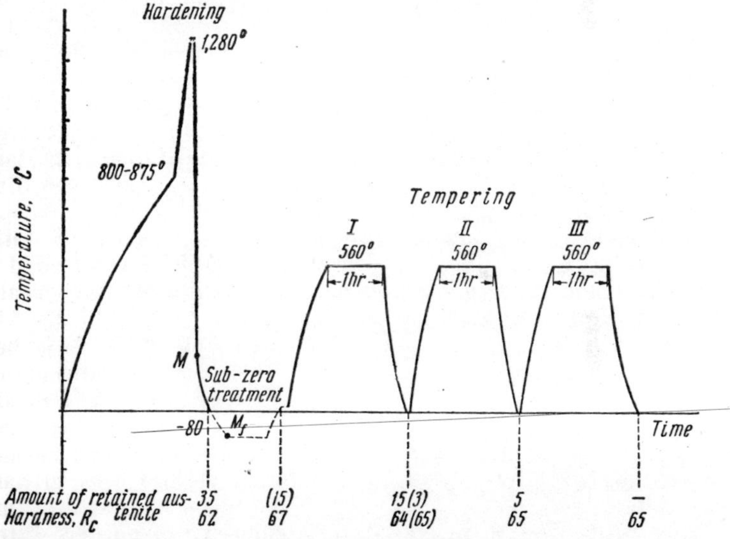

Fig. 238. Temperature vs time diagram for hardening and tempering high-speed steel (after A. P. Gulyaev)

en salt bath for tools from 10 to 15 mm in diameter (or thickness) may be taken as 8 or 9 seconds per millimetre of cross-section.

Oil is most frequently the quenching medium for high-speed steels. Martempering in molten salts (usually KNO_3), having a temperature

Table 38

Hardening Temperatures for High-Speed Steels (after Y. A. Geller)

Type of tool	Hardening temperature, °C	
	Steel P18	Steel P9
Twist drills over 15 and up to 22 mm in diameter and single-point tools	1,280-1,300	1,240-1,250
Form tools, 10 to 70 mm in diameter (milling cutters, broaches, etc.), and twist drills 5 to 15 mm in diameter	1,270-1,290	1,220-1,230
Form tools 5 to 10 mm and over 70 mm in diameter	1,260-1,280	1,210-1,230
Tools less than 3-5 mm in diameter	1,250-1,270	1,200-1,220

of 450-550°C, is also applied to reduce warping. The tools are held in the bath for 2 to 5 minutes.

Y. A. Geller has shown that austempering by holding in a bath at a temperature of 200-300°C for 30 to 60 minutes (according to the bath temperature) will reduce deformation in quenching large tools.

Directly after quenching, high-speed steel has a hardness of R_c 61-63. Its structure comprises martensite, carbides that remained undissolved in heating, and retained austenite (Fig. 236, *c*). The higher the hardening temperature, the lower the martensite points (M and M_f) will be and the larger the amount of austenite retained. Ordinarily, no more than 20 to 30 per cent austenite is retained in steel P18. Since the retained austenite lowers the cutting properties, a sub-zero treatment,* at a temperature of 70° to 80°C below zero, is often applied directly after hardening (not later than 30 to 60 minutes) to prevent stabilisation of the austenite. This treatment should be immediately followed by tempering. The main purpose of the last operation is to relieve internal stresses and to transform retained austenite into martensite.

Tools are subject to several (three) tempering operations at 550-570°C (holding time from 45 to 60 minutes); tools undergoing sub-zero treatment after quenching are tempered once or, less frequently, twice.

Two or three tempering operations are applied after martempering or after oil quenching to transform the maximum possible amount

* If the sub-zero treatment is to be carried out in liquid air, the tools must first be carefully degreased.

of retained austenite. Tempering is repeated three or four times after austempering or for tools of large cross-section (Fig. 238).

The curve in Fig. 239 shows the influence of the tempering temperature on the hardness of quenched high-speed steel. The drop in hardness in the temperature range from 150° to 400°C is probably due to the decomposition of the martensite and coagulation of carbides of the cementite type which precipitate from the martensite.

The increase in hardness at higher tempering temperatures (known as secondary hardness) is due to the transformation of the retained austenite into martensite and to the precipitation of carbides from the martensite (precipitation hardening of the martensite). The high hardness (R_c 63-65) obtained by tempering in the range from 550° to 570°C is retained when the steel is subsequently heated to 600-620°C. This constitutes the high red-hardness of high-speed steel tools.

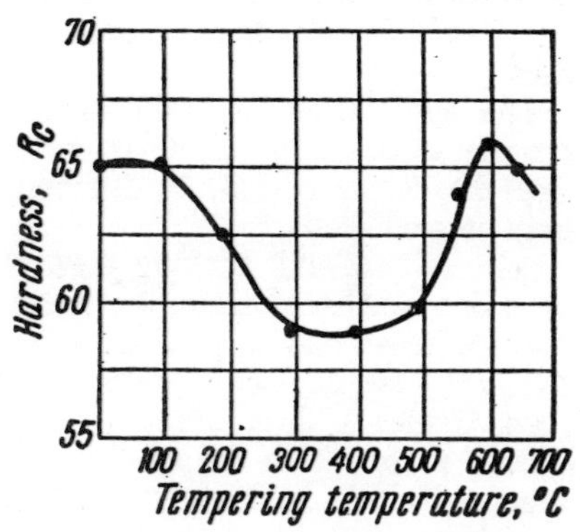

Fig. 239. Effect of tempering temperature on the hardness of high-speed steel

Low-temperature liquid cyaniding or carbonitriding (at 550-570°C) is frequently applied after heat treatment to improve the cutting efficiency of tools (see Sec. 11-4).

A steam treatment has been developed recently to increase wear resistance. This procedure consists in admitting steam into the furnace during tempering. This forms a film of Fe_3O_4 on the surface of the tool. The film has a high hardness and tightly adheres to the base metal. A certain degree of porosity of the film promotes the adsorption of the cutting fluid or lubricant and this, in turn, increases the wear resistance.

STEELS FOR MEASURING TOOLS

Steels for measuring tools should possess high hardness and wear resistance and should retain their size over a long period of service.

Such tools are most often manufactured of steels X and XΓ (1.3-1.5 per cent C, 0.45-0.7 per cent Mn, and 1.3-1.6 per cent Cr), which acquire a high hardness after quenching and low tempering.

Snap gauges, scales, rules and other flat, long measuring tools are usually made of steel plate Grades 15 and 15X. They are carburised and quenched to obtain a working surface of the required high hardness and wear resistance.

DIE STEELS

Dies for blanking, heading, and drawing operations, draw plates for wire drawing, and other tools for the cold working of metals are made of steels which possess sufficient toughness in addition to their high hardness and wear resistance (Table 39).

Small dies of simple form are frequently made of the carbon steels У10A and У12A (see Table 35) which are easily machinable in the annealed condition. This facilitates die manufacture. Larger and more complex dies are made of the alloy steels Grade X, 9XC, and others (see Table 35) which have increased hardenability and are oil-quenched.

The application of carburising, cyaniding, and chromising will provide satisfactory results in increasing the wear resistance of alloy steel dies which operate in difficult conditions but are not subject to large impact loads.

Cold cutting and working dies, requiring high wear resistance and increased toughness (complex punches and dies, deep drawing lower dies for sheet metal, profiled rolls of complex form, thread rolling dies, etc.) are made of high-chromium steel of the carbidic (cementitic) class, Grades X12, X12M, X12Ф1, and X12Ф. High-chromium steels possess a very high hardenability (up to 300-400 mm), exceptional wear resistance, and undergo almost no dimensional changes after quenching.

Cast high-chromium steel has a eutectic structure. This is broken up in subsequent hot working to form separate carbides (Fig. 223). Even with low reduction in forging these carbides become arranged in bands or lines along the direction of metal flow and this impairs the quality of the steel.

In annealed steels of the X12M and X12Ф type, the carbide phase comprises complex carbides such as $(Cr, Fe)_7C_3$. Steel X12, with an increased carbon content, also contains carbides of the cementite type $(Fe, Cr)_3C$. Annealed steel X12M may contain up to 13-16 per cent carbides.

When the steel is heated above the eutectoid temperature, the carbides dissolve and alloy the austenite with chromium. Upon heating to 1,100-1,200°C most of the vanadium (for steel X12Ф) and molybdenum (for steel X12M) is also dissolved. The stability of the supercooled austenite depends to a considerable extent upon the heating temperature in hardening. The higher the heating temperature, the more the austenite will become alloyed and the higher its stability will be.

There are two main procedures for heat treating dies of high-chromium steels:

1. Hardening (heating temperatures: steel X12Ф1—1,070-1,090°C, X12M—1,020-1,040°C, and X12—1,000-1,040°C; quenching in oil)

Table 39

Chemical Compositions of Certain Die Steels

Grade	Chemical composition, per cent							
	C	Mn	Si	Cr	W	V	Ni	Other elements
				Steels for cold-working dies				
X12Ф1	1.2-1.4	⩽0.35	⩽0.4	11-12.5	—	0.7-0.9	—	—
X12Ф	1.4-1.6	⩽0.35	⩽0.4	11-12.5	—	0.2-0.4	—	—
X12M	1.45-1.7	⩽0.35	⩽0.4	11-12.5	—	0.15-0.3	—	Mo 0.4-0.6
X12	2-2.3	⩽0.35	⩽0.4	11.5-13	—	—	—	—
4XB2C	0.35-0.44	0.2-0.4	0.6-0.9	1.0-1.3	2-2.5	—	—	—
6XB2C	0.55-0.65	0.2-0.4	0.5-0.8	1.0-1.3	2.2-2.7	—	—	—
5XBГ	0.55-0.7	0.9-1.2	0.15-0.35	0.5-0.9	0.5-0.8	—	—	—
				*Steels for hammer dies**				
5XHB	0.5-0.6	0.5-0.8	0.15-0.35	0.5-0.8	0.4-0.6	—	1.4-1.8	—
5XHCB	0.5-0.6	0.3-0.6	0.6-0.9	1.3-1.6	0.4-0.6	—	0.8-1.2	—
5XHC	0.5-0.6	0.3-0.6	0.6-0.9	1.3-1.6	—	—	—	Mo 0.15-0.3
5XHT	0.5-0.6	0.5-0.8	⩽0.35	0.9-1.25	—	—	1.4-1.8	Ti 0.08-0.15
				Steels for press-moulds				
3X2B8	0.3-0.4	0.2-0.4	⩽0.35	2.2-2.7	7.5-9.0	—	⩽0.3	Ti 0.2-0.5
4X8B2	0.35-0.45	0.2-0.4	⩽0.35	7-9	2-3	—	⩽0.3	—

* Steel 5XHM is the most suitable for hammer dies, but is rarely used due to its high cost.

followed by low tempering (150-170°C). The hardness after this treatment will be R_c 61-63.

Steel acquires high hardness (R_c 61-63) and wear resistance after this treatment. If increased toughness is required at the expense of hardness (decreased to R_c 57-59) and if the dimensions of the part are to be reduced in comparison to those obtained after quenching, the tempering temperature is raised to 200-275°C.

Dies operating under very heavy impact loads are tempered at 400-425°C.

2. Hardening (heating temperatures: steel X12M—1,115-1,130°C and X12Ф1—1,150-1,170°C; quenching in oil or molten salt at a temperature of 400-500°C) followed by repeated tempering (2 or 3 operations) at 510-520°C. The steel is not very hard (R_c 44-52), directly after quenching, due to the large amount of retained austenite. During repeated tempering, retained austenite is transformed into martensite and the hardness is increased to R_c 60-61. A single tempering at 510-520°C will prove sufficient if a sub-zero treatment at 80°C below zero is applied after quenching.

The first procedure will provide high mechanical properties and minimum deformation.

Steel, to which the second procedure has been applied, acquires high red-hardness and wear resistance. This procedure is recommended for dies operating at 400-500°C and also for cutting tools (milling cutters). It must be noted, however, that quenching from high temperatures is associated with large dimensional changes and does not ensure high mechanical properties.

Disadvantages of high-chromium steels are their poor machinability in the annealed state (hardness *Bhn* 207-269) and their lowered mechanical properties in instances when carbide inhomogeneity is sharply defined.

Steels with increased toughness, such as Grades 4XB2C, 6XB2C, 5XBГ, and others, are used in the manufacture of air-hammer chisels, trimming dies, cold-cutting shear blades, and other tools with thin working edges. These steels acquire a hardness of R_c 50-55 after quenching with subsequent tempering at 250-270°C.

Steels for hammer dies used in the hot working of metals must have sufficiently high mechanical properties at elevated temperatures; they must resist softening after repeated heating, have a high hardenability, minimum deformation upon heat treatment, and no susceptibility to temper brittleness.

Steel 5XHCB is usually employed for making large dies (length of a side$\geqslant$400 mm), steels 5XHB and 5XHCB for medium-sized dies, and steel 5XHC for small dies (with the shortest side up to 300 mm in length). These dies are hardened by heating to 840-870°C and quenching in oil (or in air). This is followed by tempering at 520-570°C to relieve the stresses and to obtain a homogeneous sorbite structure.

Table 40

Principal Defects in Alloy Steels

Defect	General description	Cause	Revealed by
1. Flakes	Thin cracks or fissures 1 to 100 mm long. The boundaries of the crack form the rounded or oval form of the flake. The surface of the crack has a metallic lustre as it passes along the grain boundaries. Found chiefly in medium- and high-carbon steels	Flakes are formed due to the presence of dissolved hydrogen in steel in quantities which cause supersaturation at temperatures below 250°C. Reduced ductility and high stresses in the hydrogen-saturated metal, in conjunction with stresses caused by non-simultaneous structural transformations under conditions of well-developed dendritic segregation, are the causes of flakes	1) Fracture 2) Macrography 3) Ultrasonic flaw detection
2. Slaty fracture	Wood-like arrangement of fibres in longitudinal fractures of steel usually revealed after hardening and high tempering. Found in medium-carbon alloy structural steels	Slatiness as well as dendritic segregation are determined by the degree of deoxidation of the steel	In the fracture of hardened and tempered (at 400-600° C) steel
3. Stone-like fracture	Dull projections, seen in the fracture of large forgings, which are the boundaries of large grains. Most frequently found in the structural steels: Cr-Al, Cr-Ni-Si, and Cr-Ni-W	Overheating during hot working and an insufficient degree of subsequent deformation	Fracture
4. Naphthalene-like fracture	Large bright spots on a field of the uniformly dull hardened steel fracture. Found in high-alloy steels of the high-speed type	Appears after hardening and tempering in cases of previous heat treatment at a higher temperature producing coarse grain and a martensite structure or when the final temperature in forging was too high	Fracture

After tempering, the hardness will be within R_c 35-45 (*Bhn* 321-430).

The *components of die-casting dies* for copper and aluminium alloys are made of steel 3X2B8; steel 4X8B2 is used for these components in die casting zinc alloys. These two grades of steel have good thermal endurance, high mechanical properties at elevated temperatures, and they are corrosion and erosion resistant. They are employed for heading dies and for moulds that operate for prolonged periods, at high temperatures.

DEFECTS IN ALLOY STEELS

The most typical defects of alloy structural and tool steels, appearing during their manufacture and subsequent treatment, are given in Table 40.

13-9. Wear-Resistant (Austenitic) Steels

Parts subject to wear under conditions of high contact pressures and impacts (crawler links, stone-crusher jaws, excavator buckets, frogs of railroad and street car tracks, etc.) are made of high-manganese austenitic steels Г13Л and Г13 containing 1.0-1.5 per cent C and 11.0-15.0 per cent Mn. As cast, these steels have an austenitic structure with excess carbides (Mn_3C) along the grain boundaries. This decreases the strength and toughness of the steel. Therefore, parts made of steel Г13Л are hardened by heating to 1,100°C and quenching in water. Cracks are prevented by heating first to 650°C at a rate of 70° to 100° per hour for small castings and 50° per hour for large complex castings. The holding time at 650°C is from one to four hours. At the hardening temperature, the holding time is 0.5 to 4 hours, in accordance with the size of the casting.

These steels are very sensitive to decarburisation which causes the formation of martensite. Therefore, thin articles must be protected against decarburisation by metal coating or special pastes.

After properly conducted heat treatment, these steels have an austenitic structure and the following mechanical properties:

Tensile strength σ_B, kg/mm²	63-70
Yield point σ_s, kg/mm²	31-35
Relative elongation δ, per cent	25-15
Reduction in area ψ, per cent	30-20
Hardness, *Bhn*	170-228

The steel is readily work-hardened in the course of operation and acquires high wear resistance. If the wear is purely abrasive and the steel is not subject to impact loads which would cause work hardening, the wear resistance will not be very high.

13-10. Stainless and Acid-Resistant Steels

The destruction of metals and alloys by the action of the surrounding medium is called corrosion. Corrosive destruction begins at the surface, i.e., at the interface between the metal and the medium, and gradually penetrates into the metal. At this, metals and alloys lose their metallic lustre and become coated with the corrosion products (rust). The mechanical properties of corroded steel are sharply reduced even when no visible change is observed on the metal surface.

One effective method of protecting steel against corrosion is to add elements that will form a stable protective film on the surface of the metal to prevent destruction of the base metal.

Stainless steels. Steels resistant to atmospheric corrosion are called stainless steels. In the Soviet Union, the most extensively used steels of this type (GOST 5632-51) are the chromium steels 1X13, 2X13, 3X13, and 4X13 containing from 0.1 to 0.4 per cent C and 12-14 per cent Cr (Table 41) and the high-carbon steel X18 (0.9-1.0 per cent C and 17-19 per cent Cr) (Table 41).

Grades 1X13 and 2X13 are hypoeutectoid steels while grades 3X13 and 4X13 are eutectoid and hypereutectoid steels, respectively. The austenite formed in heating these steels is transformed into martensite upon quenching. High-carbon steel X18 is of the cementitic class. Heat treating procedures and applications of chromium stainless steels are listed in Table 41.

Chromium stainless steels have a high corrosion resistance in air, water, and alkali or acid solutions. This resistance is due to the formation of a thin oxide film (Cr_2O_3) on the surface of the steel. The corrosion resistance is most pronounced when all of the chromium is in the solid solution and the steel has a single-phase structure. The formation of carbides is associated with the depletion of chromium from the solid solution. This lowers the corrosion resistance.

Consequently, the higher the carbon content, the lower the corrosion resistance will be. The corrosion resistance of a part is increased by heat treatment and polishing. Machine parts made on automatic screw machines and thereby requiring improved machinability (screws, nuts, and other threaded parts) are made of steel X14 ($\leqslant$0.15 per cent C, 13-15 per cent Cr, and 0.2-0.4 per cent S) which contains more sulphur than the other grades. Its resistance to atmospheric corrosion is quite satisfactory.

Acid-resistant steels are highly resistant to corrosion under the action of various aggressive media. They include ferritic steels with a high chromium content (X17, X25, and X28) and austenitic chromium-nickel steels (Table 42).

The most important of the acid-resistant steel series are the austenitic chromium-nickel steels listed in Table 42. Their structure in the equilibrium state consists of two phases [austenite + carbides

Table 41

Heat Treatment and Applications of Chromium Stainless Steels

Grade	Heat treatment	Applications
1X13	Hardening: heat to 1,000-1,050°C, quench in water (or oil), and temper at 700-790°C	Parts requiring high ductility and subject to impact loads (turbine blades, valves of hydraulic presses, cracking plant fittings, bolts, nuts, household appliances, etc.). Good resistance in air, in water and in steam
2X13	Hardening: heat to 1,000-1,050°C quench in water (or oil), and temper at 660-770°C	The same parts but with higher hardness (3X13 has a hardness of R_c 48)
3X13	Hardening: heat to 1,000-1,050°C, quench in oil, and temper at 200-300°C	Higher resistance in air, as well as in water and steam
4X13	Hardening: heat to 1,050-1,100°C, quench in oil, and temper at 200-300°C	Cutting and measuring tools, surgical instruments, carburettor needles, and ball bearings (hardness—R_c 50). Corrosion resistance lower than for preceding grades
X18 *	Ditto	Ball bearings, high-quality knives, bushings, globe valves, and other parts subject to wear and requiring increased corrosion resistance

* Contains 0.9-1.1 per cent C.

$(Cr, Fe)_{23}C_6]$. Fig. 240 shows that only alloys containing less than 0.04 per cent C have structures free of excess carbides. After hardening (heating to 1,100-1,150°C and quenching in water), these steels acquire a single-phase structure (Fig. 223, *d*).

Due to the low rate of chromium diffusion at temperatures below 500°C, the austenite is stable and no carbides precipitate from it. The hardened steel has a comparatively low strength but high ductility. Its mechanical properties are: σ_u=55-58 kg/mm², σ_s=20-22 kg/mm², δ=45-40 per cent and ψ=60-55 pei cent.

Table 42

Compositions and Applications of Chromium-Nickel Acid-Resistant Steels

Grade	Chemical composition, per cent					Certain applications and characteristics
	C	Mn	Cr	Ni	Ti	
1X18H9	⩽0.14	⩽2	17-20	8-11	—	Floats and fairing of seaplanes, fuel tanks, nonmagnetic parts of ship navigation instruments, devices exposed to sea-water and weak alkalis. Heat treatment is applied, after welding and other fabricating processes involving heating, to prevent intercrystalline corrosion
2X18H9	0.15-0.25	⩽2	17-20	8-11	—	
1X18H9T	⩽0.12	⩽2	17-20	8-11	(C* — 0.03%) ×5, but ⩽0.8 per cent	Used in the following industries: nitrogen (acid tanks, absorption towers, etc.), paint and varnish (autoclaves, mixers, etc.), coal (pumps and devices operating in acidic waters), and dairy (cans and flasks)

* C—carbon content in per cent.

The strength of this steel may be increased by work hardening (Fig. 241). The hardened steel is easily formed and has good weldability.

One notable factor is that improper heat treatment (slow cooling in the range from 500° to 700°C or heating to 500-700°C), for instance in welding, will cause the austenite to decompose and chromium carbide will be precipitated along the grain boundaries. Due to the low rate of chromium diffusion, carbides are formed only by the chromium at the grain surfaces. This impoverishes the chromium content of the austenite adjacent to the grain boundaries below 12 per cent and the corrosion resistance of the steel is sharply reduced. The steel acquires a susceptibility to corrosion on the grain boundaries (intercrystalline corrosion).

Upon a further development of intercrystalline corrosion,* steel loses its metallic sound and tears will appear in flexure tests along

* For intercrystalline corrosion tests, see GOST 6032-51.

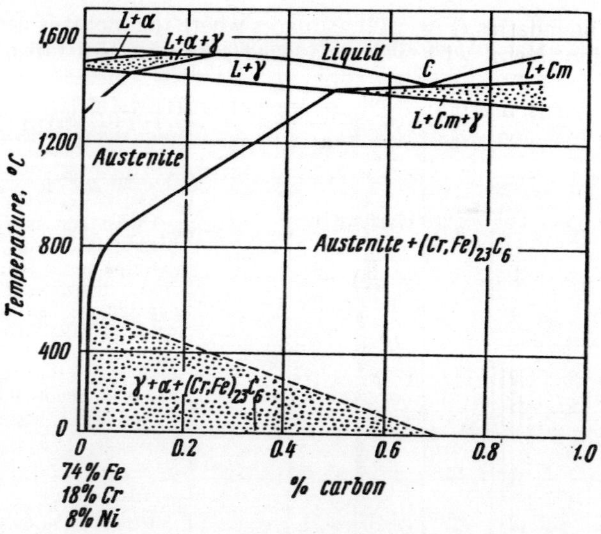

Fig. 240. Pseudobinary section of the system Fe-Cr-Ni-C. Vertical section at 18 per cent Cr and 8 per cent Ni (after E. Bain): *Cm*—(Cr, Fe)$_{23}$C$_6$

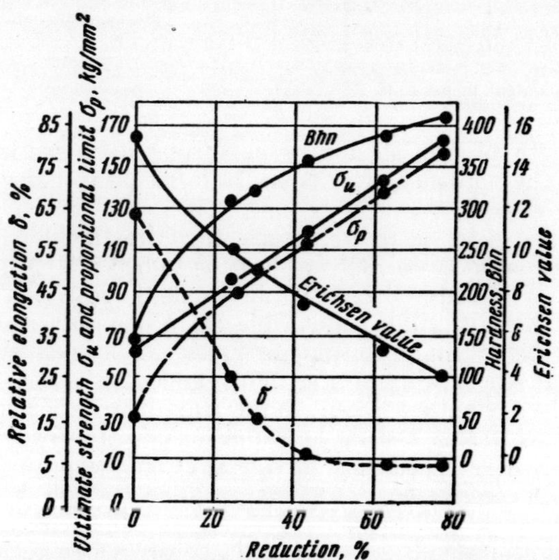

Fig. 241. Variation of the mechanical properties of steel 2X18H9 with the reduction in the cold rolling of sheets (after F. F. Khimushin)

the grain boundaries (Fig. 242) at places where the metal is destroyed by corrosion. Metal with well developed corrosion will fail at small stresses.

Two methods are employed to prevent intercrystalline corrosion: 1) the carbon content is reduced; this correspondingly reduces the

Fig. 242. Destruction of steel 1X18H9 by intercrystalline corrosion:
a—initial specimen after quenching in water from 1,150° C, b—intercrystalline destruction in the form of transverse cracks appearing after bending through 90°, c—intensive destruction over the whole surface of the specimen (after V. M. Doronin)

amount of precipitated carbides (as in steel 0X18H9) and 2) alloying elements are added to prevent the precipitation of chromium carbide. Such elements are titanium (steel 1X18H9T) and niobium (steel X18H9Б).

The addition of these elements combines all of the carbon in the stable carbides TiC and NbC thus preventing the possibility of chromium carbide precipitation. The amount of titanium or niobium is specified so that all of the carbon is combined in the carbides of these elements. The amount of titanium should be (% C—0.03)5 but$\leqslant$0.8 per cent; niobium is added up to 1.5 per cent (experimental data). Steel 1X18H9T is not susceptible to intercrystalline corrosion.

As a substitute for steel 1X18H9, steel 1X13H4Г9 is used to some extent. Here a large part of the nickel has been replaced by manganese. This substitute has lower corrosion resistance and is susceptible to intercrystalline corrosion.

13-11. Scale-Resistant and Heat-Resistant Steels and Alloys

A scale-resistant steel is one not susceptible to scale formation (gas corrosion) at high temperatures. This property of a steel is improved, chiefly, by the addition of chromium, aluminium, and silicon which, upon heating, form dense oxide films, such as $(Cr, Fe)_2O_3$, $(Al, Fe)_2O_3$ and others, that protect the base metal against oxidation. An addition of 5 to 8 per cent Cr raises the scale resistance to 700-750°C, increasing the content to 15-17 per cent will prevent scaling up to 950-1,000°C, and 25 per cent Cr will prevent scaling up to 1,100°C. It is necessary to point out that scale resistance depends only on the composition and not on the structure of the steel. Ferritic and austenitic steels with equal chromium content have practically the same scale resistance.

Table 43 lists the compositions and properties of certain scale-resistant and heat-proof steels (according to GOST 5632-51).

Steels X6C, X9C2, X6CM, and X10C2M contain large amounts of chromium and silicon. They are known as silchrome steels. These steels are of the martensitic class and require a preliminary heat treatment to improve their machinability (high tempering at 820-850°C). Final heat treatment consists of hardening by heating to 1,040-1,100°C, quenching, and then tempering at 750-850°C to stabilise the structure which is a ferrite-carbide mixture.

Silchromes are susceptible to temper brittleness and grain growth. Molybdenum may be added, as in steel X6CM, to eliminate temper brittleness.

Heat-resistant steels retain sufficient strength and scale resistance at high temperatures. Such steels must resist creep well and possess high short-time and rupture strengths (see Sec. 5-6).

Heat resistance depends upon the strength of the interatomic bonds and upon the structure of an alloy. Alloying increases the heat resistance of alloys having the same base by increasing the intensity of interatomic attraction and raising the recrystallisation temperature. These changes impede the development of plastic deformation at elevated temperatures. This resistance to plastic deformation is also increased by the formation of a heterogeneous structure and by breaking up the blocks of mosaic structure.

The heat resistance of steels and alloys is improved by alloying elements that increase their tendency to age, and thereby they are

Table 43

Compositions, Properties, and Applications of Scale-Resistant and Heat-Proof Steels

Grade	Composition	Chief properties	Typical applications
X6C	⩽0.15% C, 1.5-2% Si, 5-6.5% Cr	Scale-resistant up to 750°C	Boiler components operating under increased loads at temperatures up to 750°C, valves, etc.
X9C2	0.35-0.5% C, 2-3% Si, 8-10% Cr	Scale-resistant up to 800°C	Valves and components operating under low loads at temperatures up to 800°C
X12ЮC	0.07-0.12% C, 1.2-2% Si, 11.5-14% Cr, 1.0-1.8% Al	Scale-resistant up to 900°C	Components operating under low loads at temperatures up to 900°C
X6CM	⩽0.15% C, 1.5-2% Si, 5-6.5% Cr, 0.45-0.6% Mo	Heat-resistant up to 650°C and scale-resistant	Pipe used in cracking processes under sulphur corrosion conditions, pump components, gate valves, piston rods, etc.
X18H25C2	0.3-0.4% C, 2-3% Si, 17-20% Cr, 23-26% Ni	Scale-resistant up to 1,100°C, heat- and acid-resistant	Heavily loaded components: furnace conveyers and fastening parts subject to high temperatures and pressures
X25H20C2	⩽0.2% C, 2-3% Si, 23-27% Cr, 18-21% Ni		

hardened by the precipitation of microscopic particles that impede plastic deformation at high temperatures.

Austenitic steels are more heat resistant than ferritic grades because, in the former, such processes as recovery and recrystallisation take place at higher temperatures. Larger grain size also increases heat resistance.

Nickel-base alloys are extensively used as heat-resistant materials.

Table 44

Heat-Resistant Alloys Used in the U.S.S.R. (after F. F. Khimushin) *

Grade	Chemical composition, per cent					Applications
	C	Cr	Ni	Fe	Other elements	
Iron-base heat- and scale-resistant alloys						
ЭИ69 (4X14H14B2M)	0.45	14	14	Base	W—2, Mo—0.5	Exhaust valves of aviation engines, fastening parts (studs, bolts, etc.)
ЭИ388 **	0.4	14	7	Ditto	Mn—8, V—1.6, Mo—0.7	Gas-turbine working blades, bolts, studs and nuts
ЭИ481	0.4	13	8	Ditto	Mn—8, V—1.3, Mo—1.1, Nb—0.3	Wheels, annular parts, and fastening parts for turbines
ЭИ696	0.06	10	20	Ditto	Ti—3, B—0.015, Al—0.4	Blades, wheels, rings, sheets and other parts of gas turbines
ЭИ395	0.10	16	25	Ditto	Mo—6, N_2—0.15	Turbine wheels, studs, bolts, etc.
Nickel-tase heat-resistant alloys						
ЭИ437 *** ЭИ437А ЭИ437Б	0.08	20	Base	⩽2.0	Ti—2.6, Al—0.7, B—0.005-0.008	Gas-turbine working blades and wheels
ЭИ617	0.08	15	Base	⩽5.0	Ti—2, Al—2, Mo—3, W—7, B—0.008, V—0.3	Working and nozzle blades of gas turbines

Continued

Grade	Chemical composition, per cent					Applications
	C	Cr	Ni	Fe	Other elements	
ЭИ598	0.08		Base	⩽5.0	Ti—2.5, Al—1.5, W—6, Mo—3, B—0.008	Working blades of gas turbines
ЭИ826 **** (ЭИ617АБ)	0.08	13	Ditto	⩽5.0	Ti—2, Al—3.0, Mo—3.5, W—5, B+V—0.015	Working blades (cast)
Cobalt-base alloys						
ЛК-4	0.20	27	3	⩽5.0	Co—63, Mo—6	Cast working blades of turbo-superchargers
ЭИ416 (ВК36)	0.40	20	20	4	Co—45, Mo—4, W—4	Blades of turbo-superchargers

* F. F. Khimushin, *Heat-Resistant Alloys*, published by the Dzerzhinsky House of Scientific and Engineering Information, Moscow, 1957.

** Alloy ЭИ388 was developed by F. F. Khimushin and L. Y. Kontorovich; alloy ЭИ481—by F. F. Khimushin, K. I. Terekhov, I. M. Gavrilova, Y S Ovsepyan, and Z. A. Shevankova; alloy ЭИ616—by F. F. Khimushin, S. G. Kishkin, D Y. Lifshitz, Y. F. Trusova, and others.

*** Differ in manufacturing technique and addition of boron (ЭИ437Б), alloy ЭИ437Б has highest heat resistance of this series.

**** Increased boron and aluminium content promotes heat resistance but makes hot working more difficult.

Modern iron- and nickel-base heat-resistant alloys are given in Table 44. They find wide application in the manufacture of components for gas turbines and jet engines.

Iron-base heat-resistant alloys may be classified into the following two groups:

1. Austenitic-carbidic steels (ЭИ69, ЭИ388, ЭИ481, ЭИ590, and ЭИ374) whose high heat resistance is associated with the strengthening of the austenite by the formation of highly dispersed carbide phases in ageing. In steel ЭИ69, the carbide phase consists principally of $Cr_{23}C_6$; in steels ЭИ388 and ЭИ481, it consists of $Cr_{23}C_6$ and VC.

2. Austenitic-intermetallic steels (ЭИ696) achieve their high heat resistance by the formation of intermetallic phases Fe_2Ti and Ni_3Ti or the intermediate phase α' during ageing or operation at high temperatures.

Heat-resistant alloys ЭИ437, ЭИ617, and others (Table 44) are of the nickel-base type. The nickel forms a solid solution with the chromium, tungsten, and molybdenum. During ageing, aluminium, titanium, and boron form excess intermetallic phases. In the initial period of solid solution decomposition, an intermediate α'-phase is formed on the basis of the compound Ni_3(Al, Ti, Cr). It has a face-centred cubic lattice, coherently bound to the solid solution. The stable phases, nickel titanide (Ni_3Ti) with a hexagonal lattice and Ni_3Al are formed after prolonged ageing at high temperatures (800-900°C). Carbides and borides may also be formed in addition to these phases. Nickel-base alloys acquire their high heat resistance due to precipitation hardening, associated with the formation of intermetallic phases, and by the alloying of the solid solution with molybdenum and tungsten. These alloying elements increase the intensity of interatomic attraction and impede the softening of the alloy at high temperatures. Figs. 243 and 244 show the variation in 100-hour rupture strength with the testing temperature. The heat treating procedures given in Table 45 provide for maximum heat resistance of these alloys.

The alloys are heated to high temperatures in hardening to put the excess phases (carbides and intermetallides of the Ni_3Ti type and others) into the solid solution. Rapid cooling (quenching) fixes the supersaturated solid solution. This saturation with alloying elements considerably distorts the crystal lattice. It also promotes stress increase and breaks up the mosaic structure. The result is increased resistance to plastic deformation. During ageing, the excess phases precipitate from the solid solution. Proper ageing procedure will increase the hardness and heat resistance of the alloy.

The precipitation of excess phases improves the heat resistance only if the solid solution remains alloyed to a sufficient extent. When the solid solution is considerably impoverished and the grain of the excess phase is coarsened, the heat resistance will be lowered. Ageing

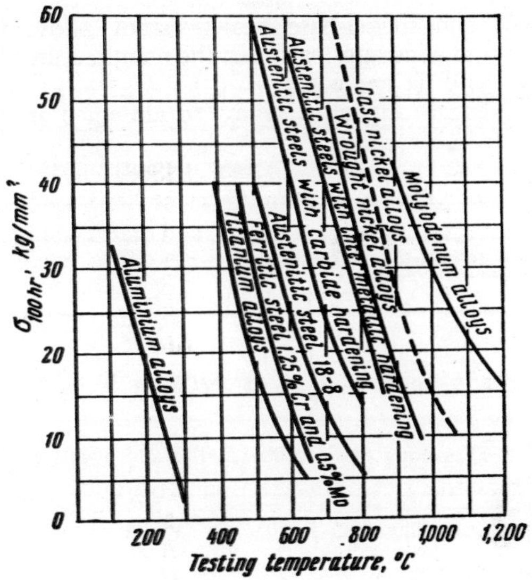

Fig. 243. Variation in the 100-hour rupture strength with the testing temperature

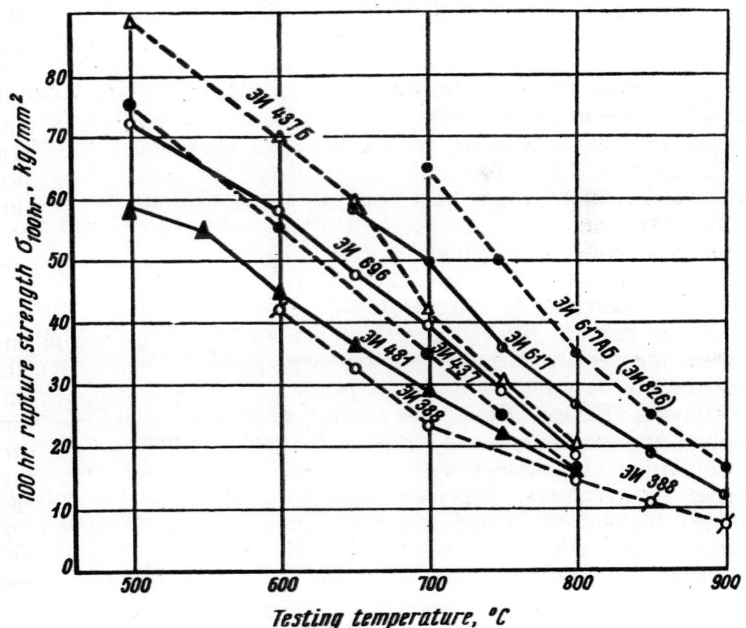

Fig. 244. Variation in the 100-hour rupture strength of heat-resistant steels and allovs with the testing temperature (after F. F. Khimushin)

Table 45

Heat Treatment of Heat-Resistant Alloys (after F. F. Khimushin)

Grade	Heat treatment
ЭИ388	Quenching from 1,180° C in water and ageing at 800° C for 8-10 hours
ЭИ481	Quenching from 1,140-1,200° C and double ageing: 1) at 690° for 16 hours and 2) at 780-800° C for 10 to 16 hours
ЭИ696	Quenching from 1,100-1,200° C and ageing at 700-800° C
ЭИ437	Quenching from 1,080° C in air and ageing at 700° C
ЭИ617	Double quenching: 1) from 1,200° C (holding time 2 hours) in air, and 2) from 1,050° C (holding time 4 hours) in air, followed by ageing at 800° C for 16 hours

must be carried out at temperatures that exceed the working temperatures of subsequent operation.

Heat-resistant alloys are very difficult to work and to machine.

Tables 46 and 47 indicate the chemical compositions and strength characteristics of certain heat-resistant alloys used in the U.S.A.*

The first group includes iron-base alloys hardened by heat treatment and "hot-cold working".

Timken alloys 16-25-6 and 19-9DL are used in the manufacture of turbo-superchargers and gas turbine rotors after a reduction of 22-25 per cent at 650-760°C (hot-cold working). This procedure increases the tensile strength and the yield point. No additional heat treatment is required by these alloys.

Alloys A-286 and Discaloy-24 which contain small additions of titanium and aluminium, are hardened by heat treatment.

Cobalt-nickel-chromium-iron alloys are intended for service at higher temperatures. They are usually employed in the solution treated or solution treated and aged condition. The strength properties of these alloys are improved by the addition of molybdenum, tungsten, and niobium (see Table 46). Titanium and aluminium,

---

* *Metal Progress*, Vol. 66, No. 1a, July 1954, Report of the ASM Committee on Heat-Resisting Alloys.

Table 46

Chemical Compositions of Heat-Resistant Alloys

Grade	Chemical composition, per cent												Other elements
	C	Si	Mn	Cr	Ni	Co	Mo	W	Nb	Ti	Al	Fe	
							Iron chromium-nickel alloys						
Wrought alloys													
A-286	0.08	1.0	1.25	14.7	25.5	—	1.25	—	—	2.1	0.35	Remainder	V—0.25
Discaloy	0.04	1.0	1.4	13.5	26.2	—	3.9	—	—	1.6	0.11	Ditto	—
Timken-16-25-6	0.10	0.7	1.35	16.0	25	—	6	—	—	—	—	Ditto	N—0.15
19-9 DL	0.30	0.6	1.10	19.0	9	—	1.25	1.2	0.4	0.3	—	Ditto	—
							Cobalt-nickel-chromium-iron alloys						
Wrought alloys													
Refractaloy 26	0.05	0.7	0.7	18	37	20	3	—	—	2.8	0.2	18	—
S-590	0.42	0.4	1.25	20.5	20	20	4	4	4	—	—	24	—
S-816	0.38	0.4	1.20	20	20	43	4	4	4	—	—	4	—
V-36	0.31	0.5	0.9	25	20	42	4	2.6	2.2	—	—	3	—
							Nickel-base alloys						
Wrought alloys													
Inco 700	0.10	0.2	0.05	15	49	28	3	—	—	2	3	0.5	—
Inconel X	0.03	0.3	0.5	15	73	—	—	—	0.6	2.3	0.9	6.5	—
Inconel X, 550	0.03	0.3	0.5	15	73	—	—	—	0.6	2.5	1.1	6.5	—

Continued

Grade	Chemical composition, per cent												Other elements
	C	Si	Mn	Cr	Ni	Co	Mo	W	Nb	Ti	Al	Fe	
M-252	0.15	0.6	1	19	Rem.	10	10	—	—	2.5	0.9	5	—
Nimonic 90	0.08	0.4	0.5	20	58	16	—	—	—	2.3	1.4	0.5	—
Nimonic 80A	0.05	0.5	0.7	20	76	—	—	—	—	2.3	1.0	0.5	—
Waspaloy	0.10	0.7	1	19.5	Rem.	13.5	4.2	—	—	2.5	1.25	2	—
Cast alloys													
Hastelloy Alloy B	0.10	0.7	0.8	1	65	—	28	—	—	—	—	5	—
Hastelloy Alloy C	0.10	0.7	0.8	16	57	—	17	4	—	—	—	5	—
Cobalt-base alloys													
Wrought alloy													
Haines Alloy 25 (L-605)	0.12	1	1.5	20	10	51	—	15	—	—	—	1.0	—
Cast alloys													
Haines Alloy 21	0.25	0.6	0.6	27	3	62	5	—	—	—	—	1.0	—
Haines Alloy 30 (422-19)	0.40	0.6	0.6	24	17	51	6	—	—	—	—	1.0	—
Haines Alloy 31 (X-40)	0.40	0.6	0.6	25	10	55	—	8.0	—	—	—	1.0	—
Haines Alloy 36	0.40	0.5	1.2	19	10	54	—	14.5	—	—	—	1.0	B—0.03

Table 47

Stress-Rupture Properties of Heat-Resistant Alloys

Grade	Stress to rupture in 100 and 1,000 hours, kg/mm²									
	650° C		732° C		816° C		871° C		982° C	
	100 hr	1,000 hr	100 hr	1,000 hr	100 hr	1,000 hr	100 hr	1,000 hr	100 hr	1,000 hr
Iron-chromium-nickel alloys										
Wrought alloys										
A-286	42.6	33.5	24.5	15.1	9.7	5.4	—	—	—	—
Discaloy 24	38.5	28.7	22.4	14	10	—	—	—	—	—
Timken 16-25-6	31.5	23.8	17.5	11.9	9.5	6.3	—	—	—	—
19-9 DL	36.4	26.6	19.6	13.3	11.9	7	—	—	—	—
Cobalt-nickel-chromium-iron alloys										
Wrought alloys										
Refractaloy 26	56	44.1	35.7	26.6	18.9	12.6	—	—	—	—
S-590	33.6	26.6	21.0	15.4	15.4	11.2	8.75	6.3	3.9	2.45
S-816	42	31.5	25.2	17.6	16.8	12.3	9.8	6.7	3.8	2.1
V-36	—	—	24.5	18.6	14.1	12.6	10.5	7.7	6.0	3.5
Nickel-base alloys										
Wrought alloys										
Inco 700	—	—	51.1	42	29.4	21	19.6	13	4.5	2.45
Inconel X	56	47.6	33.6	26.6	19.6	12.6	12.6	6.3	2.3	1.6
Inconel X, 550	—	—	—	—	23.8	14.7	12.6	7.2	2.3	—
M-252	—	—	36.4	24.5	20.3	12.6	11.2	7	—	—
Nimonic 80A	47.1	39.2	33.6	25.2	16.8	10.9	9.8	6	—	—
Nimonic 90	53.3	44.1	35.4	26.6	17.6	12.5	—	—	—	—
Waspaloy	—	—	40.3	—	22.1	14	13.7	—	—	—
Cast alloys										
Hastelloy Alloy B	35.7	28.4	24.5	17.9	13	9.9	—	—	—	—
Hastelloy Alloy C	34.7	29.8	22.4	17.5	13.3	10.2	9.3	6.4	—	—
Cobalt-base alloys										
Wrought alloys										
Haines Alloy 25 (L-605)	49	40.6	30.1	23.1	16.1	11.9	10.9	7.4	4.9	2.7
Cast alloys										
Haines Alloy 30 (422-19)	—	—	32.9	25.2	20	15.2	11.2	10.4	7	5
Haines Alloy 21	35.7	31	22.4	15.4	15.4	9.95	11.7	9.3	6.6	4.9
Haines Alloy 31 (X-40)	38.5	32.2	31.5	23.1	12.9	16.4	14.7	12.6	8	6.9
Haines Alloy 36	—	—	33.6	29	20.3	17.9	16.1	13	7.4	5

which form intermetallic compounds, may be added to provide greater response to heat treatment.

Development of the high-strength heat treatable nickel-base alloys such as Inconel X, 550; Inco 700; M-252; and Waspaloy was based on the necessity for new alloys to replace, in part, the cobalt-base alloys (such as S-816) since cobalt is very expensive.

The fourth group lists the cast and wrought cobalt-base alloys which have been extensively used for aircraft gas turbine guide vanes and buckets.

Cast alloys, which contain more carbon and have a coarser grain structure, are more heat resistant than the wrought alloys.

Alloys of molybdenum with titanium, niobium, vanadium, and cobalt have exceptionally high heat-resisting properties. When reliable methods are found for protecting molybdenum alloys from gas corrosion, they will lead among materials designed for high-temperature service.

Scale-resistant alloys with high electrical resistivity. This group includes scale-resistant alloys and steels with high electrical resistivity intended for making round and flat heating elements of electrical appliances (for example, furnaces). Ferritic steels, containing chromium and aluminium which sharply increase the scale resistance and electrical resistivity (X13IO4-fechral, 1X17IO5, OX17IO5, OX25IO5, and 1X25IO5),* are extensively employed for this purpose, as well as chromium-nickel alloys (of the nichrome type) such as X15H60 and X20H80. The resistivity of these alloys ranges from 1.15 (for X20H80) to 1.3 (for X13IO4) ohm-sq mm per m and they have a scale-resistance from 850°C (for X13IO4) to 1,100-1,150°C (for 1X25IO5 and X20H80).

13-12. Magnetic Steels and Alloys

Magnetic steels and alloys may be classified into the two following general groups: 1) permanent-magnet types for making permanent magnets (as the name implies); such steels and alloys must possess high coercive force (H_c) and residual induction (B_r) and these magnetic properties must be stable in the course of time, and 2) soft-magnet types intended for transformer cores, armatures and pole pieces of electrical machines, as well as for the cores of chokes, relays, measuring instruments, and other components of magnetic circuits. Soft-magnet materials must have a low coercive force, high permeability (μ) and low hysteresis losses.

Permanent-magnet steels and alloys. Table 48 lists the compositions and magnetic properties (after heat treatment) of steels intended for permanent magnets (according to GOST 6864-54).

* Aluminium coarsens the grain of high-chromium steel to a great extent and imparts high brittleness. Therefore, its content should never exceed 5-7 per cent.

Table 48

Compositions and Properties of Steels for Permanent Magnets

Grade	Chemical composition, per cent					Magnetic properties	
	C	Cr	W	Co	Mo	Minimum residual induction B_r, gauss	Minimum coercive force H_c, oersteds
EX	0.95-1.1	1.3-1.6	—	—	—	9,000	58
EX3	0.9-1.1	2.8-3.6	—	—	—	9,500	60
E7B6	0.68-0.78	0.3-0.5	5.2-6.2	—	—	10,000	62
EX5K5	0.9-1.05	5.5-6.5	—	5.5-6.5	—	8,500	100
EX9K15M	0.9-1.05	8-10	—	13.5-16.5	1.2-1.7	8,000	170

Chromium steels are easily machined and worked and may be applied for manufacturing magnets of complex form.

Maximum values of H_c and B_r are obtained after hardening. Subsequent tempering (ageing), which relieves the quenching stresses, reduces the coercive force to some extent but provides stability of magnetic properties in operation.

The optimum heat treating technique for chromium steels, intended for permanent magnets, consists of heating to 820-840°C and quenching in oil (or water). This is followed by ageing for 10 to 24 hours at 100°C to stabilise the magnetic properties.

Steel E7B6 is most widely used in Soviet practice. After quenching from 820-860°C in oil (or water) and ageing at 100°C, it has higher magnetic properties than the chromium types.

If chromium steels (EX and EX3), and particularly the tungsten types (E7B6) are held at 700-900°C, for example in annealing, to soften the steel or in heating for forging, they will be spoiled.

Magnetic spoiling is caused by the coagulation of carbides at high temperatures and the formation of the stable carbides WC and W_2C (in steel E7B6) which dissolve with extreme difficulty in the austenite. After quenching, such steel will consist of martensite, impoverished in carbon and alloying elements, and coagulated carbides. This leads to a substantial deterioration of the magnetic properties (coercive force).

Spoiled magnet steel can be restored by normalising (hardening), holding for a short time at high temperatures (steel EX at 1,000-1,050°C and E7B6 at 1,180-1,200°C). This puts the carbides into the γ solid solution. Steel E7B6 is softened by high tempering at 660-680°C followed by cooling in air.

Cobalt steels possess higher magnetic properties. Cobalt intensively increases H_c with no appreciable reduction in B_r. As a result.

the energy product ($B_r \times H_c$) of the magnet is sharply increased. Cobalt steels are highly susceptible to magnetic spoiling. This must be taken into consideration because cobalt steels are very hard and require high tempering to soften them (EX5K5 requires 700-750°C and EX9K15M requires 800-820°C).

Since a certain degree of magnetic spoiling is inevitable in the manufacture and treatment of cobalt steels (tempering for softening and heating to bend the magnets), the finished magnets undergo a special three-step treatment:

1) hardening by air-quenching from 1,200-1,220°C (1,180-1,200°C for steel EX5K5);

2) tempering at 700°C followed by cooling in air;

3) hardening by heating to 1,030-1,050°C (950°C for steel EX5K5) and quenching in oil.

The principal purpose of this three-step treatment is to bring the carbides into a state in which they will go over into the solid solution more completely during the final hardening operation. The first hardening forms highly alloyed austenite which decomposes during subsequent tempering to produce highly dispersed carbides. The latter easily dissolve in the austenite during the final hardening.

The highest magnetic energy product is possessed by Fe-Al-Ni-Co alloys. The best of these is an alloy called *magnico* (14 per cent Ni, 8 per cent Al, 24 per cent Co and 3 per cent Cu) which is characterised by complex transformations in the decomposition of the solid solution during casting and heat treatment. This alloy is subject to a peculiar heat treatment which consists in heating to 1,300°C followed by cooling at a rate of 2° per second in a magnetic field having an intensity from 2,000 to 3,000 oersteds. The magnetic field is arranged along the direction most important for a magnet of the given configuration. After this treatment, the magnico alloy acquires the following magnetic properties: B_r=11,500-12,500 gauss and H_c= =600-650 oersteds.

Magnets are made of magnico by casting since the alloy is too hard and brittle for working or machining. The only possible subsequent processing is to grind the castings with carborundum wheels flooded with coolant.

Soft-magnet steels and alloys. Minimum coercive force and maximum magnetic permeability may be obtained only if the ferromagnetic material is as free as possible of all impurities and inclusions, i.e., it must have a homogeneous structure (pure metal or solid solution) and must be completely recrystallised to eliminate all internal stresses caused by work hardening. Pure iron may be used as a soft-magnet material. The purer the iron, the higher its magnetic properties will be (Table 49).

The most harmful impurities in this case are carbon (as Fe_3C) oxygen, and sulphur. The coarser the grain of the ferrite, the higher

Table 49

Magnetic Properties of Iron

Type of iron	Carbon content, per cent	Magnetic properties			
		Initial permeability μ_0, gs/oersted	Maximum permeability μ_{max}, gs/oersted	Residual induction B_r, gauss	Coercive force H_c, oersteds
Carbonyl iron	0.005	2,000-3,000	20,000-21,000	5,500-6,000	0.08
Electrolytic iron	0.02	500-600	15,000	10,500	0.36
Armco-type iron	0.06	250	3,500-4,000	—	0.8-1.2

the magnetic permeability (μ) will be. Even the smallest degree of work hardening will considerably reduce μ and increase H_c.

Commercial iron is used to make solid cores through which a magnetic flux of constant value is to pass, when losses from eddy currents may be disregarded. High-purity carbonyl iron is used to manufacture high-frequency cores which are compacted of iron powder and an insulating bond. They are called magnetodielectrics.

Low-carbon alloys of iron and silicon are widely employed as soft-magnet materials.

By forming a solid solution with the iron, silicon increases the electrical resistivity. Consequently, it reduces eddy current losses and increases magnetic permeability. It slightly lowers the coercive force and hysteresis losses because it coarsens the grain, has a graphitising action, and deoxidises the steel to a greater extent.

The silicon content of electrical sheet steel ranges from 0.8 to 4.8 per cent with a minimum carbon content.

Table 50 lists the characteristics of certain electrical steels (according to GOST 808-58).

It must be mentioned, however, that silicon reduces the induction in strong fields and increases the brittleness of the alloy. This becomes especially notable for a silicon content of 3-4 per cent. Silicon steels are available as sheet material from 0.35 to 1.0 mm thick. The average magnetic properties of steel Э41 are: $\mu_0=300$ gauss per oersted, $\mu_{max}=$ $=6{,}000$ gauss per oersted, and $H_c=0.45$ oersted. The specific losses $P_{10/50}=1.55$ W per kg and $P_{15/50}=3.5$ W per kg.*

* $P_{10/50}$ and $P_{15/50}$ denote the losses with inductions of 10,000 and 15,000 gauss respectively, and a field frequency of 50 cps.

Table 50

Compositions and Applications of Electrical Steels

Grade *	Si, per cent	Typical applications
Э11 Э12 Э13 Э1100 Э1200	0.8-1.8	Armatures and poles of d-c electrical machines, rotors and stators of asynchronous motors for industrial frequency with an output up to 100 kW, and magnetic circuits of apparatus and instruments. High ductility
Э21 Э22	1.8-2.8	Stators and rotors of asynchronous motors for industrial frequency with an output from 100 to 400 kW and rotors of synchronous motors from 1,000 to 10,000 kW. Good ductility
Э31 Э32 Э320 Э330A	2.8-3.8	Rotors and stators of asynchronous motors from 400 to 1,000 kW output, low-rating power transformers. Fair ductility
Э41 Э42 Э43 Э48	3.8-4.8	Magnetic circuits of large electrical machines and power transformers. Low ductility

* The figures and letters denote: letter Э — electrical steel, first digit (1 to 4, inclusive) — silicon content in per cent, second digit (1 to 8, inclusive) — guaranteed electrical and magnetic properties. Here, 1, 2 and 3 indicate the specific losses when magnetised at high magnetic densities in alternating fields of 50 cps frequency (characterised as 1 — normal specific losses, 2 — lowered losses, and 3 — low losses); letter A indicates especially low specific losses, 4 — specific losses when magnetised at medium magnetic densities in alternating fields of 400 cps frequency, 5 and 6 — magnetic permeability in weak fields from 0.002 to 0.008 A/cm (5 — with normal magnetic permeability, 6 — with increased permeability), 7 and 8 — magnetic permeability in medium fields from 0.03 to 10 A/cm (7 — with normal permeability, 8 — with increased permeability). The third integer 0 indicates that the steel is cold rolled and texturised; if both third and fourth integers are 0 (00), the steel is cold rolled but texturised only to a small extent.

Electrical sheet steel undergoes a special annealing process for recrystallisation, grain coarsening and removal of harmful impurities (burning out the carbon). Best results are obtained by annealing in hydrogen (dissociated ammonia) or in a vacuum at 1,100-1,200°C for 20 to 40 hours. However, heat treatment after rolling is frequently restricted to annealing at 850-900°C for a long period of time.

At the present time, steels Э310 and Э320 are being more and more extensively employed. They have a definite crystallographic texture, known as "cube texture" in which the crystals are so oriented in the direction of rolling as to provide easy magnetisation. Therefore, the magnetic permeability of such steels will be considerably higher in the direction of rolling than in a transverse direction.

Steel Э310 has the following average properties: μ_0=500 gauss per oersted, μ_{max}=16,000 gauss per oersted, and H_c=0.2 oersted. The specific losses are $P_{10} \leqslant 0.80$ W per kg and $P_{15} \leqslant 1.75$ W per kg.

Various apparatus in the communications industries (telephone, radio, etc.) require alloys with an especially high initial permeability (μ_0) since they must become strongly magnetised in very weak magnetic fields. Fe-Ni alloys of the Hiperm and Permalloy types are most frequently employed for this purpose (Table 51).

Table 51

Magnetic Properties of Certain Alloys

Alloy	μ_0, gs/oersted	μ_{max}, gs/oersted	H_0, oersted
Hiperm 50 (50% Ni and 0.3% Mn)	3,400	70,000	0.0028-0.03
Permalloy (78.5% Ni, 3.5% Cr, 2% Mn, and 3.5% Cu)	30,000	85,000	0.05
Supermalloy (79% Ni and 5% Mo)	100,000	800,000	0.004

High initial permeability is attained after homogenising at 1,200°C and reheating to 600°C followed by cooling at a specified rate. Even higher values may be obtained by cooling in a magnetic field.

The alloy *Alsifer* (9.6 per cent Si, 5.4 per cent Al, and the remainder Fe) is applied for magnetic shields, instrument housings, and components of magnetic circuits operating under constant or slowly changing fluxes. This alloy has the following magnetic characteristics: $\mu_0 \approx 20{,}000$-$35{,}000$, $\mu_{max} \approx 115{,}000$, $B_r \approx 3{,}000$, and $H_c \approx 0.025$. The alloy is brittle and hard (R_c 45-50) and is used in the cast form. Alsifer powder is used for making high-frequency compacted cores.

Nonmagnetic steels. Many components of magnetic instruments must be made of nonmagnetic materials since the presence of a ferromagnetic material may be the cause of errors in readings of the instrument. Previously, nonferrous metals (brass or bronze) were used for this purpose. At the present time, austenitic steels (μ=1.05-2 gs per oersted) have replaced nonferrous metals to a great extent.

Steel H24X2 and the less expensive austenitic steels, for example, 55Г9H9, ЭИ269 (0.5-0.6 per cent C, 4.0-5.5 per cent Mn, and 18.5-21.5 per cent Ni) and others are often used as nonmagnetic grades.

13-13. Alloys with Definite Expansion and Elasticity Properties

Fe-Ni alloys are customarily used as materials having a definite linear coefficient of thermal expansion which varies according to a complex relationship for these alloys (Fig. 245). Alloys with different linear coefficients of thermal expansion may be obtained by varying the nickel content. An alloy containing 36 per cent nickel (H36) and called *Invar* is employed to make length standards, precise gauges, and precision mechanisms.

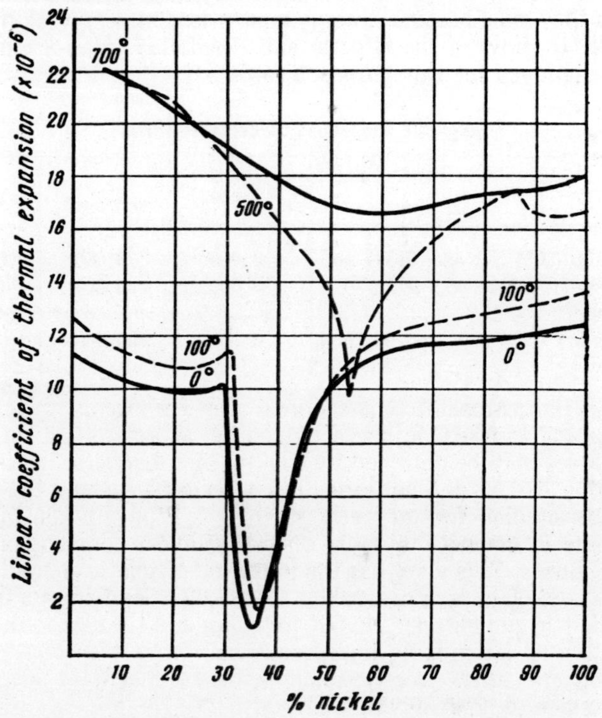

Fig. 245. Linear coefficient of thermal expansion for Fe-Ni alloys

It is evident from Fig. 245, that Invar has the lowest linear coefficient of thermal expansion ($\alpha=1.2\times10^{-6}$) which remains practically constant in the temperature range from 0 to 100°C. An alloy containing 42 per cent Ni, *Platinite* (H42), has a thermal expansion which is the same as that of glass and platinum. It is used as electrodes in incandescent lamps, lens mountings, radio components, etc.

In many instances, a metal is required with a modulus of elasticity that does not depend on the temperature. A typical representative of this group of alloys is *Elinvar* (0.7-0.8 per cent C, 2-3 per cent Mn, 33-35 per cent Ni, 7-9 per cent Cr, and 2-4 per cent W) which is used to make watch springs, chronometers, and components of measuring instruments.

Chapter 14

CAST IRON AND ITS HEAT TREATMENT

14-1. Grey Cast Iron

Cast iron is one of the most important of engineering materials. Machine tool beds, cylinder blocks, gears, piston rings, and many other parts are made of cast iron.

The properties that make cast iron such a valuable metal industrially are its high castability, fair mechanical properties, excellent machinability, and its lack of sensitivity to the quality of surface finish.

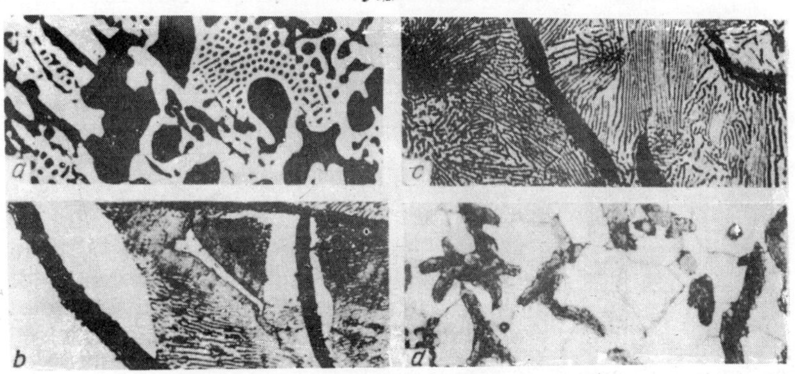

Fig. 246. Microstructure of cast iron, × 550:
a—white cast iron, *b*—pearlitic grey cast iron, *c*—ferrito-pearlitic grey cast iron, *d*—ferritic grey cast iron

All cast irons may be classified according to their structure into:

1. *White cast irons*. Here, practically all of the carbon is in the combined state as cementite. The structure of hypoeutectic white cast iron consists of pearlite and ledeburite (Fig. 246, *a*). The fracture has a dull white appearance.

2. *Grey cast irons*. In the structure of grey cast iron, a large part or all of the carbon is in the form of flakes or nodules of graphite (Fig. 246,

b, c, and *d*). Graphitic cast iron has a dark grey or almost black fracture.

Upon small degrees of supercooling, graphite is formed when the cast iron solidifies from its liquid state. Slow cooling promotes graphitisation. Rapid cooling partly or completely suppresses graphitisation and leads to the formation of cementite.

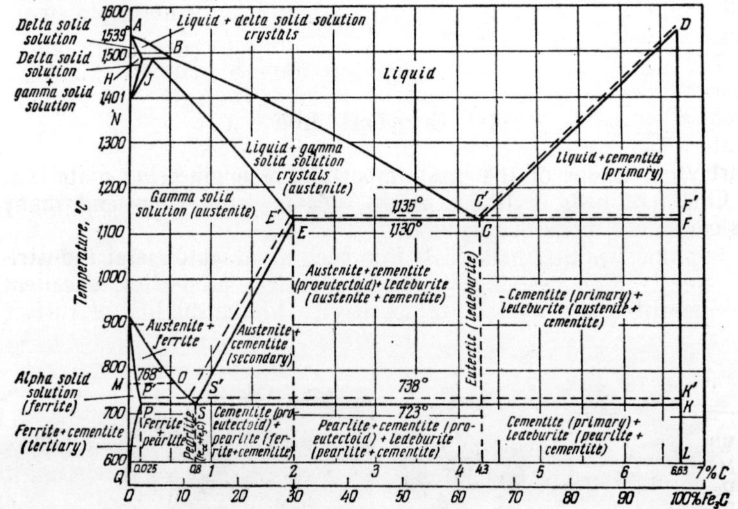

Fig. 247. The Fe-C equilibrium diagram

In the iron-carbon equilibrium diagram shown in Fig. 247, the dashed lines indicate the formation of graphite (stable system) while the solid lines denote the metastable state in which cementite is formed.

Eutectic graphite precipitates along the line *E'C'F'* (graphitic system) while excess (secondary) and eutectoid graphite are formed along the lines *S'E'* and *P'S'K'*, respectively.

The secondary and eutectoid graphite builds up on the previously formed inclusions, increasing their size.

When cast iron, in which the carbon is in the form of iron carbides, is heated to a high temperature and held for considerable time, graphitisation will occur, i.e., the cementite will be transformed into graphite. In this process, inclusions of the stable phase of graphite nucleate and grow while cementite crystals dissolve. After a certain period of time, the metastable phase, cementite, disappears.

The mechanism of graphitisation has not been made sufficiently clear as yet. According to K. P. Bunin, this process comprises the following: 1) dissolving of the cementite in the austenite, 2) diffusion of carbon atoms through the solid solution lattice toward the centres of graphitisation, 3) iron atoms leave regions of the solid solution where graphite inclusions nucleate and grow (this is called self-diffusion of the iron), and 4) carbon precipitates from the solution, i.e., graphite inclusions are formed. The rate of graphitisation is limited by the slowest component process, which is self-diffusion of the iron, i.e., the process in which the iron atoms leave the places where graphite is formed.

The movement of iron atoms in austenite (or ferrite) is considerably slower than the diffusion of carbon atoms. Therefore, graphitisation frequently fails to occur when iron-carbon alloys are cooled in the solid state while cementite forms comparatively easily since it requires only carbon diffusion and a slight rearrangement of the iron atoms.

Cast iron classification, based on the amount of the carbon combined as carbides, is given in Table 52.

Table 52

Types of Cast Irons

Name	Amount of carbon combined as Fe_3C, per cent	Structural components
White	all	Ledeburite + pearlite (Fig. 246, *a*)
Mottled	over 0.8	Ledeburite + pearlite + graphite
Pearlitic	0.7-0.8	Pearlite + graphite (Fig. 246, *b*)
Ferrito-pearlitic	0.7-0.1	Pearlite + ferrite + graphite (Fig. 246, *c*)
Ferritic	none	Ferrite + graphite (Fig. 246, *d*)

The chemical composition of a cast iron has a large influence on its graphitisation.

Alloying elements added to cast iron may be divided into graphite-forming elements (Si, Al, Ni, C, and others), which promote cementite decomposition, and carbide-forming elements (Mn, Cr, W, V, Mo, and others) which dissolve in the cementite and impede its decomposition.

Various degrees of cementite decomposition may be obtained by changing the amount of graphite- and carbide-forming elements in the cast iron. This is most easily accomplished in actual practice by varying the silicon content. In any given composition of cast iron, its structure will depend on the rate of cooling (wall thickness) of the casting. The structural diagram, given in Fig. 248, may be used to determine the structure obtained, depending on the composition (Si + C content) and the cooling rate (which practically may depend on the wall thickness of the casting).

The relation between the properties of cast iron and its structure is much more complicated than that for steel. Grey cast irons consist

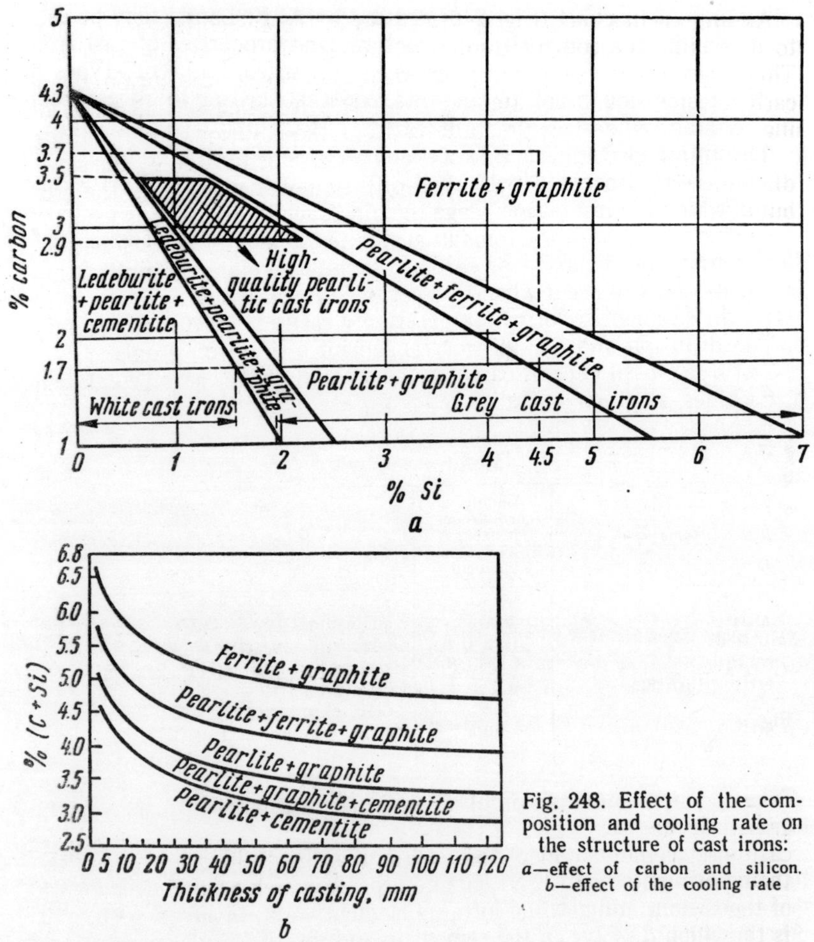

Fig. 248. Effect of the composition and cooling rate on the structure of cast irons: *a*—effect of carbon and silicon, *b*—effect of the cooling rate

of a metallic matrix in which graphite inclusions are disseminated. Therefore, the properties of cast irons will be determined by the structure of the matrix and the character of the graphite inclusions. The effect of the latter on the properties can only be qualitatively evaluated. The larger the inclusions and the less they are isolated, the weaker the cast iron will be with the same matrix.

Graphite inclusions in cast iron may be regarded as cracks or internal notches which violate the continuity of the metal.

The metallic matrix of common foundry cast iron consists of pearlite and ferrite. An increase in pearlite in the structure with the same form of graphite precipitation will improve the mechanical properties.

An alignment chart (Fig. 249) devised by A. F. Landa may be used to determine the composition, structure, and properties of cast iron. This chart gives the relation between the silicon content, types of carbon (combined, graphite, and total content), casting thickness (cooling rate), tensile strength, and structure for unalloyed cast irons.*

This alignment chart is a combination of three diagrams. Two diagrams, at the right and left, have a common axis of ordinates but different axes of abscissas (the left axis is the thickness of the

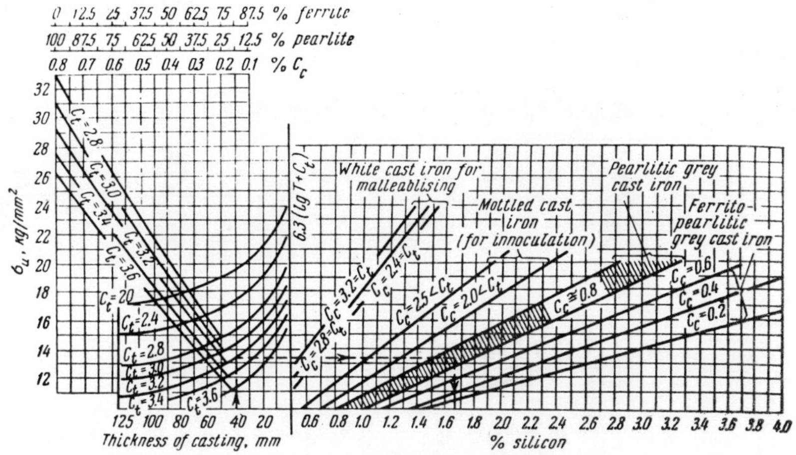

Fig. 249. Alignment chart for determining the composition, structure, and properties of unalloyed cast iron cast in sand moulds (after A. F. Landa)

casting, the right axis is the silicon content). On the left-hand diagram, the curved lines characterise the total carbon (C_t) for various cast irons; on the right-hand diagram the straight lines characterise the combined carbon (C_c). In the diagram at the extreme left, values of the tensile strength (σ_u) are ordinates while the abscissa (upper axis) is the amount of combined carbon C_c and the amounts of pearlite and ferrite.

The so-called pearlitic cast irons are most extensively employed in the engineering industries. Their structure consists of a pearlitic matrix in which graphite flakes or lamellae are distributed. The amount of flakes must be as small as possible and they must be isolated from each other. A network structure of graphite (Fig. 250) is undesirable since here the graphite is usually surrounded by ferrite and it is difficult to obtain a purely pearlitic matrix that provides high mechanical properties.

* *Physical Metallurgy and Heat Treatment*, VNITOMASH, Mashgiz, Moscow, 1955.

According to GOST 1412-54, grey cast iron is identified by the Russian letters СЧ (which represent the first letters of the Russian words "серый чугун" — meaning grey cast iron). These letters are followed by numbers. The first number indicates the average value of the tensile strength and the second number, the bending strength in kg per sq mm.

All grey cast irons used at the present time may be divided into the following groups:

Fig. 250. Graphite network structure, × 170

1. *Low-strength cast irons*. They have a tensile strength from 12 to 21 kg per sq mm and a bending strength from 28 to 40 kg per sq mm (СЧ00, СЧ12-28, СЧ15-32, and СЧ18-36). Their approximate composition is: 3.2-3.6 per cent C, 1.7-3.0 per cent Si, $\leqslant$0.5 per cent Mn, $\leqslant$0.5 per cent P, and $\leqslant$0.12 per cent S. They have the "ferrite + graphite" or "pearlite + ferrite + graphite" structures shown in Fig. 251. The graphite is in the form of large flakes.

These cast irons are applied for less responsible machine parts subject to low loads in operation.

2. *Medium-strength cast irons* (СЧ21-40, СЧ24-44, СЧ28-48, СЧ32-52, СЧ35-56, and СЧ38-60). These grades are applied in casting the beds of powerful machine tools and mechanisms, pistons, cylinders, etc.

This group includes the pearlitic (steely) cast irons. About 10 to 30 per cent steel scrap is added to the foundry cupola charge when melting these cast irons. They have a lower carbon content which enables a pearlitic matrix to be obtained with a small amount of graphite inclusions. Their approximate composition is: 2.8-3.0 per cent C, 1.5-1.7 per cent Si, 0.8-1.0 per cent Mn, $\leqslant$0.3 per cent P, and $\leqslant$0.12 per cent S.

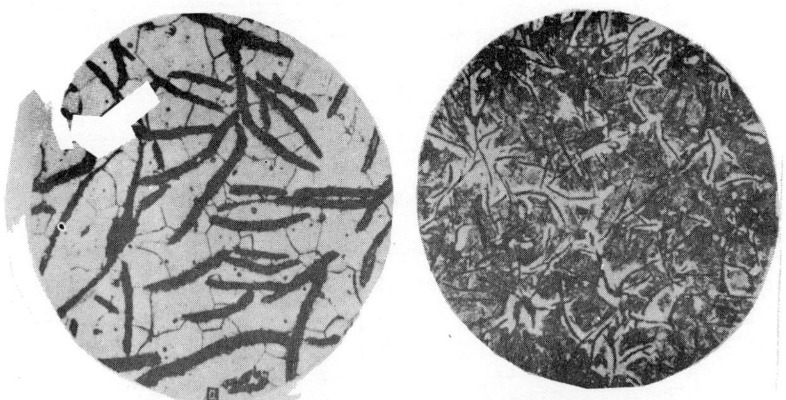

Fig. 251. Microstructure of cast iron, × 150:
a—cast iron СЧ00, *b*—cast iron СЧ18-36

The silicon content of these cast irons must be sufficient to prevent chilling.

The same group also includes *inoculated cast irons* obtained by special additions, called *inoculants* (ferrosilicon with 75 per cent Si, calcium silicide, etc.), to the liquid cast iron, made before pouring the moulds.

The principal purposes of inoculating are to prevent the formation of eutectic and secondary cementite, to eliminate graphite network structure, and to obtain a pearlitic matrix with fine graphite flakes. As a rule, inoculation is applied to low-carbon cast irons, containing a comparatively small amount of silicon and an increased amount of manganese. Without the inoculants, they would acquire the structure of mottled cast iron, i.e., pearlite + graphite + cementite. In this type of cast iron the lower carbon content decreases the amount of graphite, and the low silicon and carbon content and the increased amount of manganese facilitate the formation of pearlite.

The approximate composition of the cast iron is: 2.9-3.2 per cent C, 1-1.5 per cent Si, 0.8-1.2 per cent Mn, $\leqslant$0.3 per cent P, and $\leqslant$0.12 per sent S. Ladle additions of inoculants promote a reduction in the degree of supercooling, deoxidation of the cast iron, and provision of an abundance of nuclei. By reducing the degree of supercooling, chilling and network structure of the graphite are avoided. Inoculated cast irons are used for machine parts subject to wear, such as gears, brake drums, steam engine cylinders, etc.

A factor to be remembered in selecting the grade of cast iron is that the mechanical properties are lowered when the cooling rate is decreased (by increasing the thickness of the walls of the casting) (Table 53).

Table 53

Relationship Between Mechanical Properties of Cast Iron and Cooling Rate

Grade	Diameter of casting, mm							
	30		50		100		200	
	σ_u, kg/mm²	*Bhn*	σ_u, kg/mm²	*Bhn*	σ_u, kg/mm²	*Bhn*	σ_u, kg/mm²	*Bhn*
СЧ24-44	24	187-217	23	187-217	22	163-207	20	143-187
СЧ32-52	32	170-241	30	170-241	28	170-229	26	170-229
СЧ38-60	38	197-255	36	197-255	32	197-255	27	197-255

This is due to the reduction in the degree of dispersity of the pearlite in thick sections and the formation of a considerable amount of ferrite in addition to pearlite.

3. *High-strength* (*spheroidal graphite*) *cast irons* (ВЧ). They are also called nodular graphite or ductile cast irons and may be obtained from ordinary grey pearlitic cast iron by a ladle addition to the liquid metal (at 1,400-1,450°C) of magnesium in an amount from 0.3 to 1.2 per cent of the weight of the cast iron. The larger the casting to be made, the more magnesium is added.

Magnesium causes the graphite, which precipitates in the structure of cast iron during solidification, to take a strictly rounded spheroidal form (Fig. 252) instead of the flakes found in grey cast iron. Graphite in this form weakens the matrix to a lesser extent thus providing for high mechanical properties.

Magnesium not only changes the form of the graphite but increases the strength of the matrix as well. In addition, it increases the capacity of cast iron to be supercooled and, consequently, to be chilled. Chilling can be prevented by a double addition: ferrosilicon to promote graphitisation and magnesium to obtain spheroidal graphite.

Table 54 lists the mechanical properties of various grades of high-strength cast iron (according to GOST 7293-54). The first number in the grade designation indicates the tensile strength while the second is the relative elongation in per cent,

Table 54

Properties of High-Strength Cast Irons

Grade	σ_u, kg/mm²	σ_s, kg/mm²	δ, per cent	a_k, kg-m/cm²	*Bhn*
ВЧ45-0	45	36	0	—	187-255
ВЧ50-1.5	50	38	1.5	1.5	187-255
ВЧ60-2	60	42	2	1.5	197-269
ВЧ45-5	45	33	5	2	170-207
ВЧ40-10	40	30	10	3	156-197

High-ductility cast iron ВЧ45-5 has a structure comprised of pearlite + ferrite + spheroidal graphite (see Fig. 252) while grade ВЧ40-10 has ferrite + graphite. The formation of ferrite is desirable in this case since the cast iron acquires ductility without a loss in strength.

The presence of spheroidal graphite greatly increases the effect of heat treatment.

Low-alloy and high-alloy cast irons find certain applications in industry. In low-alloy grades the amount of alloying elements does not exceed 3 per cent.

The chief purposes of alloying are to increase the strength and the wear-, scale-, heat-, and corrosion-resistances, to obtain required magnetic properties, and others.

Combined alloying, in which sets of carbide-forming and graphitising elements are added (Cr and Ni; Cr, Mn, and Si, etc.), has proved most effective.

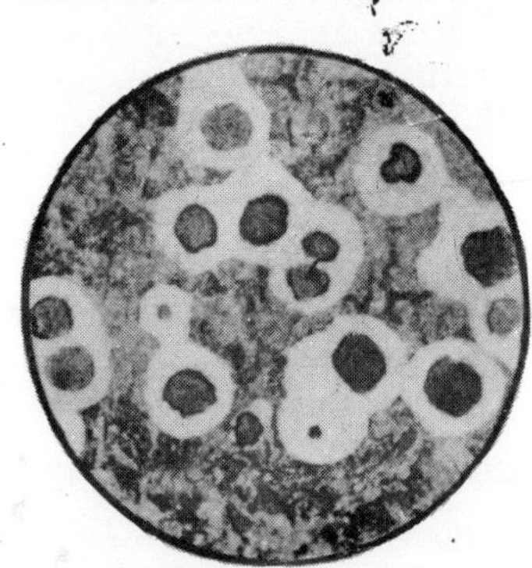

Fig. 252. Microstructure of high-strength (spheroidal graphite) cast iron, × 100

14-2. Heat Treatment of Cast Iron

One of the principal types of cast iron heat treatment is heating to relieve internal stresses, which are inevitable in castings, and to improve the mechanical properties. Heat treatment is also frequently applied to eliminate the nonuniformity in the chemical composition arising in the cooling of castings.

The heat treatment of cast iron involves the same transformations that occur in steel. It must be noted, however, that graphitisation, which takes place when cast iron is heated, has a large influence on the structure and final properties.

Annealing to relieve internal stresses. Internal stresses are often relieved by natural ageing, i.e., storing the castings in still air from 6 to 15 months. This treatment relieves about 30 to 50 per cent of the stresses. A faster method, used at the present time, more completely relieves the stresses and consists of annealing the castings at 500-550°C for 6 to 8 hours. For this process, a heating rate of 75° to 150°C per hour is recommended; the cooling rate in the range from 500° to 200°C should be 30-60° C per hour. This treatment almost completely eliminates internal stresses.

Annealing to soften cast iron. Annealing to soften the chilled layer of parts cast in sand or metal moulds involves heating to 850-900°C and holding for from 30 minutes to 5 hours. Graphitisation, which takes place during this treatment, reduces the hardness.

It may prove expedient to raise the annealing temperature to 1,050-1,150°C to speed up this process. Graphitisation is completed in several minutes at these temperatures if the treatment is accomplished in a molten salt bath. Induction heating can reduce the annealing time to 3-5 seconds.

Hardening and tempering cast iron is applied to improve its properties. This treatment involves heating the cast iron slightly above the eutectoid transformation temperature (760° to 790° for ordinary cast irons) followed by quenching. After quenching, the structure consists of martensite with graphite inclusions.

Cast iron is ordinarily heated to 860-900° C for hardening. Oil is used as the quenching medium while castings of complex form are air-quenched. Subsequent tempering relieves internal stresses.

Both high and low tempering are applied to quenched cast iron. After tempering at a low temperature (200-250°C for several hours), cast iron will have a hardness of *Bhn* 350-400 and can be machined only by grinding. High tempering (300-500°C) will reduce the hardness (a sorbite matrix is obtained) but the tensile strength will be raised. Highest tensile strength values are obtained by tempering at 300-350° C. It is not advisable to raise the tempering temperature above 500° C as this leads to additional graphitisation of the eutectoid cementite. Cast iron acquires excellent machinability after high tempering.

Heat treatment (hardening) does not affect the size and distribution of the graphite inclusions and is little effective. Therefore (and also because of the danger of quenching cracks), it is not applied to castings of unalloyed cast iron to any practical extent.

On the other hand, new research work has shown that good results are obtained by isothermal quenching (austempering) which consists in heating to 850-900°C and quenching in a bath at 500-550°C with a holding time of 30 to 45 minutes. This treatment considerably increases the resistance to wear. Induction surface hardening procedures may also be applied to cast iron.

Heat treatment of spheroidal graphite cast iron. All types of heat treatment are applicable to magnesium cast irons. Most widely used are:

a) Normalisation (cooling in air from 900-920°C), which is applied if the casting has a chilled layer (structure: ledeburite + pearlite + spheroidal graphite surrounded by ferrite). As a result of the graphitisation of the eutectic cementite, cast iron will acquire a structure of sorbitic pearlite + spheroidal graphite after this treatment.

If the cast iron is not sufficiently hard, it is heated to 850-880°C (to dissolve the graphite in the austenite) and cooled in air.

b) Annealing at 650-720°C may be applied to increase the toughness of cast iron having a pearlitic or ferrito-pearlitic matrix. Graphitisation of the eutectoid cementite, in this case, will give a ferrite + spheroidal graphite structure and increased ductility.

c) Good results are obtained, insofar as increasing the structural strength and wear resistance of high-strength cast iron is concerned, by austempering and by induction hardening. Magnesium cast iron is very seldom hardened, however.

14-3. Malleable Cast Iron

Similar to grey cast iron, the malleable type also comprises a metallic matrix (pearlite or ferrite) and graphite.

In grey cast iron, graphite is precipitated as flakes formed in solidification and further cooling. *Malleable cast iron*, on the other hand, contains graphite in the form of small rounded nodules (called *temper carbon*), and this circumstance is responsible for the considerable increase in the mechanical properties of the cast iron. Malleable cast iron or, simply, *malleable iron* is obtained as a result of a special annealing process (*malleablisation*) from white cast iron.

Two types of white cast irons are suitable for malleable castings: low-carbon type (1.7-2.7 per cent C, 1.4-1.6 per cent Si, 0.4-0.7 per cent Mn, up to 0.2 per cent P, and up to 0.12 per cent S) and the ordinary cupola composition (2.8-3.2 per cent C, 0.5-0.8 per cent Si, 0.4-0.7 per cent Mn*, up to 0.2 per cent P, and up to 0.12 per cent S).

In accordance with the cast iron composition, one of two principal annealing procedures is applied:

a) Castings of low-carbon white cast iron are annealed in a neutral (inert) medium. The malleable iron produced has a characteristic dark fracture surrounded by a clear-cut grey border and a structure comprising ferrite and temper carbon (graphite) (Fig. 253). This is often called ferritic malleable iron.

b) Castings of ordinary cupola white cast iron, annealed in an oxidising medium, have a light-grey fracture and a structure made

* Not to exceed 0.4 per cent to produce black heart malleable iron.

up of pearlite, ferrite, and graphite (Fig. 253). This variety is often called ferrito-pearlitic malleable iron.

Annealing in a neutral medium is done by packing the castings in boxes (or pots) filled with sand. This reduces deformation and oxidation of the castings.

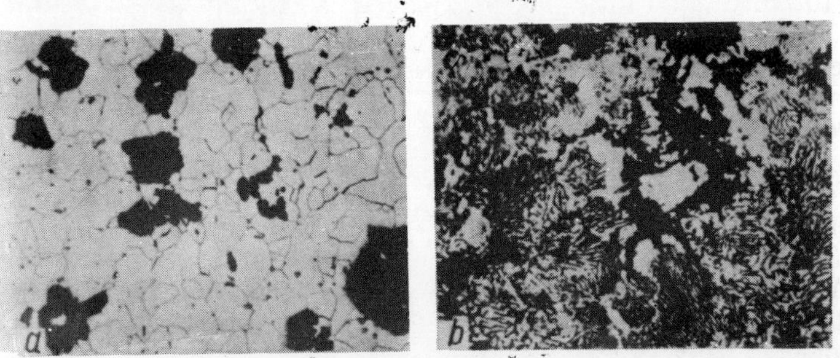

Fig. 253. Microstructure of malleable iron, × 200:
a—ferritic malleable iron, b—ferrito-pearlitic malleable iron

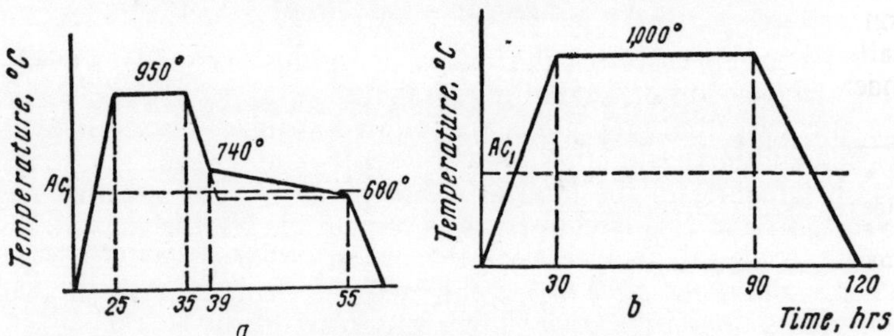

Fig. 254. Malleablising procedures:
a—for obtaining ferritic malleable iron, b—annealing in an oxidising medium

Packing the castings in boxes filled with iron ore is practised for annealing in an oxidising medium.

Annealing procedure, used to obtain ferritic malleable iron, follows the diagram given in Fig. 254. Holding at high temperature (950-1,000°C) graphitises the cementite in the ledeburite eutectic. This is the first stage of graphitisation. Holding at 720-730°C or slow cooling in the range from 740° to 680°C (Fig. 254) is required to decompose the cementite in the pearlite (second stage of graphitisation). In the intermediate stage (790-950°C), secondary cementite is precipitated and graphitised.

The annealing process may be shortened by preceding it with a hardening operation consisting of heating to 900-920°C and quenching in oil.

Malleablising may also be shortened by carrying out the first stage of graphitisation in molten chlorides (75 per cent $BaCl_2$+25 per cent NaCl) at 1,030-1,050°C for 0.5-2.0 hours. The second stage is carried out in a box-type electric furnace according to the following procedure: 1) heating with the furnace to 710-720°C, 2) holding for 8 hours at 710-720°C, and 3) cooling with the furnace to 600°C and further in air. The structure after this treatment will comprise temper carbon and spheroidised pearlite. The mechanical properties will be: σ_u=40-50 kg per sq mm, δ=2-5 per cent, and *Bhn*=170-230*.

Pearlitic malleable iron of this type is obtained by applying only the first stage of graphitisation (decomposition of eutectic cementite).

Ferritic malleable iron is classified in accordance with its tensile strength and relative elongation into the following grades: КЧ37-12, КЧ35-10, КЧ33-8, and КЧ30-6** (according to GOST 1215-41).

Malleable iron, obtained after annealing castings (of cupola white cast iron) in an oxidising medium for a long period, may be of the following grades: КЧ40-3, КЧ35-4, and КЧ30-3.

Due to the fact that malleablisation is accompanied by decarburisation, and not only graphitisation, it is impossible to obtain uniform properties in castings of large cross-section. As a rule, malleable iron castings of this type should not be over 20 mm in thickness. Malleable iron castings are used to make machine parts which function under impact and vibration loads.

* This procedure was developed by the TSNITMASH Institute and has found wide application in the Moscow Borets Works.

** The Russian letters КЧ indicate malleable iron, the first number is the tensile strength, in kg per sq mm, while the second is the relative elongation in per cent.

Chapter 15

COPPER AND ITS ALLOYS

15-1. Copper

Copper is one of the most extensively used nonferrous metals in industry. It has a crystal lattice of the face-centred cubic type with a lattice constant of 3.607 Å. Copper has a melting point of 1,083°C and a specific weight of 8.93 grammes per cu cm. The mechanical properties of rolled and annealed copper are characterised by the following data: σ_u=23-25 kg per sq mm, δ=40-50 per cent, and *Bhn*=30-40.

Cold working considerably increases the tensile strength of copper and reduces its relative elongation at the same time.

Cold worked copper acquires the following mechanical properties: σ_u=40-43 kg per sq mm, δ=1-2 per cent, and *Bhn*=100-120.

Copper, with its high ductility, can easily be worked both hot (at 550-750°C) and cold (with intermediate annealing at 500-600°C).

The microstructure of pure annealed copper (99.26 per cent Cu) is illustrated in Fig. 255, *a*.

The most valuable properties of copper are its high electrical and thermal conductivity. Purest copper has an electrical conductivity of 64×10^4 ohm^{-1}-cm^{-1}. Standard copper, often employed as the reference for expressing the conductivity of other metals, has a conductivity of 58×10^4 ohm^{-1}-cm^{-1} at 20°C in the annealed state. The thermal conductivity of copper is 0.910 cal per cm-sec-degree. The linear coefficient of thermal expansion is 16.42×10^{-6} mm per mm-degree. The application of copper for electric conductors, cable, busbars, etc., is based on these properties.

The amount of impurities found in copper depends upon the method by which it is produced (Table 55).

Table 55

Impurities in Copper of Various Grades

Grade	Cu content, per cent	Impurities, per cent (max)		
		Bi	Pb	Total impurities
M0	99.95	0.002	0.005	0.05
M1	99.9	0.002	0.005	0.1
M2	99.7	0.002	0.01	0.3
M3	99.5	0.003	0.05	0.5
M4	99.0	0.005	0.3	1.0

All impurities, and particularly those in solid solution (Al, Fe, P, Ni, Sn, Zn, Ag, Cd, As, and others) reduce the electrical and thermal conductivity of copper but increase its hardness and strength. They have practically no effect on its formability.

Bismuth and copper (Fig. 256, *a*) form a eutectic with a low melting point which is arranged in a film along the grain boundaries when the copper solidifies (Fig. 256, *b*). Copper containing bismuth will easily fail in cold working due to the brittleness of this boundary film. This

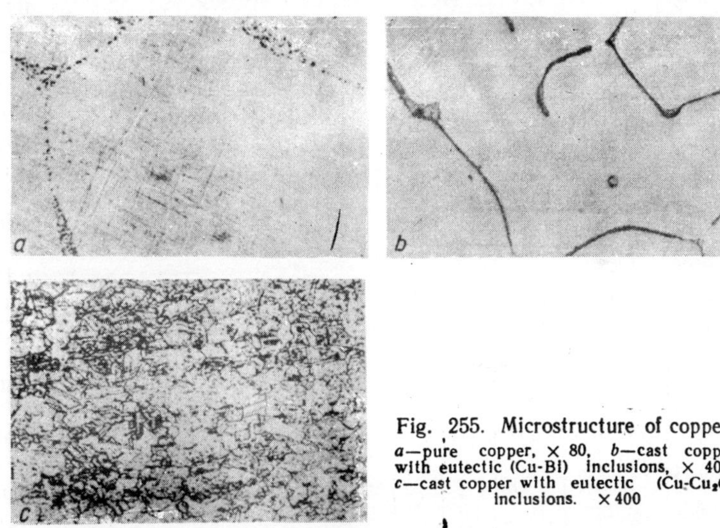

Fig. 255. Microstructure of copper:
a—pure copper, × 80, *b*—cast copper with eutectic (Cu-Bi) inclusions, × 400, *c*—cast copper with eutectic ($Cu-Cu_2O$) inclusions. × 400

type of brittleness is often called cold shortness. In hot working, the eutectic melts and separates the grains; this also leads to failure. It is impossible to hot work copper which contains even hundredths of one per cent of bismuth. All copper-bismuth alloys are red-short.

Lead, which forms a eutectic with a low melting point, in the same manner as bismuth, also promotes red-shortness in copper (Fig. 256, *b*). It is evident, however, from Table 55 that lead may be present in larger amounts than bismuth in copper without deteriorating its quality.

Sulphur and oxygen form the brittle chemical compounds Cu_2S and Cu_2O with copper. These compounds (Fig. 255, *c*) form a film or network on the grain boundaries, in the form of a eutectic, and interfere with cold working, thus making the copper cold-short.

Copper, if it contains cuprous oxide, is susceptible to *hydrogen attack* (formation of flakes).

The essence of hydrogen attack is the following: if copper containing Cu_2O is heated in the presence of hydrogen, the latter diffuses into the metal. As a result of the reaction $Cu_2O + H_2 \rightarrow 2Cu + H_2O$, water vapour is formed. This vapour creates a high pressure, within the metal, in attempting to escape to the surface. This results in microcracks or flakes which may lead to complete failure of the part involved.

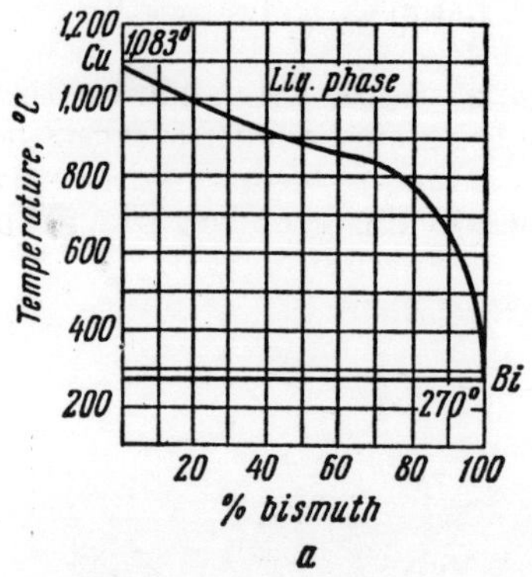

Of other impurities found in copper, antimony is of note. Antimony completely dissolves in copper but, nevertheless, must be considered among the harmful impurities since it interferes with working operations.

Copper has excellent resistance to atmospheric corrosion due to

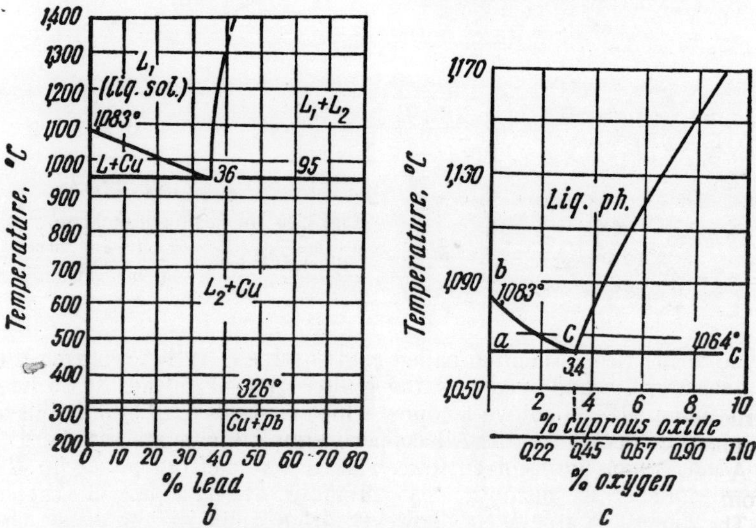

Fig. 256. Equilibrium diagrams of copper and its main impurities:
a—Cu-Bi, *b*—Cu-Pb, *c*—$Cu\text{-}Cu_2O$

the formation of a protective film, consisting of $CuSO_4 \cdot 3Cu(OH)_2$, on its surface.

Pure copper has almost no applications as a structural material since its mechanical properties are very low. Copper Grades M0 and M1 are used for electric conductors and as a constituent of alloys with high properties. Grades M2 and M3 are used for wrought alloys and grade M4—for cast bronze and less important alloys. Alloys of copper with zinc, tin, aluminium and others have much higher mechanical and fabrication properties than pure copper and have, therefore, found widespread applications in industry.

15-2. Brasses

The alloys of copper and zinc are generally called *brasses*. The copper-zinc equilibrium diagram is shown in Fig. 257. As the diagram shows, copper containing up to 39.0 per cent Zn has a single-phase

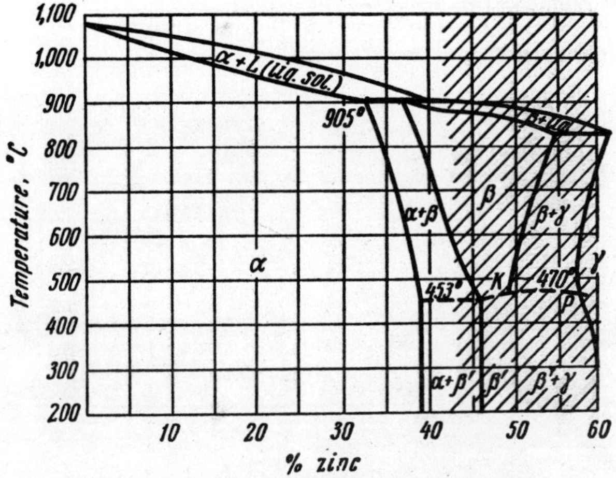

Fig. 257. The Cu-Zn equilibrium diagram

structure consisting of crystals of the solid solution of zinc in copper (α-phase).

A new structural component (β-phase) appears in alloys containing from 39 to 50 per cent zinc.

The β-phase is also a solid solution. The solvent in this case, however, is not the pure metal but an electron compound CuZn, having a space-centred cubic lattice with a $^3/_2$ ratio of the number of valence electrons to the number of atoms (see Sec. 6-5).

The β-phase has a disordered distribution of atoms at high temperatures. Upon lowering the temperature below 453-470°C, the copper and zinc atoms in the solid solution acquire an ordered structure which is designated as β'.

As is evident in Fig. 257, the beta solid solution has a wide range of homogeneity at high temperatures. It decomposes as the temperature is lowered and crystals of the excess α- or γ-phase begin to precipitate from the solution.

The γ-phase, formed in high zinc alloys of copper, is a solid solution on the basis of the electron compound Cu_5Zn_8 $\left(\frac{21}{13}\right)$, having a complex cubic lattice. The γ-phase is a very brittle component and, therefore, brasses, intended for room temperature service consist either completely of the α-phase (α-brasses) or of the mixture $\alpha + \beta'$ ($\alpha + \beta'$-brasses).

Photomicrographs of extensively used α- and $\alpha + \beta'$-brasses are given in Fig. 258.

Fig. 259 shows the variations in the mechanical properties of copper and zinc alloys with their composition.

Small additions of zinc raise the tensile strength and increase the relative elongation.

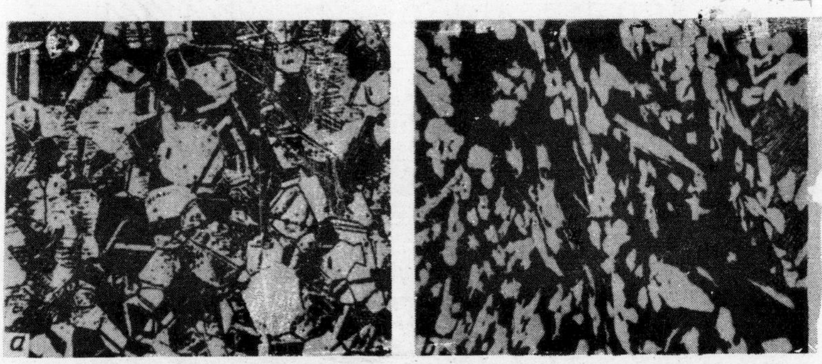

Fig. 258. Microstructure of brass, × 200:
a—α-brass, b—$\alpha+\beta'$-brass

Brass acquires maximum relative elongation at a zinc content from 30 to 32 per cent which approximately coincides with the boundary between the α-phase and $\alpha+\beta'$-phase regions (Fig. 257). The α-phase is the most ductile structure. When the β'-phase appears in the structure, the relative elongation is reduced and reaches very low values when the zinc content corresponds to the point at which the β'-phase structure is formed. Upon increasing the zinc content to 45-47 per cent, the tensile strength first grows and then sharply falls.

The α-brasses have good workability (undergo plastic deformation easily) at room temperature. They are much less ductile at temperatures from 300° to 700°C.

Brasses are strain-hardened by cold working (rolling, drawing, etc.). Their ductility may be restored by annealing at 500-700°C. In ductility, the $\alpha+\beta'$-brasses are inferior to the α-brasses. For hot working, the $\alpha+\beta'$-brasses are heated to the range of homogeneous β solution which possesses high ductility.

Brass with internal stresses, appearing, for example, after working, is susceptible to intercrystalline corrosion if it is stored for a long period and, in particular, when exposed to a reactive corrosive medium. This leads to disintegration and failure of the metal and is known as *season cracking*.

Season cracking may be eliminated by annealing strain-hardened brass at 200-300°C for several hours.

The considerable toughness of brass will not allow an even surface without scoring to be obtained in machining. A small amount of

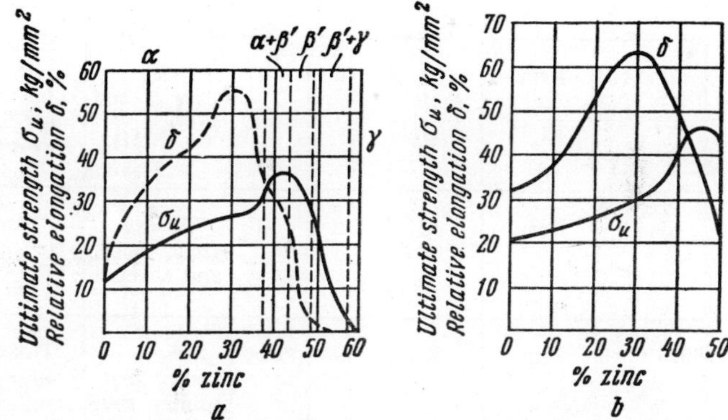

Fig. 259. Variation in mechanical properties of brass with the zinc content: *a*—cast brass, *b*—rolled and annealed brass

lead is frequently added to improve the machinability. Such alloys of copper and zinc to which lead has been added are called *leaded* or *free cutting brasses*.

The lead* in the brass forms small inclusions located chiefly within the grain and facilitates chip forming in machining operations.

Table 56 lists the commercial grades of brass widely used in the U.S.S.R.

Grade Л70 α-brass, intended for deep drawing, has the highest ductility; the α+β′-brass ЛС59-1 has the lowest. The latter can easily be machined, however, and may be hot worked.

Brasses Л68 and Л62 are chiefly applied in the sheet or strip form for articles fabricated by deep drawing, as well as for radiators, sealing rings, connection pipes, tubing, etc.

* Lead is added to α+β′-brasses which undergo the α⇄β transformation in heating or cooling. As a result of this transformation the lead is arranged, not along the grain boundaries, as in pure copper or α-brass, which do not undergo transformation in the solid state, but within the grains.

Table 56

Wrought Brasses

Grade	Chemical composition, per cent		Average mechanical properties			Applications	Fabricating methods
	Cu	Zn	σ_u, kg/mm^2	δ, per cent	*Bhn*		
ЛТ96 Tombac	95-97	Rem.	35	—	—	Radiator tubes	Drawing
Л80 Semi-tombac	78-81	Rem.	30	40	—	Sheet, strip, and wire	Hot and cold rolling, and drawing
Л70	69-71	Rem.	30	40	60	Tubing, strip, wire, and sheet	Ditto
Л62	60.5-63.5	Rem.	30	40	60	Strip, sheet, bar, tubing, rod, and wire	Hot and cold extrusion, rolling, and drawing
ЛС59-1	57-60	0.8-1.9 Pb, rem. Zn	35	15	75	Sheet, strip, rod, and tubing	Extrusion, rolling, and drawing

Note: Brasses are designated by the Russian letter Л, which is followed by a number indicating the mean copper content, in per cent.

Brasses Л59 and ЛС59-1 are used as rolled or extruded bar stock for making machine parts on machine tools (bushings, nuts, rings, cocks, etc.).

In addition to plain brasses, i.e., copper and zinc alloys and *Muntz metal* (an alloy of the copper-zinc-lead system), special brasses find wide applications in engineering. These are complex alloys that differ from ordinary brasses in structure and properties. This is due to the addition of various alloying elements, such as tin, silicon, manganese, aluminium, and iron.

The special or alloy brasses may be divided into three groups: 1) brasses with increased corrosion resistance, 2) high-strength brasses, and 3) antifriction brasses.

It should be noted that certain grades of alloy brasses may have all three properties at the same time.

Table 57 lists the compositions, properties, and applications of certain grades of alloy brasses.

Wrought brasses contain a comparatively small amount of alloying additions and are copper-base homogeneous solid solutions. This explains their high plasticity and malleability. Cast brasses, which do not require high ductility, contain larger amounts of special additions which improve their castability. Copper-zinc alloys (ПМЦ36, ПМЦ48, and ПМЦ54), as well as copper-zinc-silver alloys (ПСр10, ПСр25, ПСр70, and others) *, are used to a great extent as brazing solders (their melting point ranges from 700° to 850°C).

15-3. Bronzes

Alloys of copper with all other elements except zinc are called *bronzes*. Zinc may be a constituent of bronze but only together with other components and in a relatively small amount.

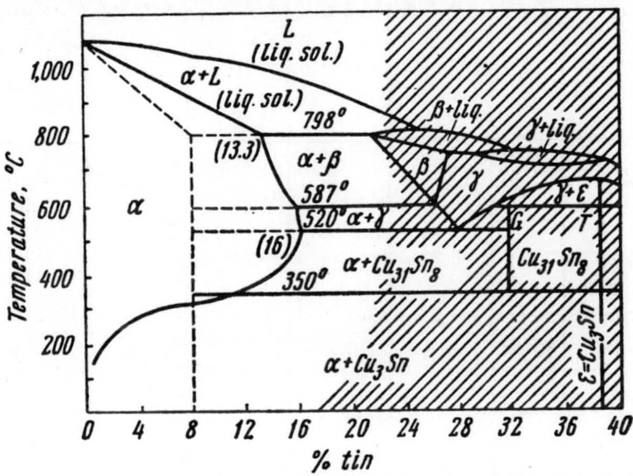

Fig. 260. The Cu-Sn equilibrium diagram

Tin bronzes. The Cu-Sn equilibrium diagram is shown in Fig. 260. In the Cu-Sn system, the α-phase is a solid solution of tin in copper. Electron compounds of the type $Cu_{31}Sn_8$, having a $\frac{21}{13}$ ratio of valence electrons to the number of atoms, and Cu_3Sn $\left(\frac{7}{4}\right)$ are also formed in this system.

* The Russian letter П denotes solder, Cp denotes silver while the number (10, 50, 70, etc.) indicates the silver percentage.

Special (Alloy)

Grade	Chemical composition, per cent					As applied
	Cu	Fe	Al	Ni	Mn	
						a) With increased corrosion
ЛА67-2.5	66-68	—	2-3	—	—	Permanent mould castings Sand mould castings
ЛА77-2	76-79	—	1.75-2.5	—	—	Drawn tubing
						b) High-strength (and
ЛАЖМц70-6-3-1	68-72	2.5-3.5	5-6	—	0.5-1.5	Permanent mould castings Sand mould castings
ЛМцНЖ52-2-2-1	50-54	0.5-1.5	—	—	1.5-2.5	Permanent mould castings Sand mould castings
ЛМцЖ52-4-1	50-54	0.5	—	—	3.0-4.5	Permanent mould castings Sand mould castings
ЛАН59	57-60	—	2.5-3.5	2-3	—	Tubing, bar
						c) Anti
ЛКС80-3-3	79-81	—	1.5-4.5% Pb	1.5-4.5% Si	—	—
ЛМцС58-2-2	57-60	—	1.5-2.5% Pb	—	1.5-2.5	—

Note: Special (alloy) brasses are designated by the Russian letter Л (denoting brass) Н—nickel, С—lead, etc. The first two figures following the letters indicate the average indicate the average content of the alloying elements. The zinc content is the remainder.

Table 57

Brasses

Average mechanical properties			Applications
σ_u, kg/mm²	δ, per cent	*Bhn*	
resistance			
40	15	90	Castings for ship-building and general engineering
30	12		
38	23	—	Tubing for manufacturing parts in naval and general engineering
corrosion-resistant)			
55	5.0	150	Castings for heavily-loaded components (large worms)
50	7.0	120	
55	15	130	Large critical castings in ship-building
50	15	100	
50-40	15-12	100-80	Load-carrying components of aircraft; unimportant bearings and fittings
45-35	12-8	100-70	
43-55	25-12	—	Heavily-loaded components in automobile manufacture and ship-building
friction			
—	—	—	Castings
—	—	—	Bearings; bushings

followed by letters wh ch denote: A — aluminium Ж — iron, K — silicon, Мц — manganese, copper content in per cent; all other figures, separated from each other by hyphens,

The diagram (Fig. 260) shows that the Cu-Sn system has several peritectic transformations and two eutectoid transformations in the solid state. At 587°C, the β-phase crystals undergo a eutectoid decomposition and form a mixture of the α-and γ-phases, while at 520°C, γ solid solution crystals decompose into a mixture of the α and $Cu_{31}Sn_8$ phases.

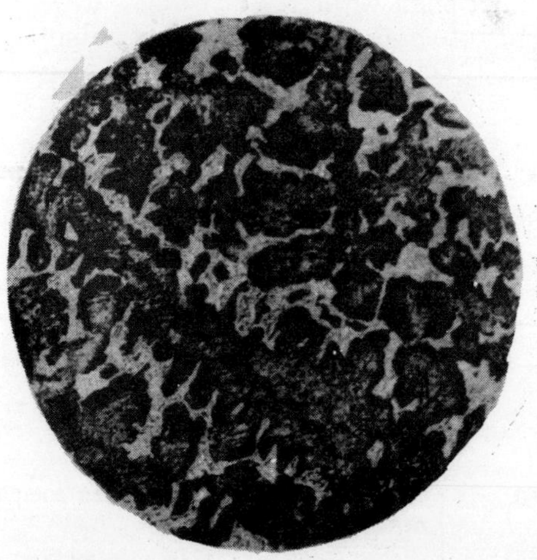

Fig. 261. Structure of cast tin bronze, containing 10 per cent Sn (after Rumyantsev), × 300

At a temperature of about 350°C, the electron compound $Cu_{31}Sn_8$ decomposes into the alpha solid solution and the compound Cu_3Sn. This transformation occurs only in a state of complete equilibrium; under actual cooling conditions the structure of bronze consists of the phases $\alpha+Cu_{31}Sn_8$.

Only the left-hand side of the diagram is of practical importance. It corresponds to the formation of the alpha solid solution of tin in copper.

Due to the wide solidification temperature range, tin bronzes are susceptible to segregations and a sharply defined dendritic structure is always obtained upon insufficiently slow cooling. Bronzes containing from 5 to 6 per cent Sn practically consist only of the alpha solid solution. At higher tin content, the eutectoid $\alpha+Cu_{31}Sn_8$ appears in addition to the α solution (Fig. 261). After working and annealing, bronze acquires a polyhedral structure.

The curves given in Fig. 262 show the variation in the mechanical properties with the tin content for cast bronzes.

The tensile strength of bronze increases until the tin content reaches 20 per cent. It drops sharply upon further increases in the tin content due to the presence of a considerable quantity of eutectoid which contains the brittle compound $Cu_{31}Sn_8$. The relative elongation increases up to a tin content of 5-6 per cent and then drops sharply when the eutectoid begins to appear in the structure of cast bronzes (Fig. 262). Consequently, only bronzes containing up to 5-6 per cent Sn are workable.

Tin bronzes have good castability and are widely used in the foundry. A special feature of bronzes is their antifriction properties. This is due to the well defined dendritic structure formed in the solidification of tin bronzes (Fig. 261). Here the axes of the dendrites contain less tin than the spaces between them. In the operation of such a bronze as a bearing, this combination of hard and soft components is expedient since the hard areas support the shaft while the soft areas, which are subject to wear from friction, promote the formation of minute channels through which the lubricant circulates.

Zinc may be added to tin bronzes. This allows the tin content to be reduced and is done frequently to reduce the cost and improve the castability.

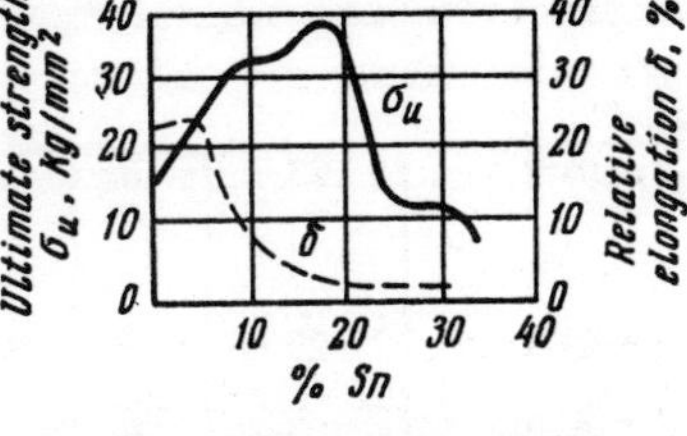

Fig. 262. Effect of tin on the mechanical properties of bronze

Machinability of bronze may be improved by adding lead. Phosphorus acts as a deoxidiser when added to bronze, as well as an element for strengthening the alloy. An addition of about 1 per cent P improves castability and increases the mechanical and antifriction properties.

Wrought tin bronzes undergo heat treatment consisting of annealing (at 600-650°C), as an intermediate operation during cold working or as a final operation to provide the required properties in the semi-finished product (sheet or strip).

Bronzes containing a large quantity of expensive tin are being replaced to a great extent at the present time by cheaper alloys, low-tin bronzes (according to GOST 613-50) to which zinc, lead, nickel, and other metals have been added (БрОЦСН-3-7-5-1, and БрОЦС3-12-5 used for fittings, and БрОЦС-5-5-5 and БрОЦС4-4-17 used for antifriction components).

Table 58 lists the properties and applications of several tin bronzes.

Tin-free bronzes. As tin is very expensive, various grades of tin-free bronzes have found extensive application in industry in addition to the tin bronzes. Their characteristics are listed in Table 59.

Most tin-free bronzes have higher mechanical properties than tin bronzes, better corrosion resistance, excellent antifriction properties, etc., and have therefore found widespread applications in industry. Their machinability and castability, however, are inferior to those of tin bronzes.

Aluminium bronzes are extensively employed. It is evident from the Cu-Al equilibrium diagram (Fig. 263) that all aluminium bronzes, applicable in industrial practice, consist of the homogeneous alpha solid solution and the phase mixture $\alpha+\gamma'$. The eutectoid constituent $\alpha+\gamma'$, formed as a result of decomposition of the β-phase, is found, however, in the structure of aluminium bronzes containing 6-8 per cent Al upon rapid cooling under actual conditions. The effect of aluminium addition on the mechanical properties of copper is illustrated

Table 58

Properties of Tin Bronzes

Grade	Mechanical properties			Typical applications
	σ_u, kg/mm²	δ, per cent	Bhn	
Cast bronzes				
БрОЦ10-2	23	5	75	Gears, bushings, and bearings operating at high contact pressures and high speeds; has high antifriction and corrosion-resistant properties
БрОФ10-1	25	3	90	Same applications; ductility somewhat less than for БрОЦ10-2
БрОС10-10	20	5	55	Various bushings for high contact pressures and speeds, and sealing discs
БрОЦС6-6-3	18	4	60	Bushings, bearings and other components subject to friction. Fittings
БрОЦСН3-7-5-1	18	8	60	Fittings which resist corrosion in sea water and also fittings operating under hydraulic pressures up to 25 atm
Wrought bronzes				
БрОФ6.5-0.4	28	15	80	Sheet and strip in electrical and instrument industries. Gears and bushings
БрОЦ4-3	20	10	70	Springs, contacts (spring type) in electrical devices; membranes and apparatus in the chemical industries
БрОФ6.5-0.15	50	28	150	Rod. Components subject to friction, wire for springs; bar and strip

Notes 1. Chemical compositions are not given since they are sufficiently clearly indicated by the grade designations. The Russian letters Бр indicate that the alloy is called a bronze. The figures following the letters and separated from each other by hyphens indicate the average content of the alloying elements. These figures are in the same order as the Russian letters which indicate the presence of various elements (О—tin, Ц—zinc, С—lead, Ф—phosphorus, etc.). 2. Mechanical properties of cast bronzes refer to sand mould castings (GOST 493-54).

by the curves in Fig. 264. The drop in relative elongation is due to the appearance of the brittle γ'-phase in the structure. The character of these property variation curves determines the procedures and techniques of working these bronzes.

Bronzes containing up to 6-8 per cent Al can be successfully worked either hot or cold. The mechanical properties of these bronzes are considerably improved by strain hardening.

Bronzes containing 10 per cent Al can only be worked at high temperatures but are easily cast and are much used as foundry materials.

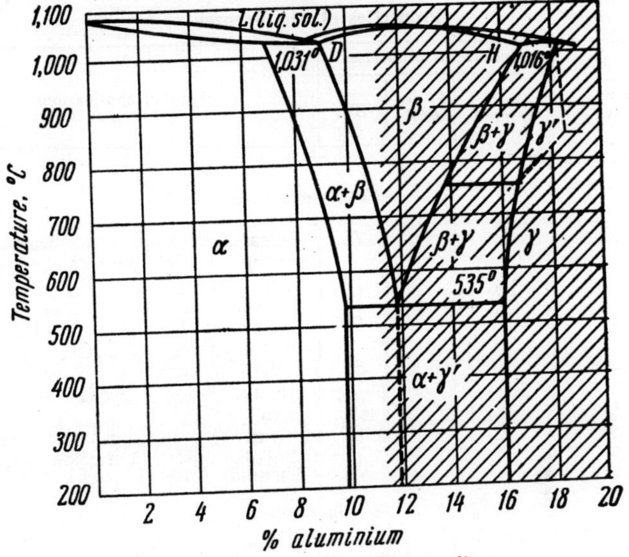

Fig. 263. The Cu-Al equilibrium diagram

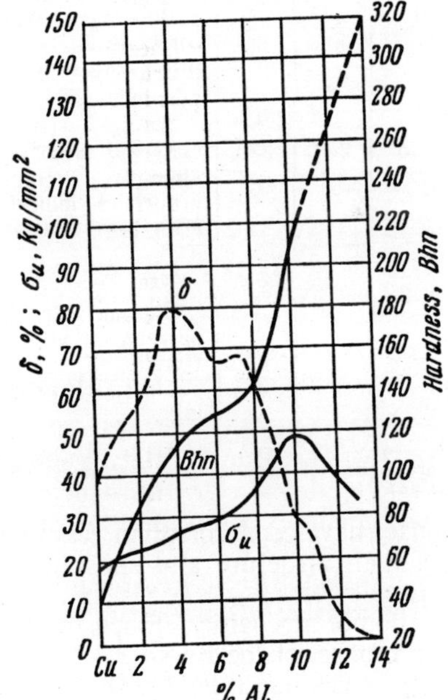

Fig. 264. Effect of aluminium on the mechanical properties of copper

Table 59

Properties of Tin-Free Bronzes

Grade	Mechanical properties			Typical applications
	σ_u, kg/mm²	δ, per cent	*Bhn*	
		Aluminium bronzes		
БрА5	60	3-4	—	Sheet, strip, rod, and wire (mechanical properties are given for strain-hardened sheets)
БрА10	52	13	110	Rod, sheet, and small castings
БрАЖ9-4	40	10	100	Fittings exposed to steam or sea water; gears, bushings, and other components subject to friction (and corrosion). Permanent and sand mould castings
БрАЖН11-6-6	60	2	250	Components subject to friction at high temperatures (valve seats); bushings exposed to sea water
		Silicon bronzes		
БрКМц3-1	80-90	1-2	—	Welded components and spring-type contacts. High corrosion resistance and mechanical properties
БрКН1-3	50	15	—	Bushings, valves, bolts, tubing, and components exposed to sea water (high corrosion resistance)
		Beryllium bronzes		
БрБ2	up to 150	2-3	370	Spring-type contacts and high-strength components (mechanical properties indicated here are acquired after heat treatment)
		Leaded bronzes		
БрОС5-25	12	4	—	Various bushings and axle-box bearings of high-power steam locomotives
БрС30	6	4	25	Liners of main and crankpin bearings

Note: Mechanical properties of cast bronzes are given for sand mould castings (according to GOST 493-54).

All aluminium bronzes have good corrosion resistance. An addition of iron to them refines their structure and substantially improves their mechanical properties.

A better combination of mechanical properties may be obtained for components made of bronzes of the БрАЖ9-4 type by applying normalisation or hardening, consisting in heating to 650°C followed by water quenching.

The high mechanical properties developed at room temperature by aluminium bronzes with nickel additions are retained at elevated temperatures as well. A very desirable addition to bronze is manganese which improves both mechanical and fabricating properties, and, in particular, formability.

Silicon bronzes. In certain applications, as for spring-type components, tin bronzes may be expediently replaced by cheaper silicon bronzes, alloyed with manganese, which also have high mechanical properties.

Beryllium bronzes develop exceptionally high mechanical properties after heat treatment. After hardening (or, more properly, *solution heat treatment*) by heating to 800°C and quenching in water, these alloys acquire a tensile strength of 54 kg per sq mm and a relative elongation of 25 per cent. A subsequent tempering (age hardening) at 350°C for 1 or 2 hours raises the tensile strength to 150 kg per sq mm though the relative elongation falls to 2-4 per cent.

Beryllium bronze is an excellent material to use for springs due to its high elasticity and resistance to fatigue and corrosion. At the same time, it has a high hardness and wear resistance. A disadvantage is its high cost.

Leaded bronze is applied for the bearing liners of machinery operating at high speeds. In the solid state lead and copper are almost insoluble in each other (Fig. 256, *b*) and the structure of these bronzes consists of crystals of copper and soft lead. The excellent antifriction properties are due to this structure.

Leaded bronzes give satisfactory service as bearings subject to high specific pressures and speeds, since their fatigue limits reach high values and their resistance to impact loads exceeds that of other bearing materials. A disadvantage of leaded bronzes is their tendency to segregation. To avoid this, special casting procedures are required.

A copper-base alloy with nickel (MH19), called cupro-nickel, is applied in industry as a corrosion-resistant material for service at elevated temperatures. It can also be used for the manufacture of netting, surgical instruments, coins, kitchen utensils, and various ornaments. Cupro-nickel with 30 per cent Ni (НЖМц-30-08-1) is widely employed for naval condenser tubes due to its good resistance to corrosion by sea water.

A copper alloy containing 40 per cent Ni and 1.5 per cent Mn (МНМц-40-1.5), called constantan, is used for rheostats, thermocouples, and heating devices operating at moderately high temperatures. The main feature of constantan is its high electrical resistivity which remains almost constant upon variations in temperature.

Chapter 16

NICKEL AND ITS ALLOYS

Nickel has a melting point of 1,452°C. Its specific gravity is 8.7 grammes per cu cm. Nickel has a face-centred cubic crystal lattice with a lattice constant of 3.512 Å. The mechanical properties of nickel are comparatively high (σ_u=50 kg per sq mm, δ=50 per cent); it can easily be worked both hot and cold, possesses high corrosion resistance and is ferromagnetic. Commercial nickel contains from 99.5 to 97.5 per cent Ni and 0.6-1.0 per cent Co.

Table 60

Properties of Certain Nickel Alloys

Properties	Monel metal НМЖМц28-2.5-1.5	Alumel НМцАК2-2-1	Chromel НХ9.5
Tensile strength σ_u, kg/mm²	50-60	56	60-70
Relative elongation, δ, %	40	36	35-48
Electrical resistivity, ohm-mm²/m	0.42	0.32	0.674
Temperature coefficient of electrical resistivity, ohm/cm-degree $\times 10^{-4}$	18.0	26.0	4.90
Thermoelectric emf vs platinum at 100° C, millivolts	2.5	−1.2	+3
Thermal conductivity, cal/cm-sec-degree	0.06	—	—
Linear coefficient of thermal expansion, α, mm/mm-degree $\times 10^{-6}$	14.0	—	12.8
Maximum service temperature, °C	—	1,000	1,000

In its commercially pure form (not less than 99 per cent Ni), nickel is available as strip, wire, and tubing used for various purposes in the instrument and electrical engineering industries.

The chief admixtures contained in nickel alloys are: Co, Fe, Si, C, S, and Cu.

Cobalt, iron, silicon, and copper form solid solutions with nickel and somewhat increase its electrical resistivity, hardness, and strength. Carbon, in quantities exceeding 0.1-0.15 per cent, forms graphite inclusions in nickel and lowers its workability. The low melting point of the eutectic (Ni-Ni_3S_2) complicates hot working to a considerable extent.

Alloys of nickel, which forms solid solutions with many different metals, possess high electrical resistivity, low temperature coefficient of resistivity, and high corrosion and heat resistance. Nickel alloys are widely employed as heat-resistant (see page 353), electrical engineering (filaments in vacuum installations) and corrosion-resistant materials (Table 60).

The well-known alloy *Monel metal* (Table 60) has a very high resistance to corrosion in acids and alkalis. Produced as sheet, bar, rod, and castings, Monel metal finds extensive applications for components of chemical apparatus, as well as high-tension oil-filled cables, components of textile and dyeing equipment, architectural items, medical instruments, etc.

Alumel and *chromel* are employed to make compensating leads and thermocouples which operate at temperatures up to 1,000°C. The latter are known as chromel-alumel couples.

Chapter 17

ALUMINIUM AND ITS ALLOYS

17-1. Aluminium

Characteristic features of *aluminium* are its low specific gravity (2.72), low melting point (658°C), and high electrical and thermal conductivities.

The electrical conductivity of pure aluminium is 34×10^4 ohm^{-1}-cm^{-1}. It has a face-centred cubic crystal lattice with a lattice constant of 4.0414 Å.

Aluminium possesses high ductility ($\delta=40$ per cent) but low strength and hardness ($\sigma_u=8$-10 kg per sq mm and $Bhn=20$).

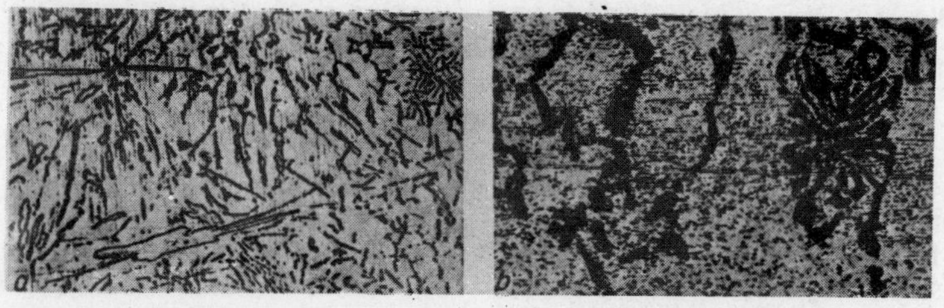

Fig. 265. Precipitation of the $FeAl_3$ phase in aluminium alloys containing 4 per cent iron:
a—×80, *b*—×400

Pure aluminium resists corrosion well. This is due to the dense, strong passivating film of aluminium oxide (Al_2O_3) which forms immediately on its surface upon exposure to the atmosphere. The purer aluminium is, the higher its corrosion resistance and electrical conductivity.

The chemical compositions and applications of commercial grades of aluminium are listed in Table 61.

The addition of iron and silicon increases the strength of aluminium to a certain extent but reduces its ductility and corrosion resistance.

Table 61

Commercial Grades of Aluminium (According to GOST 3549-58)

Grade	Chemical composition, per cent						Typical applications
	Al	Fe max	Si max	Fe+Si max	Cu max	Total impurities	
AB0000	99.996	0.0015	0.0015	—	0.0010	0.004	Aluminium of high purity for research and other special purposes
AB000	99.99	0.0030	0.0025	—	0.0050	0.010	
AB00	99.97	0.015	0.015	—	0.0050	0.03	Foil, cladding in special cases, cable and current-conducting items, special-purpose aluminium alloys, chemical industry
AB0	99.93	0.04	0.04	—	0.01	0.07	
A00	99.7	0.16	0.16	0.26	0.01	0.30	
A0	99.6	0.25	0.20	0.36	0.01	0.40	
A1	99.5	0.30	0.30	0.45	0.015	0.50	Current-conducting items, aluminium alloys, cladding, foil, aluminium powder for paints, kitchen utensils
A2	99.0	0.50	0.50	0.90	0.02	1.0	Aluminium-base and other alloys, aluminium utensils, cable and current-conducting items, special alloys
A3	98.0	1.1	1.0	1.80	0.05	2.0	Aluminium alloys and alloying material, for aluminothermic processes and consumer goods

The most harmful impurity in aluminium is iron which forms the compound $FeAl_3$ (Fig. 265) and reduces the workability.

Due to its high ductility and malleability, pure aluminium can easily be rolled into very thin sheets (foil). It can readily be die-forged or formed as well.

The strength of aluminium can be increased to a considerable extent by strain hardening (cold working).

Large reductions will raise the tensile strength to 16-18 kg per sq mm but will sharply reduce the ductility at the same time. Aluminium may be subjected to annealing at 330-360°C, after cold working, to eliminate strain-hardened condition. Annealing at higher temperatures is not recommended as coarse grains may be formed.

Annealed aluminium and strain-hardened aluminium are used in industry to produce tubing, wire, sheet, strip, rod, and other semi-fabricated products.

The application of aluminium as a structural material is limited by its low strength. It may only be employed for lightly loaded structures and for cable and current-carrying items. It is used to a far greater extent for various alloys.

17-2. Classification of Aluminium Alloys

All aluminium-base alloys may be divided into two main groups: I) wrought alloys and II) casting alloys.

The two subdivisions of the first group of wrought aluminium alloys are:

1. Alloys which do not respond to heat treatment, called non-heat-treatable alloys. These alloys consist of a homogeneous solid solution and are characterised by a comparatively low strength and high ductility. Al-Mn and Al-Mg alloys are included in this subdivision. They may be strain-hardened.

2. Heat-treatable alloys, which require heat treatment to develop optimum properties. All alloys of this type belong to systems with limited solubility in the solid state.

These alloys find widespread applications in the engineering industries.

The most typical heat-treatable aluminium alloys are *avial* (Al-Mg-Si) and the *duralumins* (based on the Al-Cu-Mg system). Wrought alloys are available as sheet, tubing, sections, wire, and forgings.

The best known of the casting alloys are the *silumin* alloys, which are Al-Si alloys with additions of Mg, Mn, and Cu. Aluminium-copper alloys are also used to some extent.

Many aluminium casting alloys are heat treatable.

17-3. Non-Heat-Treatable Wrought Aluminium Alloys

Table 62 lists the average chemical composition and the properties of wrought aluminium alloys which do not respond to heat treatment.

Alloy AMц, containing manganese, has a structure comprising the solid solution of manganese in aluminium and the compound $MnAl_6$ [(Mn, Fe)Al_6]. This alloy is not subject to heat treatment and is strengthened by strain-hardening (Table 62).

The AMr alloy series are binary alloys of aluminium and magnesium.

It is evident from the Al-Mg equilibrium diagram (Fig. 266) that alloys AMr, AMr3, and AMr5 are practically single-phase in structure and consist of the solid solution of magnesium in aluminium. Only in alloy AMr5, the excess β-phase may appear after annealing. Therefore, these alloys cannot be strengthened by heat treatment but only by strain-hardening. Only alloys containing over 8 per cent magnesium are heat treatable.

Table 62

Chemical Compositions and Properties of Wrought Aluminium Alloys

Grade	Content, per cent			Treatment	Guaranteed mechanical properties				
	Mg	Mn	Si		σ_u, kg/mm^2	σ_s, kg/mm^2	δ, per cent	δ_{-1}, kg/mm^2	*Bhn*
AMц	—	1.3	—	Annealing (M)	13	5	20	5.5	30
				Strain-hardening (П)	16	13	10	6.5	40
AMr	2.5	0.3	—	Annealing (M)	20	10	23	12.5	45
				Strain-hardening (П)	25	20	6	13.5	60
AMr3	3.5	0.4	0.7	Annealing (M)	24	12	20	—	50
AMr5	5.0	0.4	—	Annealing (M)	28	14	20	—	70

All alloys listed in Table 62 have comparatively low strength, high ductility, good weldability (especially the Al-Mn alloys) and increased resistance to corrosion.

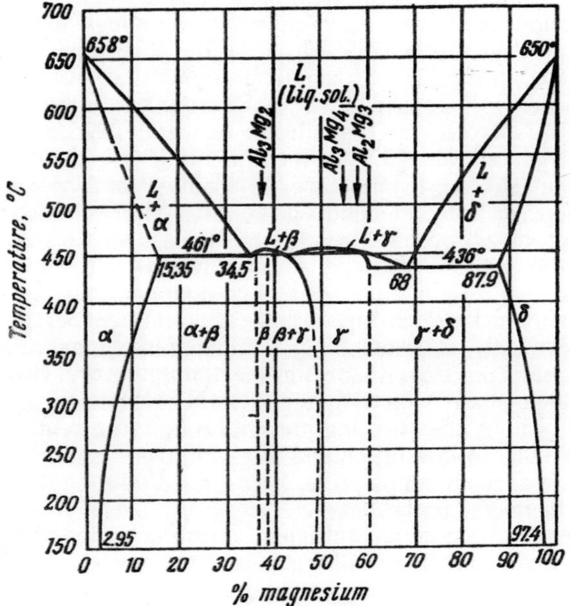

Fig. 266. The Al-Mg equilibrium diagram

These properties render them suitable for manufacturing lightly-loaded components subject to deep drawing operations, for welded parts, and also for parts exposed to corrosive agents.

17-4. Heat-Treatable Wrought Aluminium Alloys

As mentioned above, heat-treatable alloys belong to systems having limited solubility in the solid state.

The heat treatment of aluminium alloys comprises the following stages:

1. Heating the alloy to a temperature at which the excess phases are dissolved to obtain a homogeneous solid solution.
2. Rapid cooling (quenching) to fix the structure (supersaturated solid solution), obtained at the heating temperature, which is unstable at room temperature.
3. Ageing (room-temperature precipitation) which involves holding at room temperature for several days (natural ageing) or holding for fractions of an hour or for several hours at elevated temperatures (artificial ageing). In the course of ageing the solid solution approaches a more stable condition and the alloy is strengthened.

Let us examine the heat treatment of aluminium alloys in more detail, using the binary aliminium-copper alloys as an example.

A part of the equilibrium diagram for the aluminium-copper system is illustrated in Fig. 267. This diagram shows that aluminium forms a solid solution with copper in which the latter has a maximum solubility of 5.7 per cent at the eutectic temperature (548°C). The solubility is reduced when the temperature is lowered and is less than 0.5 per cent at room temperature. At a content of 54 per cent, copper forms the chemical compound $CuAl_2$. After slow cooling, alloys containing up to 0.5 per cent copper will have a single-phase structure (alpha solid solution of copper in aluminium). If the copper content ranges from 0.5 to 5.7 per cent, a two-phase structure ($\alpha + CuAl_2$) will be obtained. When any alloy, containing copper in this range (0.5-5.7 per cent), is heated above the line of maximum solubility, the excess crystals of $CuAl_2$ dissolve in the aluminium and the alloy acquires a single-phase structure, alpha solid solution.

Rapid cooling after heating prevents the $CuAl_2$ from precipitating from the alpha solid solution and this single-phase solid solution will be retained at room temperature. Such a solid solution is supersaturated and, therefore, unstable.

If, then, the atoms are sufficiently mobile at the temperature of the medium in which the alloy is held, the supersaturated solid solution will spontaneously decompose.

This process of returning the supersaturated solid solution to a more stable condition occurs in ageing as well. Lowering the temperature impedes ageing and may even completely suppress it. The work of S.T. Konobeyevsky, D. A. Petrov and other investigators shows that the decomposition of a supersaturated solid solution in ageing involves several stages and is dependent upon the ageing temperature and holding time.

No decomposition of the solid solution is observed in natural ageing since there is no precipitation of the excess phase from the supersaturated solution. What actually occurs is a congregation of the solute atoms (copper, in our case) in definite planes of the crystal lattice and the formation of zones rich in copper atoms. These zones, with increased concentration of the solute element, have the form of extremely thin discs, several atom layers thick, and can only be observed by X-ray analysis. They are known as *Guinier-Preston zones.*

This type of change within the solid solution of alloys undergoing natural ageing is accompanied by pronounced distortion of the crystal lattice and by an increase in strength. Since the solid solution does not decompose in natural ageing, the alloy will retain sufficient ductility and corrosion resistance.

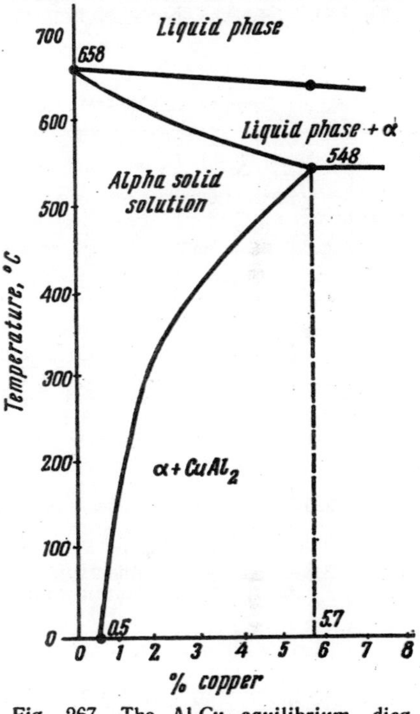

Fig. 267. The Al-Cu equilibrium diagram

When the alloy is subjected to artificial ageing at 100-150°C and higher temperatures, dispersed (lamellar) particles of an intermediate phase, which causes strengthening, are precipitated at the places where there is an increased concentration of the solute atoms. This intermediate phase in the aluminium-copper alloys is the θ′-phase. It does not differ in composition from the stable $CuAl_2$ phase but has a distorted tetragonal lattice (a=5.71 Å and c=5.80 Å).

If the alloy is held for a long period of time at the above temperatures or if the temperature is raised to 250°C and higher, coagulation of the particles, formed at the centres of decomposition, is observed and the intermediate phase becomes stable. This is called *overageing.*

For example, upon overageing of Al-Cu alloys, the intermediate phase θ′ is transformed into the stable phase $CuAl_2$ having a tetragonal lattice with the following constants: a=6.054 Å and c=4.864 Å.

The most important and extensively employed alloys, that are strengthened by *solution treatment* and subsequent ageing, are duralumin, avial, and certain other special-purpose alloys.

Duralumin alloys. Duralumin was developed by A. Wilm in 1909. Table 63 lists the average chemical compositions and typical mechanical properties of duralumin alloys employed in the U.S.S.R. at the present time (according to GOST 4784-49). Alloy Д1 is normal duralumin while Д6 and Д16 are high-strength alloys. The latter have higher strength properties than the normal grade but a lower ductility and, therefore, present difficulties in hot and cold working. The high strength is obtained by increasing the copper, magnesium, and manganese content (for Д6) or by increasing the magnesium content alone (Д16).

Table 63

Chemical Compositions and Mechanical Properties of Duralumin Alloys

Grade	Average content per cent			Treatment	Average mechanical properties			
	Cu	Mg	Mn		σ_s, kg/mm²	σ_u, kg/mm²	δ_{10}, per cent	*Bhn*
Д1	4.3	0.6	0.6	Annealing	11	21	18	42
				Solution treatment followed by natural ageing (sheet, rod, pipe)	24	42	18	100
Д6	4.9	0.8	0.8	Annealing (rod)	11	21	18	42
Д16	4.4	1.5	0.6	Solution treatment followed by natural ageing (rod)	30-32	46-47	15-17	105
Д3П	3.0	0.5	0.5	Annealing (wire)	8	17	20	—
				Solution treatment followed by natural ageing	21	34	20	80
Д18П	2.6	0.35	—	Annealing	6	16	24	38
				Solution treatment followed by natural ageing	17	30	24	70

Alloys Д3П and Д18П, which belong to the low-alloy class of duralumin, are used for rivets. These alloys are characterised by relatively low strength properties but high ductility and malleability which facilitate the rivetting operation. The most malleable and convenient for rivetting is alloy Д18П, which may be rivetted even after solution heat treatment and natural ageing.

As shown in Table 63, all duralumin alloys are based on the Al-Cu-Mg system with Si and Mn also present.

The equilibrium diagram of the Al-Cu-Mg system, given in Fig. 268, indicates the positions of the duralumin-type alloys.

The mechanical properties of duralumin are most sharply increased after heat treatment by the S-phase (Al_2CuMg), to a lesser extent by $CuAl_2$, and very little by the T-phase (Al_6CuMg_4).

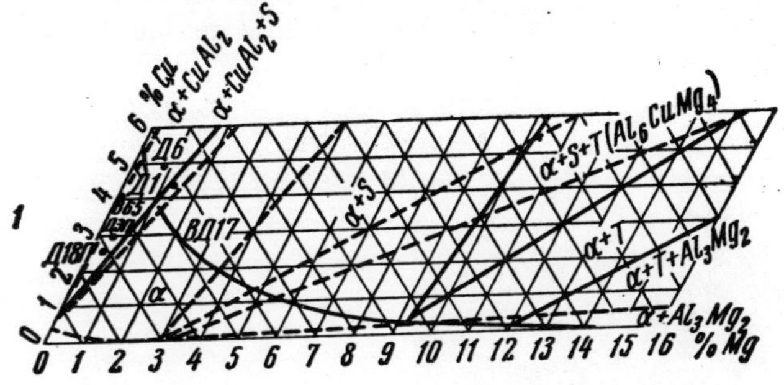

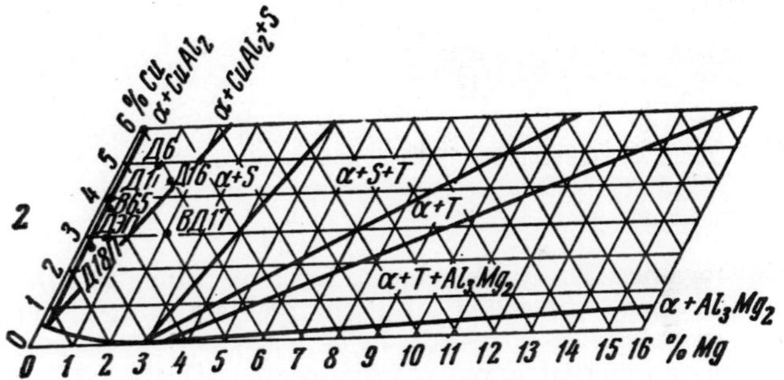

Fig. 268. The Al-Cu-Mg equilibrium diagram:

1—location of duralumin-type alloys in reference to the projection of the surface of maximum solubility of Cu and Mg in Al at the eutectic temperature, 2—location of these alloys on an isothermal section of the aluminium corner of the Al-Cu-Mg equilibrium diagram at 200° C (after D. A. Petrov)

A necessary component of all duralumin alloys is manganese which improves the corrosion resistance of the alloy.

Iron and, to some extent, silicon are impurities of duralumin alloys. Iron is a harmful impurity. By forming the compound (Mn, Fe)Al_6, iron considerably reduces the strength and ductility of duralumin. Another iron compound, combining both copper and aluminium, reduces the strengthening effect of ageing. Therefore, iron content should never exceed 0.5-0.6 per cent. To a certain extent, silicon eliminates the harmful effect of iron by fixing it in the α-phase (Al, Fe, Si) which can more easily be broken up in working. If the iron content is low, however, silicon has a detrimental effect on natural ageing of Д16 type alloys.

Duralumin, intended for the manufacture of machine parts and for structures, is available as sheet, strip, rod, shapes, and wire.

Sheet duralumin is usually cladded to protect it against corrosion. *Cladding* consists in coating with aluminium having a purity not below 99.5 per cent. Cladded duralumin is obtained by placing the duralumin ingot between two sheets of pure aluminium and rolling this pack or

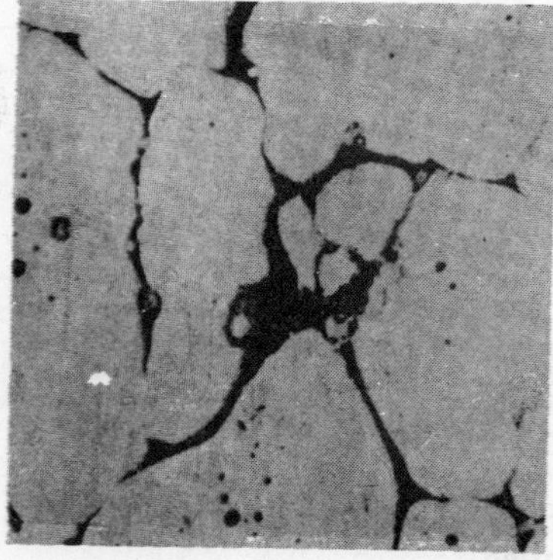

Fig. 269. Burning of duralumin: right—microstructure, × 250, left—external appearance

"sandwich" together. This pack is hot reduced by rolling to a thickness of 6 mm and then cold rolled with intermediate process annealing operations.

Duralumin and similar alloys may be comparatively easily rolled, both hot and cold, forged, formed, and have good machinability. Intermediate annealing operations at 350-370°C are required in cold rolling.

The heat treatment for strengthening duralumin comprises the following operations:

1) Solution treatment followed by water quenching (heating to 505-510°C for alloy Д1 and to 495-503°C for alloys Д6 and Д16). Heating to a lower temperature will not ensure maximum mechanical properties while heating to higher temperatures will burn the duralumin, i.e., it will cause oxidation and partial melting of the metal along the grain boundaries (melting of the complex eutectics) (Fig. 269). This leads to a sharp reduction in strength and ductility (Fig. 270). Visual evidence of burning is the bubbles appearing on the surface of the duralumin (Fig. 269). After solution treatment, a considerable part

of the strengthening phases is in the solid solution and the structure consists of the super-saturated solid solution and insoluble iron compounds.

2) Natural ageing of duralumin. Duralumin is frequently aged at room temperature. The metal develops high mechanical properties and good corrosion resistance after ageing. The influence of the ageing time on the tensile strength of aluminium is shown in Fig. 271. Lowering the temperature slows down the ageing process. The rate of the process may be accelerated by raising the temperature but optimum mechanical properties cannot be obtained and the corrosion resistance is reduced.

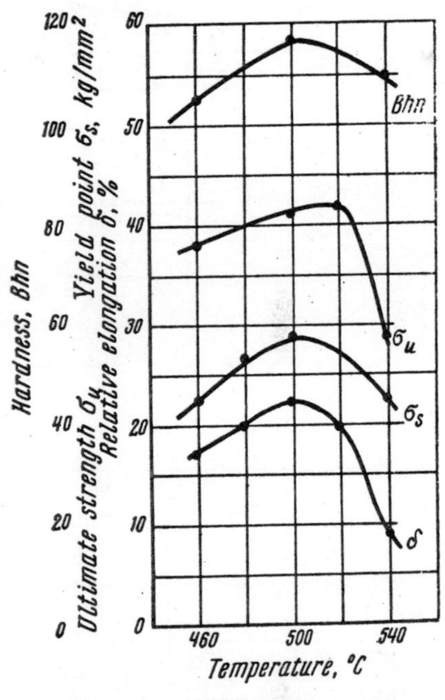

Fig. 270. Effect of the solution treatment temperature on the mechanical properties of duralumin

Heating duralumin for a short time (20-40 seconds) at 250-270°C, after natural ageing, will restore the properties of the as-quenched condition (Fig. 272). This phenomenon is called recovery and is applied in practice in place of a second quenching during cold forming operations. Recovery may be repeated many times but after each application, however, the mechanical properties are decreased to some extent.

Strain hardening may be applied in some cases to additionally strengthen duralumin after solution treatment and ageing procedures. Cold working (1 to 5 per cent reduction) substantially increases the yield point. Strain hardening directly after quenching may lower the mechanical properties in comparison with those of unworked duralumin which has undergone natural ageing.

Avial (AB) has the following composition: 0.45-0.9 per cent Mg, 0.5-1.2 per cent Si, and 0.2-0.6 per cent Cu. This alloy is heat-treatable. The procedure consists in heating to 530-540°C, quenching in water, and subsequent natural ageing or artificial ageing at 150-160°C for 12 to 16 hours. Ageing must be conducted directly after quenching, otherwise the strengthening effect is reduced. Various semi-finished products, such as sheets, pipe, shapes, etc., are made of avial.

High-strength aluminium alloy B95 (according to GOST 4783-49). Alloy B95, which has found widespread application in the Soviet Union during the last few years, contains: 6 per cent Zn, 2.3 per

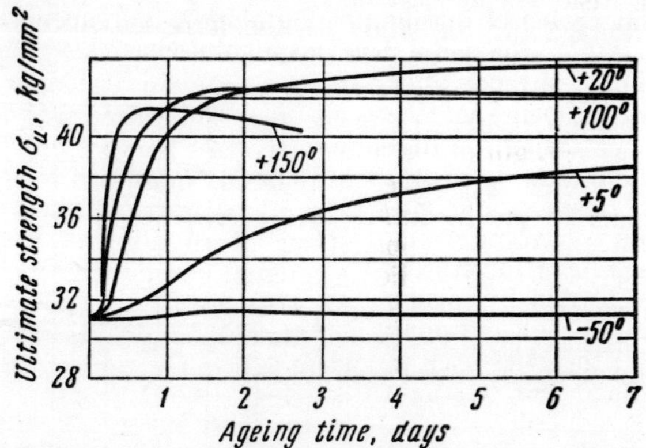

Fig. 271. Effect of the ageing temperature on the ultimate strength of duralumin (after S. M. Voronov)

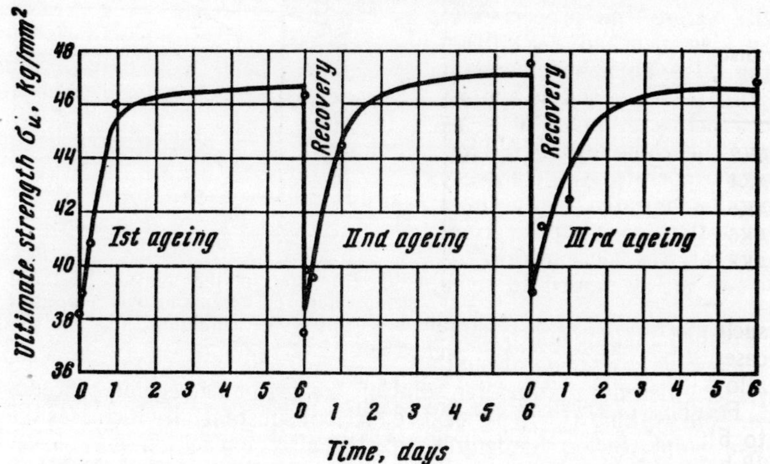

Fig. 272. Ageing curves following recovery to the as-quenched condition, produced by heating for a short time at 230° C (after D. A. Petrov)

cent Mg, 1.7 per cent Cu, and 0.4 per cent Mn. After solution treatment, by heating to 465-475°C and water quenching, and subsequent artificial ageing for 24 hours at 120-125°C, this alloy develops the following properties: σ_u=50-52 kg per sq mm, σ_s=38-41 kg per sq mm, and δ=7-5 per cent.

This alloy has satisfactory ductility in both the as-annealed and as-quenched conditions but low machinability which may be improved by solution treatment and ageing.

A disadvantage of this alloy is the increased susceptibility to corrosion of stressed material.

The strengthening phases in the Al-Zn-Mg-Cu system are $MgZn_2$ and the *T*-phase (Al-Zn-Mg). As the zinc and magnesium content is increased, the strength of the alloys increases likewise but their ductility and corrosion resistance are reduced. Additions of manganese and chromium to the Al-Zn-Mg-Cu alloys increases their corrosion resistance and strength. Applications of an alloy of even higher strength, B96, are being studied at present.

Aluminium forging alloys (GOST 4784-49). The chemical compositions and mechanical properties of aluminium alloys, used to make forgings, are given in Table 64. These alloys are used to manufacture

Table 64

Chemical Compositions and Mechanical Properties of Aluminium Forging Alloys

Grade	Chemical composition, per cent						Mechanical properties		
	Cu	Mg	Mn	Ni	Fe	Si	σ_u, kg/mm²	δ, per cent	*Bhn*
AK2	3.5-4.5	0.4-0.8	—	1.8-2.3	0.5-1.0	0.5-1.0	34-36	4-5	95
AK4	1.9-2.5	1.4-1.8	—	1.0-1.5	1.1-1.6	0.5-1.2	36-38	3-4	100
AK5									
AK6	1.8-2.6	0.4-0.8	0.4-0.8	—	—	0.7-1.2	34-38	5-8	95
AK8	3.9-4.8	0.4-0.8	0.4-1.0	—	—	0.6-1.2	44-46	8-10	130

such machine parts as pistons (AK2 and AK4), screw blades, crank-cases, rings, discs, covers, and others. The most ductile of these are alloys AK5 and AK6.

Forgings of these alloys are subjected to heat treatment (heating to 510-520°C, water quenching, and ageing at 150-160°C for 12 to 18 hours). Most aluminium alloys are forged at temperatures between 470° and 430°C.

17-5. Aluminium-Base Casting Alloys

The following subdivisions of the group of aluminium casting alloys may be indicated:

1) Alloys based on the Al-Si system (АЛ2 and АЛ4). These are called *silumin alloys*. A typical silumin is the eutectic alloy АЛ2 (Fig. 273) which is distinguished for its high castability (high flu-

idity and low shrinkage), good corrosion resistance, high ductility and low specific gravity (2.6 grammes per cu cm). Alloy АЛ2, which is not heat-treatable, is strengthened by modification. This consists in adding about 0.01 per cent Na to the melt just before casting.

Prior to modification, Al-Si alloys have a coarse acicular eutectic (α + Si) and primary Si crystals (Fig. 274, *a*).

The eutectic becomes fine-grained after being modified (Fig. 274, *b*), and hypereutectic alloys, containing from 12 to 14 per cent Si, acquire a hypoeutectic structure with primary dendrites of the α solution (aluminium). The mechanical properties are improved at the same time.

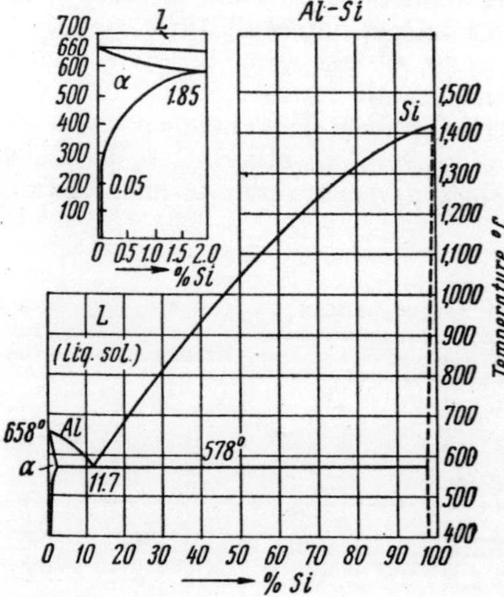

Fig 273. The Al-Si equilibrium diagram

The modifying effect is probably associated with supercooling and shifting the eutectic line downward while concentrations, corresponding to the eutectic point, are shifted to the right (Fig. 273). Certain investigators explain this refining of the eutectic and tendency to supercool by the formation of special films that envelop the silinco crystals and impede their growth.

Silumin is modified by adding a mixture of 2/3 sodium fluoride and 1/3 sodium chloride to the molten alloy. The amount added is 2 per cent of the melt by weight. Recently, other modifying mixtures have been proposed. One of them, consisting of 60 per cent NaF, 25 per cent NaCl, and 15 per cent Na_3AlF_6, degasifies the alloy and removes oxide inclusions as well.

A disadvantage of silumin as a casting alloy is the porosity of the castings often obtained. Porosity is avoided by casting under pressure in autoclaves.

The addition of magnesium and copper, which form the compounds Mg_2Si and $CuAl_2$, enables silumin alloys to be heat treated. A typical heat-treatable silumin is alloy АЛ4 which possesses excellent casting properties but has a tendency to form foam (slags) and porosity. This alloy is strengthened by modifying and subsequent heat treatment (Table 65).

Alloy АЛ4 can be easily machined and has a high corrosion resistance. It is used for large castings which are to operate under heavy-duty load conditions.

2) Alloys based on the Al-Mg system (АЛ8). The best known of these alloys is АЛ8 which has comparatively poor castability. Special foundry procedures, such as the application of special fluxes and additions to the moulding sand as well as degasification of the melt, are required in casting this alloy.

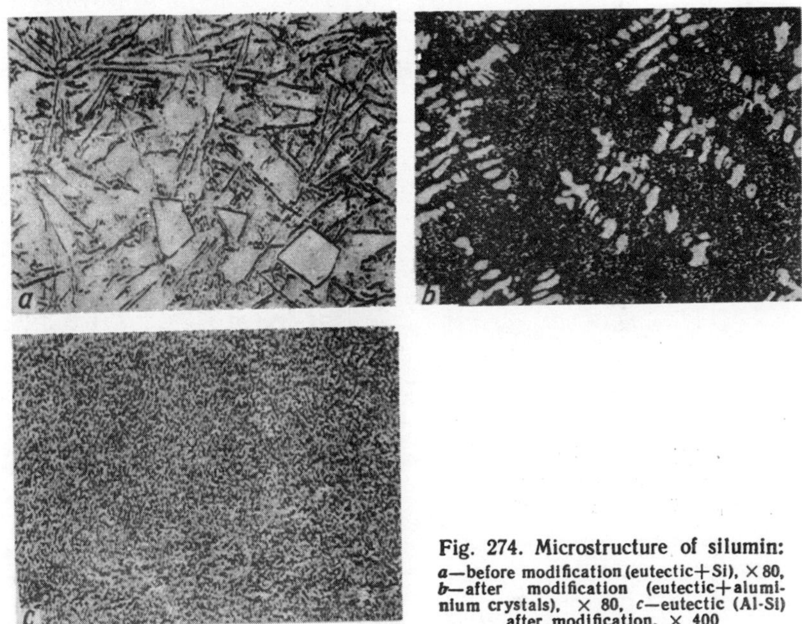

Fig. 274. Microstructure of silumin:
a—before modification (eutectic+Si), ×80, b—after modification (eutectic+aluminium crystals), ×80, c—eutectic (Al-Si) after modification, ×400

However, due to its high mechanical properties (Table 65), good corrosion resistance, and low weight, alloy АЛ8 is employed for castings of relatively simple form, subject to dynamic loads in operation and exposed to corrosive action. This alloy is used only after heat treatment (Table 65).

Upon solution treatment, the excess β-phase (Al_3Mg_2) (see Fig. 266) passes into the solid solution. Quenching fixes a homogeneous solid solution and the alloy acquires high mechanical properties and resistance against corrosion.

3) Alloys based on the Al-Cu system (АЛ7 and АЛ12). Alloy АЛ7 possesses comparatively low castability (susceptibility to hot cracking and the formation of shrinkage cavities), but develops good mechanical properties after heat treatment (Table 65). Alloy АЛ12 is used without heat treatment. It has much better casting properties and a

Table 65

Chemical Compositions and Mechanical Properties of Aluminium Casting Alloys

Grade	Chemical composition, per cent				Treatment	Mechanical properties*		
	Si	Cu	Mg	Mn		σ_u, kg/mm²	δ, per cent	*Bhn*
				Al-Si alloys (silumin alloys)				
АЛ2	10-13	—	—	—	Modified	15	4	50
АЛ4	8-10.5	—	0.17-0.3	0.25-0.5	Modified	15	2	50
					Aged at 175° C	20	1.5	70
					Solution-treated and aged at 175° C	23	3	65
				Al-Mg alloys				
АЛ8	—	—	9.5-11.5	—	Solution-treated (quenching from 435° C)	28	9	60
				Al-Cu alloys				
АЛ7	—	4-5	—	—	Solution-treated	20	6	60
					Solution-treated and aged at 150° C	22	3	70
АЛ12	—	7-14			As-cast	11	—	50
				Al-Si-Cu alloys				
АЛ5	4.5-5.5	1.0-1.5	0.35-0.6	—	Aged	16	—	65
					Solution-treated and aged at 180° C	20	—	70
АЛ6	4.5-6	2-3	—	—	Aged at 290° C	15	1	45
				Aluminium alloys containing other components				
АЛ1	—	3.75-4.5	1.25-1.75	(Ni—1.75-2.25)		20	0.5	95
АЛ11	6-9	—	—	(Zn—10-14)	As-cast	20	2	80

* Mechanical properties are given for sand-mould castings.

high strength which is retained up to 250-300°C, but is insufficiently ductile.

4) Alloys based on the Al-Si-Cu system (АЛ5, АЛ6, and АЛ10). Aluminium-silicon-copper alloys have good foundry characteristics. АЛ6, typical of this group, is used for sand-mould, permanent-mould, and die castings. Dense, air-tight castings are obtained which may be welded if required.

The heat-treatable alloy АЛ5 (see Table 66) is extensively used in the U.S.S.R. Its casting properties are good and it gives excellent service for large severely-loaded machine parts, as well as for parts operating at elevated temperatures.

Table 66

Heat Treatment Procedures for Aluminium Casting Alloys

Grade	Solution treatment			Ageing	
	Heating temperature, °C	Holding time, hrs	Quenching medium	Heating temperature, °C	Holding time, hrs
АЛ1	$515^{\pm5}$	2-4	Boiling water or still air	$220^{\pm10}$	2-4
АЛ1	$515^{\pm5}$	2-4	Boiling water or still air	Natural ageing	—
АЛ4	$535^{\pm5}$	2-6	Water at 60-80° C	$175^{\pm5}$	15
АЛ5	525	4	Boiling water	$180^{\pm5}$ $180^{\pm5}$	6-7 5
АЛ6	Annealing $290^{\pm5}$	3	Still air	$230^{\pm5}$	5
АЛ7	$515^{\pm5}$	10-15	Water at 50-100° C	$150^{\pm5}$	2-4
АЛ7	$515^{\pm5}$	10-15	Water at 50-100° C	Natural ageing	—
АЛ8	$435^{\pm5}$	15-20	Water at 60-80° C	—	—

5) Zinc silumin alloys (АЛ11). Alloy АЛ11 was proposed and investigated in detail by A. A. Bochvar.

This alloy has excellent castability and can be modified to improve its mechanical properties. It may be recommended for castings of complex form requiring high mechanical properties. A disadvantage of this alloy is its somewhat increased specific weight (2.93 grammes per cu cm). Alloy АЛ1 has relatively poor casting properties but its distinguishing feature is its increased creep resistance (a result of the addition of nickel).

Chapter 18

MAGNESIUM AND ITS ALLOYS

18-1. Magnesium

The melting point of *magnesium* is 650°C. It is the lightest of structural metals, having a specific weight of 1.74 grammes per cu cm. Its crystal lattice is of the compact hexagonal type with the constants 3.203 Å and 5.2002 Å. This explains its considerably lower ductility in comparison with aluminium, having a cubic lattice. Commercial magnesium is available in two grades in the Soviet Union, Mr1 (99.9 per cent Mg) and Mr2 (99.75 per cent Mg). Pure magnesium has very low mechanical properties (σ_u=9 kg per sq mm, σ_s=2 kg per sq mm, and δ=2 per cent) and its corrosion resistance is poor. At high temperatures it oxidises easily and burns with a bright white flame. Pure magnesium is not employed as a structural material.

18-2. Magnesium Alloys

The most extensively applied in engineering are the alloys of magnesium with aluminium and zinc to which manganese is added and, in some cases, titanium and beryllium.

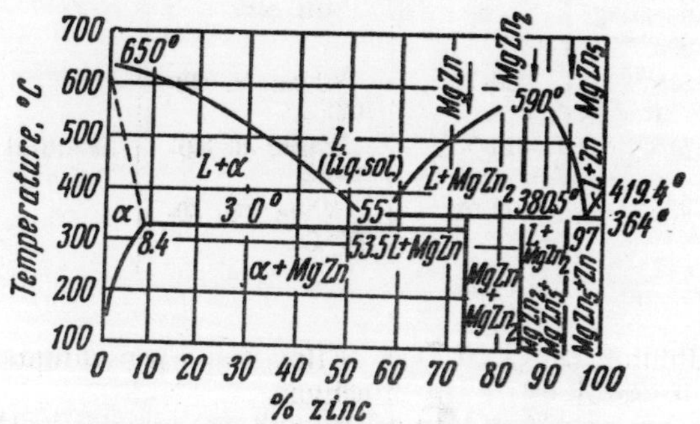

Fig. 275. The Mg-Zn equilibrium diagram

Aluminium and zinc dissolve in the magnesium (Figs. 266 and 275) and raise its mechanical properties to a considerable extent. An addition of manganese increases the corrosion resistance of the alloy. Titanium refines the grain of magnesium alloys while beryllium reduces the oxidability of the molten alloy by forming an oxide film on the bath. An addition of 0.6 to 0.85 per cent zirconium considerably refines the grain and increases the mechanical properties. Cerium increases the heat conductivity.

The most harmful impurity of magnesium alloys is iron which reduces their corrosion resistance.

In respect to their application in industry, magnesium alloys are divided into two groups.

1) Wrought alloys, available as sheet (MA1), rod, and forgings (Table 67).

Table 67

Chemical Compositions and Mechanical and Fabrication Properties of Wrought Magnesium Alloys

Grade	Average chemical composition, per cent			Mechanical properties				Fabrication properties	Applications
	Al	Zn	Mn	σ_u, kg/mm²	σ_s, kg/mm²	δ, per cent	*Bhn*		
MA1	up to 0.1	up to 0.1	1.3-2.5	21	12	8	45	High ductility, good weldability, increased corrosion resistance	Sheets, lightly loaded machine parts
MA2	3.5	0.5	0.25	26	16	8	55	High ductility, when heated, good machinability, lowered corrosion resistance	Forgings of complex form
MA3	6	1.0	0.3	30	22	8	60	Satisfactory ductility when heated, lowered corrosion resistance	Forgings

Wrought magnesium alloys have more limited applications than aluminium alloys and are used chiefly for making machine parts by hot-hammer and die-forging techniques. At normal temperatures, plastic deformation of magnesium may proceed only along the base plane of the hexagonal lattice. New slip planes appear on heating and, therefore, magnesium alloys are worked after heating to 275-

450°C. The alloys MA1, MA2, and MA3 have structures consisting of solid solution crystals.

2) Magnesium casting alloys (Table 68). Magnesium alloys are widely used as a foundry material even though the casting procedure is complicated by the necessity of protecting the melt against oxidation. The castability of magnesium alloys is inferior to that of aluminium alloys.

Table 68

Chemical Compositions and Mechanical and Fabrication Properties of Magnesium Casting Alloys

Grade	Average chemical composition, per cent			Mechanical properties			Fabrication properties	Applications
	Al	Zn	Mn	σ_u, kg/mm²	σ_s, kg/mm²	δ, per cent		
МЛ3	3	1.0	0.3	18	5.5	10	Poor castability (large shrinkage, low fluidity), excellent machinability	Parts of simple form (gasoline and oil fittings)
МЛ4	6	2.5	0.3	18	9	4	Satisfactory castability, tendency to form micro-porosity, excellent machinability, good corrosion resistance (after oxidising)	Loaded parts of aircraft: brake drums, steering wheels, trusses, chassis, instrument housings, etc.
MA5	8.4	0.5	0.3	17	9	3	High castability, excellent machinability, satisfactory corrosion resistance	Heavily-loaded machine parts subject to impact and vibration loads
MA6	10	2	0.3	25	11	8	High castability, lowered corrosion resistance	Moderately loaded machine parts

The finer the grain of magnesium casting alloys, the higher their mechanical properties will be. Grain refinement is accomplished by overheating the alloy to 850-900°C before pouring the melt or by modifying, i.e., by adding up to 1 per cent of such materials as chalk, magnesite, ferrous chloride and others to the melt.

The mechanical properties of magnesium casting alloys (МЛ4, МЛ5, and МЛ6) may be increased by a heat treatment consisting in holding for a long period (10-16 hrs) at 380°C (for МЛ4) or 415°C (for МЛ5 and МЛ6), cooling in still air, and subsequent ageing at 175°C for 15-16 hours.

Solution treatment raises the tensile strength and, if followed by ageing, the yield point as well.

Magnesium alloys are easily machined but require the observation of certain safety measures. Disadvantages of magnesium alloys in comparison to aluminium alloys are their low corrosion resistance and low modulus of elasticity ($E=4,300$ kg per sq mm).

The corrosion resistance of magnesium alloys is increased, as a rule, by treatment in special baths (for example, in a solution of $K_2Cr_2O_7+$ $+HNO_3+H_2O$). This produces a protective film on the surface. The parts are usually varnished after this treatment.

Magnesium alloys may be recommended in all cases where a strong light-weight material is required and, first of all, in the aircraft industry, for transportation vehicles, for radio apparatus, portable machinery and instruments (typewriters, adding machines, optical instruments, etc.).

Magnesium alloys should not be applied for heavily-loaded components or machine parts exposed to corrosive agents (water, steam, etc.).

Chapter 19

TITANIUM AND ITS ALLOYS

Titanium has a melting point of 1,725°C. Its specific weight is 4.5 grammes per cu cm. Titanium has two allotropic modifications: up to 880°C, it exists as α-titanium with a hexagonal lattice ($a=2.953$ Å and $c=4.729$ Å); at higher temperatures it exists as β-titanium, with a body-centred cubic lattice ($a=3.327$ Å).

The compositions and properties of commercial titanium, used in the U.S.S.R., are given in Table 69.

Table 69

Compositions and Properties of Commercial Titanium

Grade	Maximum impurities	Mechanical properties		
		σ_u, kg/mm²	δ, per cent	ψ, per cent
ВТ1Д-1	0.1% C, 0.15% O_2, 0.04% N, 0.3% Fe, 0.15% Si, and 0.015% H_2	45-60	25-30	50
ВТ1Д-2		55-75	20-25	45

Nitrogen, carbon, and oxygen are among the harmful impurities of titanium. Nitrogen and oxygen sharply reduce its ductility. Carbon in excess of 0.1-0.15 per cent lowers the formability, machinability, and drastically reduces the weldability of titanium and its alloys. Hydrogen is also a harmful impurity which sharply increases the notch sensitivity of titanium and its alloys (so-called hydrogen embrittlement of titanium).

Titanium with an increased hydrogen content (over 0.015 per cent) is subjected to vacuum annealing at 800°C (at 700-750°C for titanium alloys) for several hours to reduce the hydrogen content to the minimum (0.003-0.005 per cent). Titanium may be satisfactorily worked, both hot and cold and has good weldability (by helium-arc or argon-arc processes) but its machining behaviour is much inferior to steel. Due to the formation of a stable oxide film on the surface of titanium, it has a high corrosion resistance to fresh and sea water, and to certain acids.

The wear resistance of titanium may be raised by a chemical heat treatment (case-hardening). Good results are obtained, for instance, by nitriding in partially dissociated ammonia at 800-900°C, boron impregnation, as well as surface saturation with molybdenum, oxygen, and other elements.

Its high mechanical properties, low specific weight, and outstanding corrosion resistance have found titanium wide applications in recent years as a structural material.

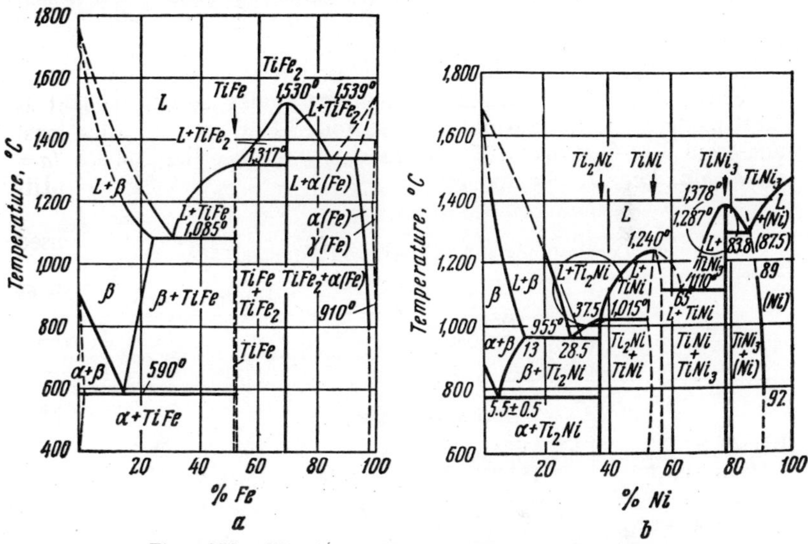

Fig. 276. Titanium alloy equilibrium diagrams:
a—Ti-Fe, b—Ti-Ni

The alloys of titanium are of special interest since they have higher mechanical properties than pure titanium, higher creep resistance, higher fatigue limits, and good corrosion resistance.

Titanium alloys have the highest specific strength, i.e., ratio of strength to specific weight, of all structural materials.

The addition of nitrogen, oxygen, and aluminium to titanium raises the temperature of the allotropic transformation and enables alloys with a stabilised α-phase to be obtained. Molybdenum, vanadium, niobium, and certain other elements stabilise the β-phase.

The most widely employed are α+β-alloys, possessing high strength and ductility; α-alloys are used to a lesser extent and β-alloys are not used at all.

Equilibrium diagrams of certain binary titanium alloys are shown in Fig. 276. The diagrams for the alloys of titanium with manganese, iron, nickel, copper, as well as tungsten, cobalt and certain other elements show that there is a eutectoid decomposition of the β-phase and, in many cases, intermetallic compounds are formed. The equilibrium diagrams of titanium with molybdenum, vanadium, aluminium, and tin do not have lines corresponding to eutectoid decomposition and these systems do not contain intermetallic compounds.

The most important alloying elements for titanium are aluminium, chromium, manganese, vanadium, iron, molybdenum, and tin which considerably increase its strength. The highest strengthening effect is obtained from iron, chromium, and aluminium; the least—from vanadium and molybdenum.

Table 70 lists the compositions of certain titanium alloys which have found wide applications, both in the Soviet Union and abroad. Most modern titanium alloys contain from 3 to 6 per cent aluminium and from 3 to 6 per cent of elements that stabilise the β-phase (such as V, Mo, Fe, Cr, Ni, and Mn).

Table 70

Compositions and Properties of Titanium Alloys

Grade	Chemical composition	Mechanical properties			
		σ_u, kg/mm²	δ, per cent	ψ, per cent	Structure
	General-purpose alloys				
BT6 (U.S.S.R.)	6% Al+4% V	95	10	20	α+β
Ti-140A (U.S.A.)	2% Cr+2% Fe+ +2% Mo		12	—	α+β
Ti-314 (Gr. Brit.)	2% Al+2% Mn	63-66	15	—	α+β
	Alloys for sheets				
BT5 (U.S.S.R.)	5% Al	80	12	30	α
C-110M (U.S.A.)	8% Mn	80	12	—	α+β
A-110AT (U.S.A.)	5% Al+2.5% Sn	80	18	—	α
	Heat-resistant alloys (up to 400° C)				
BT3 (U.S.S.R)	3% Cr+5% Al	95	10	35	—
C-130AM (U.S.A.)	4% Al+4% Mn	98	10	—	α+β
T-155A (U.S.A.)	5% Al+1.4% Cr+ +1.3% Fe+ +1.4% Mo	105	10	—	α+β

Titanium alloys may be applied as heat-resistant materials (for temperatures up to 350-500°C) in place of aluminium alloys and steel to reduce the weight of the construction.

Cold working of titanium and its alloys substantially increases the strength and lowers the ductility. Strain hardening may be relieved by annealing (at 500-550°C for commercial titanium and at 650-700°C for its alloys).

Heat-resistant titanium alloys which may become brittle at elevated temperatures (300-400°C), due to the formation of the intermediate ω-phase, are subjected to a special heat treatment called stabilisation. This operation is carried out at 700-750°C for from 2 to 4 hours. Alloy BT3 and others are stabilised at 480-550°C.

The strength of titanium alloys is sometimes increased by hardening, which comprises heating to 600-950°C and quenching in water, with subsequent ageing at 480-550°C. The properties may be varied in a wide range by altering the hardening and ageing temperatures.

Chapter 20

ZINC, LEAD, TIN, AND THEIR ALLOYS

20-1. Zinc and Its Alloys

The melting point of *zinc* is 419.4°C. Its specific weight is 7.12 grammes per cu cm and it has a close-packed hexagonal crystal lattice with the constants of 2.6595 Å and 4.9368 Å. The purity of commercial zinc is within 99.96-94.1 per cent.

The principal impurities of zinc are Fe, Pb, Sn, and Cd. The addition of iron increases the hardness and at a content exceeding 0.2 per cent Fe, zinc acquires high brittleness. The addition of lead reduces the corrosion resistance of zinc. Tin, cadmium, and lead form eutectics with zinc that have low melting points and, therefore, make hot working operations difficult to perform.

Zinc, in the deformed condition, exhibits sharply defined anisotropy of its mechanical properties. The tensile strength of sheet zinc, in the direction of rolling, does not exceed 15 kg per sq mm; in the transverse direction, this value will be from 20 to 23 kg per sq mm.

The formation of a protective oxide film on its surface renders zinc resistant to corrosion when exposed to weather or sea-water attack. The purer zinc is, the higher its corrosion resistance will be.

Zinc is extensively employed in the printing industry and for the production of galvanised iron.

The zinc alloy ЦШ1 (3.5-4.5 per cent Al, 0.06-0.1 per cent Cu, 0.02-0.06 per cent Mg, and rem. Zn) is used for hand-composed type; "Alloy 3", containing 2.2-3.0 per cent Al and 1.2-1.8 per cent Mg—for type used in type-setting machines. "Alloy 6" (4 per cent Al, 2 per cent Mg, and rem. Zn) is recommended for linotype and monotype machines.

Wrought zinc alloys, ЦАМ2-5 (2 per cent Al, 5 per cent Cu, rem. Zn) and ЦМ1 (1 per cent Cu and rem. Zn), are used to some extent in extruding operations while the antifriction alloys, ЦАМ10-5 and ЦАМ-5-10, are used for bearings.

Alloys of zinc with copper, aluminium and a small amount of magnesium (0.05-0.12 per cent) have a low melting point and excellent fluidity. They are used for making various machine parts by die casting (Table 71).

Table 71

Compositions and Mechanical Properties of Zinc Die-Casting Alloys

Chemical composition, per cent				Mechanical properties	
Al	Cu	Mg	Zn	σ_u, kg/mm²	δ, per cent
3.5-4.5	2.5-3.5	0.05-0.12	Rem.	25-31	0.5-0.2
3.5-4.3	0.6-2.4	0.06-0.0	Rem.	27-33	5-2
3.5-4.3	0.0-0.6	0.06-0.0	Rem.	25-30	6-3

The chief disadvantages of zinc alloys are: 1) variation in properties and lack of dimensional stability in natural ageing, 2) sharp drop in impact strength at lowered temperatures, and 3) susceptibility to intergranular corrosion.

20-2. Lead and Its Alloys

Lead has a melting point of 327°C and a specific weight of 11.34 grammes per cu cm. Its crystal lattice is of the face-centred cubic type with a constant of 4.4927 Å. Commercially pure lead has a purity from 99.992 per cent (C0) to 99.6 per cent (C4). The main impurities are Ag, Cu, As, Sb, Sn, Zn, Fe, and others. Lead has a low hardness and very high ductility and can be worked easily. Lead is not strain-hardened by cold working since its recrystallisation temperature is below room temperature.

All admixtures increase the strength and reduce the ductility of lead. Arsenic promotes brittleness. The mechanical properties of pure lead are: $\sigma_u=2$ kg per sq mm, $\delta=45$ per cent, $\psi\sim100$ per cent, and $Bhn=4$.

Lead has a high resistance to oxidation in high-concentration sulphuric acid but is attacked by nitric acid.

Lead is employed as radiation shielding, in the storage battery industry, for lining electrolytic baths, for sheathing sulphuric acid chambers, for cable sheathing (pure Pb, Pb+0.5 per cent Sb, and others), for type metals, etc.

Type metal for machine composition should have a low melting point (240-300°C), good fluidity, and small shrinkage. It should retain its chemical composition after repeated melting and should have good corrosion resistance. The most widely used type metals are alloys of lead with antimony and arsenic. Table 72 lists the compositions of type metals employed in the U.S.S.R. for hand composition, composing machines (linotype and typograph), and monotype machines, as well as for stereotype printing blocks cast in papier-mâché (flong) mats.

Table 72

Composition of Standard Type Metals (Lead-Base Alloys)

Grade	Principal components, per cent *			Main applications
	Sb	Sn	As	
МШ1	13.5-15	—	3.5-4.5	Hand-set type, body size—6-12 points
МШ2	14-16	—	2.7-3.3	Ditto, body size—16-48 points
МШ3	14-16	—	1.5-2.5	Ditto, body size—16-48 points
МП1	9.5-10.5	—	1.0-1.5	Casting leads and spaces
МСМ1	9.5-10.5	—	2.0-2.5	Casting stereotype printing blocks and for monotype composition
МЛН1	9.5-10.5	—	1.0-1.5	Casting rules and leads on the Elrod machine
ЛН1	11-12	4.2-4.8	1.0-1.5	Casting type on a slug-casting machine, for rules and leads

* Remainder Pb.

20-3. Tin and Its Alloys

The melting point of tin is 231.9°C and its specific weight is 7.284 grammes per cu cm. Tin has two allotropic forms. β-Sn or white tin is in an equilibrium condition at temperatures above +18°C. It has a tetragonal crystal lattice with constants of 3.1753 Å and 5.3194 Å. α-Sn or grey tin is in an equilibrium condition below +18°C and has a diamond-type crystal lattice.

As a result of the change in the compactness of the crystal lattice, the β-Sn→α-Sn transformation is accompanied by a considerable volume increase (~25 per cent) and the development of large stresses which crack up the brittle α-tin and transform it into a grey crystalline powder ("*tin pest*"). The rate of the β-Sn→α-Sn transformation is very low and β-tin also has a pronounced tendency to supercool. Consequently, this spontaneous disintegration of tin occurs only after prolonged storage at low temperatures (—30° to —40°C). Commercial tin has a purity of 96.25 to 99.9 per cent.

Tin may contain the following impurities: As, Fe, Cu, Pb, Bi, and Sb. Such impurities as lead and antimony, in small amounts, sharply reduce the rate of the β-Sn→α-Sn transition. At a content exceeding 0.5 per cent, the development of "tin pest" is practically excluded. All impurities, except antimony, are insoluble in tin.

Tin has a high ductility and can be easily worked. Plastic deformation does not produce any appreciable strain hardening since the recrystallisation temperature is below room temperature. Tin has high corrosion resistance to atmospheric conditions and to organic acids but is corroded by hydrochloric and sulphuric acids, as well as alkalis (if air is present). The mechanical properties of tin are: σ_u=2.5 kg per sq mm, δ=40 per cent, $\psi \approx$100 per cent, and *Bhn*=5.

Commercial tin is used in making babbitts (see Chapter 21), soft solders, for tin plating, for making foil, pipe, etc.

Soft solders with a low melting point (180-300°C) are usually alloys of tin with lead. Table 73 lists the compositions and applications of the most frequently employed tin solders in the U.S.S.R.

Table 73

Chemical Compositions and Applications of Soft Solders

Grade	Chemical composition, per cent			Typical applications
	Sn	Sb	Pb	
ПОС90	89-90	0.10-0.15	Rem.	Soldering internal seams of utensils for food, and of medical apparatus
ПОС40	39-40	1.5-2.0	Rem.	Soldering brass iron, and copper wire
ПОС30	29-30	1.5-2.0	Rem.	Soldering brass, copper, iron, zinc, galvanised sheet metal, tin plate, radiators, radio apparatus, and flexible metallic hose, electric motors, etc.

The tensile strength (σ_u) of the soldered seam may reach 6-8 kg per sq mm and increases when the thickness of the layer of solder is reduced or when the tin content in the solder is increased.

Chapter 21

BABBITTS (ANTIFRICTION ALLOYS)

Bearing metals or *antifriction alloys (babbitts)* are employed for lining (babbitting) bearings.

The operating conditions of a bearing liner determine the principal requisites for bearing metals. The alloy must 1) have sufficient hardness but not enough to cause excess wear of the journal, 2) be capable of deforming under local stresses, i.e., have ductility, 3) retain lubricant on its surface, and 4) provide for a low coefficient of friction between the journal and the bearing.

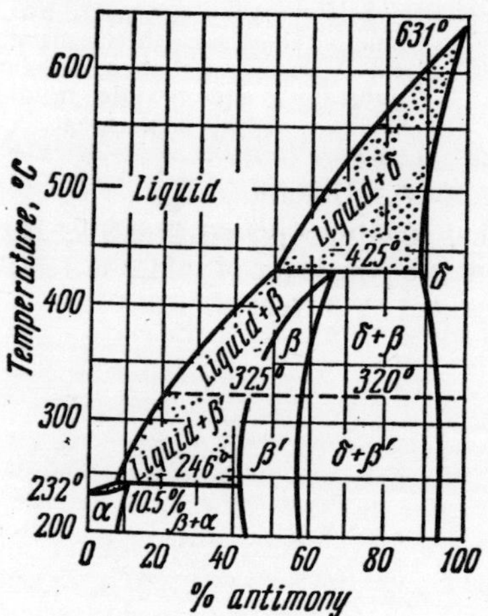

Fig. 277. The Sn-Sb equilibrium diagram

Babbitts must also conform to the following additional requirements: 1) they should not have a high melting point, 2) after being poured into the bearing, they must adhere tightly to its walls, 3) they must have good thermal conductivity and corrosion resistance, and 4) they must be cheap.

To satisfy all of these requirements, bearing metals (babbitts) must have a structure inhomogeneous in respect to hardness, i.e., they must consist of a plastic matrix into which hard particles are embedded.

In its rotation, the journal is supported by these hard, wear-resisting particles while microscopic channels are formed in the softer matrix which is more subject to wear. The lubricant circulates through these channels and carries off the products of wear.

Table 74 lists the compositions and applications of certain grades of babbitts used in the Soviet Union (tin- and lead-base babbitts).

Table 74

Chemical Compositions and Applications of Bearing Alloys (Babbitts)

Grade	Chemical composition, per cent						Applications
	Sb	Cu	Cd	Pb	Sn	Others	
Б83	10-12	5.5-6.5	—	—	Rem.	—	Babbitting bearings of highly loaded machinery (steam turbines, turbopumps, etc.)
БН	13-15	1.5-2.0	1.25-1.75	Rem.	9-11	0.75-1.21 Ni, 0.5-0.9 As	Babbitting bearings of moderately loaded machinery
Б16	15-17	1.5-2.0	—	Rem.	15-17	—	Automobile engine bearings
Б6	14-16	2.5-3.0	1.25-2.25	Rem.	5-6	0.6-1.0 As	Bearings of comparatively lightly loaded machinery
БК	0.25 max	—	—	Rem.	—	0.65-0.95 Na, 0.75-1.1 Ca	Railroad rolling stock

The tin-base babbitt, Grade Б83, is of the highest quality. Its soft matrix is the alpha solid solution of antimony in tin while the hard crystals are the β-phase, which is a solid solution based on the chemical compound SnSb (Fig. 277).

Copper is added to babbitt to eliminate segregations, due to differences in density of the compound SnSb, and forms the compound Cu_3Sn. The latter solidifies first in the form of well-branched dendrites which run through the whole of the liquid alloy and on which the crystals of the SnSb compound settle. Besides this, the Cu_3Sn crystals form hard inclusions in the babbitt, favourable for bearing service.

The microstructure of the tin-base babbitt Б83 is shown in Fig. 278, *a*. Here the darker background is the plastic matrix, consisting of the alpha solid solution, the light hard cube-shaped crystals are the compound SnSb (50.4 per cent Sb), and the star-shaped crystals are the compound Cu_3Sn which is even harder than the SnSb crystals. The SnSb and Cu_3Sn crystals constitute the projecting parts of the bearing liner surface which support the rotating journal. The soft, plastic matrix of alpha solid solution in the babbitt easily conforms to the journal after running-in. It also wears easier to form the microscopic channels between the crystals.

Babbitt, Grade Б83, contains a large amount of expensive tin and, therefore, is used only for babbitting the bearings of high-power machinery, such as steam turbines, turbo-superchargers, turbo-pumps, etc. Tin-base babbitts have a melting point in the range from 180° to 240°C and, sometimes, even higher. Their hardness is *Bhn* 30-35 when tested with a 500 kg load and a 10 mm ball indenter.

In recent years tin in babbitts is frequently replaced by cheaper lead.

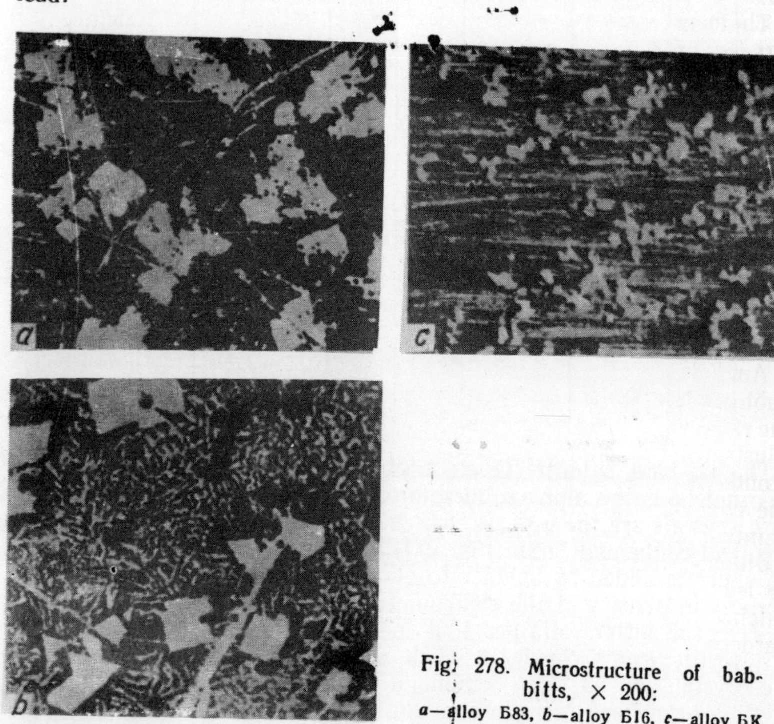

Fig. 278. Microstructure of babbitts, × 200:
a—alloy Б83, *b*—alloy Б16, *c*—alloy БК

Grade БН is used for babbitting main and crankpin bearings of internal combustion engines, crankpin and main bearings of automobile and tractor engines, radial bearings of steam turbines and steam engines of an output up to 1,200 h. p., hydraulic turbines, compressors up to 500 h.p., centrifugal pumps, etc. Babbitt Б16 has similar applications.

Table 74 gives the composition and applications of the widely used tin-lead-antimony babbitt, grade Б16.

The structure of this babbitt is illustrated in Fig. 278, *b* and consists of crystals of the solid solution of Sn in Sb (white crystals) and chemical compounds of copper with antimony which are the hard components. The eutectic Pb-Sb is the soft matrix.

Babbitt Б16 is used in the bearings of oil engines, metal-cutting machine tools, transmissions, fans, electric motors, etc.

In addition to the above-mentioned grades, a lead-base babbitt (БС) is used. Its chemical composition is: 81.75 per cent Pb, 17 per cent Sb, and 1.25 per cent Cu.

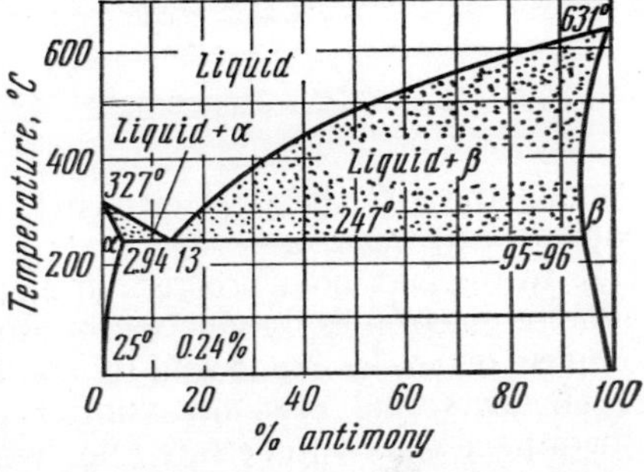

Fig. 279. The Pb-Sb equilibrium diagram

The plastic matrix in babbitt БС is the lead-antimony eutectic (Fig. 279). Crystals of β-antimony serve as the hard component.

Lead-base babbitts have a higher coefficient of friction, are softer and, at the same time, more brittle than the tin-base grades since, in them, the more brittle eutectic is the soft component instead of the plastic α-crystals.

Among the cheapest of the bearing alloys are the lead-calcium babbitts (БК in Table 74) which have found extensive application in the railroad field for babbitting the bearings of freight cars. Fig. 278, *c* illustrates the microstructure of Grade БК. The main dark background is a solid solution of sodium in lead. The hard component is the compound Pb_3Ca (lighter inclusions). This alloy has a melting point of 320-380°C. Its hardness at 20°C is *Bhn* 26-30. Calcium and sodium form a solid solution with lead. The solubility of sodium in lead is decreased, however, as the temperature falls. Therefore, calcium-sodium babbitts are subject to age-hardening (precipitation hardening).

Chapter 22

RARE METALS AND THEIR ALLOYS

Among the 102 chemical elements, about 55 are included in the group of *rare metals*.

Due to the enormous progress in electrovacuum engineering, semiconductor electronics, production of atomic energy, as well as the development of new heat- and acid-resistant alloys and cemented carbides, rare elements find ever-increasing applications in industry. Many of them have ceased to be rare (Mo, W, and others) and have entered the ranks of ordinary metals.

It is accepted practice to divide all rare metals into the following groups, in accordance with their physico-chemical properties and method of extraction*:

1) *Light rare metals*, having a low specific gravity—lithium (0.53),** rubidium (1.55), caesium (1.87), and beryllium (1.85).

2) *Refractory rare metals*—titanium (see Chapter 19), zirconium, (hafnium), vanadium, niobium, tantalum, tungsten, molybdenum, and (rhenium).

3) *Scattered metals*—indium, thallium, gallium, germanium, hafnium, selenium, tellurium, and rhenium. The metals of this group are of prime importance in semiconductor engineering. According to their properties, selenium and tellurium should be considered to be metalloids.

4) *Rare-earth metals* (Lanthanide series) which include lanthanum and fourteen elements from cerium to lutetium, inclusively.

5) *Radioactive rare metals*. These include the naturally occurring radioactive elements (polonium, francium, radium, actinium, protoactinium, thorium, and uranium), artificially produced metals (technetium, promethium, and astatine), and the transuranium elements (neptunium, plutonium, americium, and others). The radioactive properties of these metals determine their fields of application (nuclear fuel, flaw detection, etc.) and, especially, their processing techniques.

---

* Y. M. Savitsky, "Physico-Chemical Properties and Fields of Application of Rare Metals", *Physical Metallurgy and Metal Treatment*, No. 8, 1958.

** The specific gravity is given in the parentheses.

Table 75

General Characteristics of Certain Rare Metals

Metal	Specific gravity	Melting point, °C	Principal applications
Tungsten (W)	19.3	$3400^{\pm 50}$	Wire, sheets, and forgings, used in the production of incandescent lamps, X-ray apparatus, high-vacuum amplifiers, radio apparatus, etc.; also used as an alloying element in steel
Molybdenum (Mo)	10.2	$2622^{\pm 10}$	Wire and strip used in the production of incandescent lamps, in electrical high-vacuum engineering, for making electrical heating elements of high-temperature furnaces, and also as an alloying element in steel
Zirconium (Zr)	6.507	1845	Zirconium and its alloys are used in the manufacture of chemical apparatus; for cooling tubes, pumps, envelopes, fuel elements, and other devices of nuclear-power reactors; as a getter in electrical high-vacuum engineering, as well as for the production of copper- and magnesium-base alloys; deoxidant and alloying element in steel, etc.
Tantalum (Ta)	16.6	$2996^{\pm 50}$	Electrodes of vacuum tube amplifiers; anodes, grids, and other parts of electronic tubes; for chemical apparatus; in the production of cemented carbides, etc.
Niobium (Nb)	8.55	$2415^{\pm 15}$	Parts of nuclear reactors (jackets of fuel elements, cooling tubes); sheets, strips, and tubing for apparatus in the chemical industries

Continued

Metal	Specific gravity	Melting point, °C	Principal applications
Beryllium (Be)	1.84	1284	X-ray windows of X-ray tubes as a material penetrable to X-rays; sound-conducting components of acoustic apparatus; in nuclear reactors as a neutron moderator; the production of copper-base, magnesium-base, and other alloys
Germanium (Ge)	5.323	980	Semiconductor electronics

Brief characteristics follow for the more extensively used rare metals in industry.

Most interesting of these are the refractory group which are becoming more and more important in modern engineering and especially in its newest fields—nuclear and jet aircraft engineering.

Table 75 lists data on the properties and applications of the most extensively employed rare metals.

Tungsten. Characteristic features of tungsten are its high melting point, low thermal coefficient of expansion (4.44×10^{-6}), high electrical resistivity (0.055 ohm-mm^2/mm), and good corrosion resistance upon being exposed to the atmosphere and certain acids (hydrochloric, sulphuric, nitric, and others). Tungsten easily oxidises when heated above 400-500°C. The mechanical properties depend on preceding treatment. Tungsten wire has the following properties: $\sigma_u=110$-415 kg per sq mm, $\delta=0$-4 per cent, and $E=35{,}000$-38,000 kg per sq mm.

Tungsten alloys have found certain applications. An alloy, for instance, containing 20-25 per cent Mo has a linear coefficient of thermal expansion near to that of refractory glass and this alloy is used in electrical high-vacuum engineering in cases when a high electrical resistivity is required. Alloys of tungsten with copper or with silver (5 to 40 per cent), obtained by powder metallurgy techniques, display high resistance to wear and electrical erosion in conjunction with good electrical and thermal conductivity. They are widely used for electrical contacts which operate under severe conditions.

Molybdenum has a high melting point, low thermal coefficient of expansion (6.35×10^{-6}) and a high electrical resistivity (0.0517 ohm-mm^2/mm).

Depending on the purity, extraction method, and preceding working and heat treatment, the mechanical properties of molybdenum may vary in wide ranges.

The mechanical properties listed in Table 76 refer to molybdenum after undergoing cold reduction (35-90 per cent) and recrystallisation. The ductility increases with the degree of reduction.

Table 76

Mechanical Properties of Molybdenum

Production method	Mechanical properties		
	σ_u, kg/mm²	σ_s, kg/mm²	ψ, per cent
Smelting	49-54	34-37	8-72
Powder-metallurgy	53-54	38-45	8-72

Additions of nickel, cobalt, titanium, chromium, vanadium, and zirconium increase the strength of molybdenum.

Molybdenum and its alloys have a high melting point and a high recrystallisation temperature. Their recent wide applications as heat-resistant materials are due to these features (see Fig. 243).

A disadvantage of molybdenum is its lack of oxidation resistance at temperatures above 500-600°C. By alloying molybdenum with nickel (~15 per cent), and cobalt (~2.5 per cent), or with chromium (~25 per cent) and titanium (~5 per cent), its resistance to gas corrosion can be increased to a considerable extent. Recently great progress has been made towards solving the problem of protecting molybdenum against gas corrosion by means of special coatings.

Wrought molybdenum and tungsten are produced by the powder-metallurgy method and, to a lesser extent, by a vacuum arc-melting process with consumable electrodes.

Zirconium. Pure zirconium is soft; it has high ductility and good resistance to attack in a number of corrosive media. Impurities and absorbed gases make zirconium brittle. Zirconium can be worked, both hot and cold, with ease, its machinability* is fair, and it can be arc-welded in an argon atmosphere. The mechanical properties may vary in wide ranges depending on the method of production and previous treatment (σ_u=20-45 kg per sq mm and δ=20-30 per cent). Zirconium alloys with Co, Nb, Si, Ti, V, and others have higher strengths (σ up to 60-100 kg per sq mm and δ=1-24 per cent).

Zirconium and its alloys have a low thermal-neutron-capture cross-section. This, in conjunction with their high corrosion resistance to radioactive coolants, is the reason why they have become recognised as a valuable structural material for atomic-power reactors.

* Zirconium chips may ignite spontaneously.

Additions of Sn, Ni, Fe, and Cr, which have no practical influence on its thermal-neutron-capture cross-section, increase the corrosion resistance of zirconium in water at high temperatures and pressures.

Niobium. Because of its high melting point, good corrosion resistance and low thermal-neutron-capture cross-section (1.1-1.2 Barns), niobium has found extensive applications as a structural material in nuclear engineering.

Niobium has the following mechanical properties (for as-annealed sheet): σ_u=34-42 kg per sq mm, δ=20-30 per cent, and E=16,000 kg per sq mm.

Tantalum resembles niobium very closely in its physico-chemical, mechanical, and fabrication properties.

Ductile niobium and tantalum are produced either by powder metallurgy or by an arc-melting process. They are worked cold with intermediate annealing in a high vacuum.

The most important, so far, of the scattered rare metals are *germanium*, with semiconductor properties, and *rhenium*. The latter's high electrical resistivity, high corrosion resistance, and good thermoelectric properties provide opportunities for extensive applications of rhenium in electrical engineering and radio electronics.

In all probability, the rare-earth metals of the *cerium group* will soon be widely employed for alloying (modifying) steel to increase its strength, ductility, hardenability, etc. For this purpose ferrocerium is added to steel (0.3 to 2.5 kg per ton) or a mixture of rare earths, called Misch metal, is introduced.

Cerium is used together with magnesium for inoculating cast iron. Data are available on the improvement of the heat-resistant properties of nichrome and magnesium alloys by the addition of rare-earth metals.

Recently, there has been a trend towards the application of rare-earth metals as special-property materials. Cerium alloys, for example, are used as a pyrophoric material for cigarette-lighter flints while the rare-earth elements gadolinium, samarium, promethium, europium, and dysprosium are used as control rods of nuclear reactors due to their high thermal-neutron-capture cross-sections.

None of the light rare metals nor alloys based on them have yet found applications with the exception of beryllium (Table 75).

ORDINARY-QUALITY CARBON STEELS

In accordance with the USSR State Standard GOST 380-71, ordinary-quality structural carbon steels are classified into three groups.

Group A (Table 1) includes steels of ordinary quality, which are supplied with their mechanical properties, but not their chemical composition, guaranteed. These steels are identified by the Russian letters Ст (the abbreviation of the Russian word for steel) followed by a number: 0, 1, 2, 3, 4, 5 or 6. The larger the number, the higher the carbon content and, consequently, the strength, and the lower the ductility. Group A steels are used in the as-delivered condition without undergoing subsequent hot working or heat-treatment.

Group Б (Table 2) comprises steels which are supplied with their chemical composition guaranteed. They are identified by the letter Б, followed by the letters Ст and by a number: 1, 2, 3, 4, 5 or 6. The larger the number, the higher the carbon content.

Group B covers steels which are supplied with both their chemical composition and their mechanical properties guaranteed. The composition of the group B steels corresponds to that of similar steels of group Б (see Table 2), and their mechanical properties, to those of similar steels of group A (see Table 2). They are identified by the letter B, followed by the letters Ст and by a number: 1, 2, 3, 4 or 5.

Steels of groups Б and B can be hot-worked and heat-treated.

Depending upon the specified characteristics the steels of each group are further divided into categories: group A into categories 1, 2 and 3; group Б into categories 1 and 2; and group B into categories 1, 2, 3, 4, 5 and 6.

The specified characteristics are as follows: (1) group A: first category – σ_u and δ, second category – σ_u, δ and σ_b; third category – σ_u, $\sigma_{0.2}$, δ and σ_b; (2) group Б: first category – percentage of C, Mn, Si, S, P, As and N; second category – percentage of C, Mn, Si, S, P, As, N, Cr, Ni and Cu; (3) group B: first category – chemical composition and σ_u, δ and σ_b; second category – chemical composition and σ_u, $\sigma_{0.2}$, δ

and σ_b; third category – chemical composition and σ_u, $\sigma_{0.2}$, δ, σ_b and a_n at +20℃; fourth category – chemical composition and σ_u, $\sigma_{0.2}$, δ, σ_b and a_n at –20℃; fifth category – chemical composition and σ_u, $\sigma_{0.2}$, δ, σ_b and a_n at –20℃ and after ageing; and sixth category – chemical composition and σ_u, $\sigma_{0.2}$, δ, σ_b and a_n after ageing.

Steels of all groups with the numbers 1 through 4 are available in the rimming (кп), killed (сп) and semikilled (пс) conditions; those with numbers 5 and 6 may be in the semikilled (пс) and killed (сп) conditions.

To designate the category, its number is added at the end of the grade designation (the first category is not indicated), for example, Ст 3 пс-2, Б Ст 5- 2, ВСт 3кп-4, etc.

Ordinary-quality steels are used for less critical purposes, such as commercial-grade hot-rolled stock: beams, bars, channels, and angles, as well as plates, tubes and forgings. They find application in building and other welded, riveted and bolted structures (beams, trusses in cranes, bodies of vessels and apparatus, dredges, etc.), as well as for machine components not subject to high stresses (axles, shafts, gears, bolts, nuts, etc.), which are not to be heat-treated or are supplied in the heat-treated condition. Many parts made of these steels (wrist pins, pushers, gears, worms, etc.), which are subject to wear and do not require a core of high strength, undergo carburization.

Weldments are made of killed and semikilled steels of groups Б and В.

Steels of improved quality (group В) are employed mainly for special purposes (bridge and ship building, agricultural machinery, etc.) and are frequently supplied to comply with special requirements.

Table 1

Mechanical Properties of Ordinary-Quality Steels, Group A*

Grade	σ_u	$\sigma_{0.2}$	δ_{10},%	Grade	σ_u	$\sigma_{0.2}$	δ_{10},%
	kgf/mm²				kgf/mm²		
Ст0	31	–	23-20	Ст 3пс, Ст 3сп	38-49	25-21	26-23
Ст 1кп	31-40	–	25-32	Ст 4кп	41-52	26-23	25-22
Ст 1, пс, Ст 1сп	32-42	–	34-31	Ст 4пс, Ст 4сп	42-54	27-24	24-21
Ст 2кп	33-42	22-19	33-30	Ст 5пс, Ст 5сп	50-64	29-26	20-17
Ст 2пс Ст 2сп	34-44	23-20	32-29	Ст 5Гпс	46-60	29-26	20-17
Ст 3кп	37-47	24-20	27-24	Ст 6пс, Ст 6сп	60	32-30	15-12

* The thicker the rolled stock, the less σ_u, $\sigma_{0.2}$ and δ will be.

Table 2

Chemical Composition (%) of Ordinary-Quality Carbon Steels, Group Б

Grade	C	Mn	Si
Б Ст 0*	0.23	—	—
Б Ст 1кп	0.06-0.12	0.25-0.50	0.05
Б Ст 1сп	0.06-0.12	0.25-0.50	0.12-0.30
Б Ст 2кп	0.09-0.15	0.25-0.50	0.07
Б Ст 2сп	0.09-0.15	0.25-0.50	0.12-0.30
Б Ст 3кп	0.14-0.22	0.20-0.60	0.07
Б Ст 3сп	0.14-0.22	0.40-0.65	0.12-0.30
Б Ст 4кп	0.18-0.27	0.40-0.70	0.07
Б Ст 4сп	0.18-0.27	0.40-0.70	0.12-0.30
Б Ст 5пс	0.28-0.37	0.50-0.80	0.05-0.17
Б Ст 5сп	0.28-0.37	0.50-0.80	0.15-0.35
Б Ст 5Гпс	0.22-0.3	0.80-1.20	0.15
Б Ст 6пс	0.38-0.49	0.50-0.80	0.05-0.17
Б Ст 6сп	0.38-0.49	0.50-0.80	0.15-0.35

* The content of other elements in the Б Ст 0 grade steel: not more than 0.07% P; and in other steel grades: not more than 0.04% P, 0.05% S, 0.3% Cr, 0.3% Ni, 0.3% Cu and 0.08% As.

BIBLIOGRAPHY

1. M. A. Aronovich and Y. M. Lakhtin, *Osnovy metallovedeniya i termicheskoi obrabotki*, Metallurgizdat, 1952 *(Fundamentals of Engineering Physical Metallurgy and Heat Treatment).*

2. N. V. Ageyev, *Khimiya metallicheskikh splavov*, U.S.S.R. Acad. of Sciences, 1941 *(The Chemistry of Metal Alloys).*

3. *Avtomobilniye konstruktsionniye stali*, VNITOMASH-NAMI, Mashgiz, 1951 *(Structural Steels for the Automobile Industry).*

4. A. D. Assonov, Y. M. Lakhtin, and M. N. Kunyavsky, *Tekhnologiya termicheskoi obrabotki*, Mashgiz, 1952 *(Heat Treating Procedures).*

5. A. A. Bochvar, *Osnovy termicheskoi obrabotki splavov*, Metallurgizdat, 1940 *(Fundamentals of Alloy Heat Treatment).*

6. A. A. Bochvar, *Metallovedeniye*, Metallurgizdat, 1956 *(Physical Metallurgy).*

7. M. Y. Blanter, *Metodika issledovaniya metallov i obrabotka opytnykh dannykh*, Metallurgizdat, 1952 *(Metals Research Methods and Data Processing).*

8. M. L. Bernstein, *Stali i splavy dlya raboty pri vysokikh temperaturakh*, Metallurgizdat, 1956 *(Steels and Alloys for High-Temperature Applications).*

9. S. Z. Bokstein, *Struktura i mekhanicheskiye svoistva legirovannoi stali*, Metallurgizdat, 1954 *(Structure and Mechanical Properties of Alloyed Steels).*

10. N. F. Bolkhovitinov, *Metallovedeniye i termicheskaya obrabotka*, Mashgiz, 1954 *(Engineering Physical Metallurgy and Heat Treatment).*

11. Edgar S. Bain, *Functions of the Alloying Elements in Steel*, Pitsburg, Pa., U.S.A., 1939.

12. K. P. Bunin, *Zhelezouglerodistiye splavy*, Mashgiz, 1949 *(Iron-Carbon Alloys).*

13. K. P. Bunin and Y. N. Malinochka, *Vvedeniye v metallografiyu*, Metallurgizdat 1954 *(An Introduction to Metallography).*

14. I. N. Bogachov, *Metallografiya chuguna*, Mashgiz, 1952 *(The Metallography of Cast Iron).*

15. V. G. Vorobyov, *Termicheskaya obrabotka stali pri temperature nizhe nulya*, Oborongiz, 1954 *(Sub-Zero Treatment of Steel).*

16. *Voprosy fiziki metallov i metallovedeniya*, U.S.S.R. Acad. of Sciences, Kiev, 1955 *(Problems of Metal Physics and Engineering Physical Metallurgy).*

17. E. Houdremont, *Handbuch der Sonderstahlkunde*, Berlin, 1956.

18. A. P. Gulyaev, *Metallovedeniye*, Oborongiz, 1956 *(Engineering Physical Metallurgy)*.

19. A. P. Gulyaev, *Termicheskaya obrabotka stali*, Mashgiz, 1953 *(Heat Treatment of Steel)*.

20. A. P. Gulyaev, Y. M. Lakhtin, and A. I. Tarusin, *Termicheskaya obrabotka stali*, Mashgiz, 1946 *(Heat Treatment of Steel)*.

21. N. V. Geveling, *Aviatsionnoye metallovedeniye*, Part 1, ONTI, 1938 *(Engineering Physical Metallurgy in the Aircraft Industry)*.

22. Y. A. Geller, *Instrumentalniye stali*, Metallurgizdat, 1955 *(Tool Steels)*.

23. A. I. Gardin, *Elektronnaya mikroskopiya stali*, Metallurgizdat, 1954 *(Electron Microscopy of Steel)*.

24. *Zaochniye kursy po metallovedeniyu i termicheskoi obrabotke* (VNITOMASH, Komitet Metallovedeniya i Termicheskoi Obrabotki), Mashgiz, 1949-52 *(Correspondence Courses in Engineering Physical Metallurgy and Heat Treatment)*.

25. V. A. Delle, *Legirovannaya konstruktsionnaya stal*, Metallurgizdat, 1953 *(Alloyed Structural Steel)*.

26. I. N. Kidin, *Termicheskaya obrabotka stali pri induktsionnom nagreve*, Metallurgizdat, 1950 *(Induction Heating in Steel Heat Treatment)*.

27. G. A. Kashchenko, *Osnovy metallovedeniya*, Metallurgizdat, 1949 (*Fundamentals of Engineering Physical Metallurgy*).

28. Y. M. Lakhtin, *Fizicheskiye osnovy protsessa azotirovaniya*, Mashgiz, 1948 *(Physics of the Nitriding Process)*.

29. Y. M. Lakhtin, *Metody poverkhnostnogo uprochneniya detalei mashin*, Mashgiz. 1951 *(Surface Hardening Methods for Machine Parts)*.

30. G. Masing, *Lehrbuch fur Metallkunde*, 1952

31. *Metallovedeniye i teoriya metallurgicheskikh protsessov*, Metallurgizdat, 1955 *(Engineering Physical Metallurgy and the Theory of Metallurgical Process)*.

32. *Metallovedeniye i termicheskaya obrabotka*, Spravochnik, Metallurgizdat, 1956 (*Engineering Physical Metallurgy and Heat Treatment*, Handbook).

33. A. N. Minkevich, *Khimiko-termicheskaya obrabotka stali*, Mashgiz, 1950 *(Chemical Heat Treatment of Steel)*.

34. *Metallovedeniye i termicheskaya obrabotka*, VNITOMASH, Mashgiz, 1955 *(Engineering Physical Metallurgy and Heat Treatment)*.

35. L. S. Moroz and S. S. Shurakov, *Problema prochnosti tsementovannoi stali*. Published by TSNII, Mintransmash, Leningrad, 1947 *(Strength of Carburised Steel)*.

36. *Novoye v metallovedenii*, VNITOMASH , Mashgiz, 1948 *(New Developments in Engineering Physical Metallurgy)*.

37. B. S. Natapov and N. A. Blagoveshchensky, *Termicheskaya obrabotka metallov*, Metallurgizdat, 1955 *(Heat Treatment of Metals)*.

38. G. I. Pogodin-Alekseyev, Y. A. Geller and A. G. Rakhshtadt, *Metallovedeniye*, Oborongiz, 1956 *(Engineering Physical Metallurgy)*.

39. *Problemy konstruktsionnoi stali*. Transactions of LONITOMASH, Vol. 12, Mashgiz, 1948 *(Research in Structure Steels)*.

40. *Problemy metallovedeniya i fiziki metallov*, issue I. Metallurgizdat, 1949 *(Problems of Engineering Physical Metallurgy and Metal Physics)*.

41. *Problemy metallovedeniya i fiziki metallov*, issue III, Metallurgizdat, 1952 *(Problems of Engineering Physical Metallurgy and Metal Physics)*.

42. *Problemy metallovedeniya i fiziki metallov*, issue IV, Metallurgizdat, 1955 *(Problems of Engineering Physical Metallurgy and Metal Physics)*.

43. *Povysheniye ustalostnoi prochnosti detalei mashin poverkhnostnoi obrabotkoi*, Mashgiz, 1952 *(Increasing the Fatigue Strength of Machine Parts by Surface Treatment)*.

44. W. T. Read Jr., *Dislocations in Crystals*, McGraw-Hill Book Co., Inc. New York, 1953.

45. V. D. Turkin, M. V. Rumyantsev, *Struktura i svoistva tsvetnykh metallov*, Metallurgizdat, 1947 (Structure and Properties of Nonferrous Metals).

46. *Termicheskaya obrabotka metallov*, URALNITOMASH, Mashgiz, 1952 *(Heat Treatment of Metals)*.

47. *Termicheskaya obrabotka*, VNITOMASH, Mashgiz, 1948 *(Heat Treatment)*.

48. *Termicheskaya obrabotka stali*, URALNITOMASH, Mashgiz, 1950 *(Heat Treatment of Steel)*.

49. *Trudy Instituta Fiziki Metallov*, U.S.S.R. Acad. of Sciences, issue 12, 1949 *(Transactions of the Institute of Metal Physics)*.

50. *Trudy Uralskogo Filiala AN SSSR*, issue 10 *(Transactions of the Ural Branch of the U.S.S.R. Acad. of Sciences)*.

51. *Trudy Nauchno-Tekhnicheskogo Obshchestva Chernoi Metallurgii*, Vol. III, Metallurgizdat, 1955 *(Transactions of the Scientific and Engineering Society for Ferrous Metallurgy)*.

52. Y. S. Umansky, A. K. Trapeznikov, and A. I. Kitaigorodsky, *Roentgenografiya*, Mashgiz, 1951 *(Radiography)*.

53. Y. B. Fridman, *Mekhanicheskiye svoistva metallov*, Oborongiz, 1951 *(Mechanical Properties of Metals)*.

54. Y. S. Umansky, B. N. Finkelstein, M. Y. Blanter, S. T. Kishkin, N. S. Fastov, and S. S. Gorelik, *Fizicheskoye metallovedeniye*, Metallurgizdat, 1955 *(Physical Metallurgy)*.

55. Y. I. Frenkel, *Vvedeniye v teoriyu metallov*, OGIZ, 1946 *(Introduction to the Theory of Metals)*.

56. M. A. Filyand and Y. I. Semyonova, *Svoistva redkikh elementov*, Spravochnik, Metallurgizdat, 1953 *(Properties of Rare Elements*, Handbook).

57. *Fazoviye prevrashcheniya v zhelezouglerodistykh splavakh*, Mazhgiz, 1950 *(Phase Transformations in Iron-Carbon Alloys)*.

58. S. S. Steinberg, *Metallovedeniye*, Vol. I, Metallurgizdat, 1952 *(Physical Metallurgy)*.

59. S. S. Steinberg, *Izbranniye statyi*, Mashgiz, 1950 *(Sélected Articles)*.

60. A. A. Shmykov, *Spravochnik termista*, Mashgiz, 1952 *(Heat Treatment Handbook)*.

61. N. A. Shaposhnikov, *Mekhanicheskiye ispytaniya metallov*, Mashgiz, 1954 *(Mechanical Tests on Metals)*.

62. Entsiklopedichesky spravochnik "*Mashinostroeniye*", Vols, 2, 3, 4, 7 and 14 *(Mechanical Engineering Encyclopaedia)*.

63. W. Hume-Rothery, *Atomic Theory for Students of Metallurgy*, Institute of Metals, London, 1952.

64. *Sovremenniye metody ispytanii materialov v mashinostroyenii*, a survey, Mashgiz, 1955 *(Modern Materials Testing Methods in the Engineering Industries)*.

65. *Metallovedeniye i sovremenniye metody termicheskoi obrabotki chuguna*, a survey, Mashgiz, 1955 *(The Physical Metallurgy and Modern Heat-Treating Procedures for Cast Iron)*.

66. S. G. Cope, *Heat Treatment of High Speed Steel, Metal Treatment and Drop Forging*, 1954, Nos. 100-107, incl.

67. B. G. Livshitz, *Fizicheskiye svoistva metallov i splavov*, Mashgiz, 1957 *(Physical Properties of Metals and Alloys)*.

68. A. J. Forty, *Direct Observations of Dislocations in Crystals, Advances in Physics* (A quarterly supplement of the Philosophical Magazine, Jan. 3, 1954, No. 9).

69. P. P. Petrosyan, *Termicheskaya obrabotka stali kholodom*, Mashgiz, 1957 *(Sub-Zero Treatment of Steel)*.

70. W. Hume-Rothery, J. W. Christian, and W. P. Pearson, *Metallurgical Equilibrium Diagrams*, Institute of Physics, London, 1952.

SUBJECT INDEX

TO THE READER

MIR Publishers would be grateful for your comments on the content, translation and design of this book. We would also be pleased to receive any other suggestions you may wish to make. Our address is: 2, Pervy Rizhsky Pereulok, Moscow, U.S.S.R.

Printed in the Union of Soviet Socialist Republics